Engineering Materials and Processes

Springer
*London
Berlin
Heidelberg
New York
Hong Kong
Milan
Paris
Tokyo*

Series Editor

Professor Brian Derby, Professor of Materials Science
Manchester Science Centre, Grosvenor Street, Manchester, M1 7HS, UK

Other titles published in this series:

Fusion Bonding of Polymer Composites
C. Ageorges and L. Ye

Probabilistic Mechanics of Composite Materials
M.M. Kaminski
Publication due 2002

Intelligent Macromolecules for Smart Devices
L. Dai
Publication due 2002

Deborah D.L. Chung

Composite Materials

Functional Materials for Modern Technologies

With 64 Figures

Springer

Deborah D.L. Chung, PhD
Composite Materials Research Laboratory, University at Buffalo, State University of New York, Buffalo, New York 14260-4400, USA
ddlchung@buffalo.edu
http://www.wings.buffalo.edu/academic/department/eng/mae/cmrl/

British Library Cataloguing in Publication Data
Chung, Deborah D. L.
   Composite materials : functional materials for modern technologies
   1.Composite materials
   I.Title
   620.1'18
   ISBN 185233665X

Library of Congress Cataloging-in-Publication Data
Chung, Deborah D.L.
   Composite materials : functional materials for modern technologies / Deborah D.L. Chung.
     p. cm.
   ISBN 1-85233-665-X (alk. paper)
   1. Composite materials. I. Title.
   TA418.9.C6 C534 2002
   620.1'18--dc21   2002026861

Apart from any fair dealing for the purposes of research or private study, or criticism or review, as permitted under the Copyright, Designs and Patents Act 1988, this publication may only be reproduced, stored or transmitted, in any form or by any means, with the prior permission in writing of the publishers, or in the case of reprographic reproduction in accordance with the terms of licences issued by the Copyright Licensing Agency. Enquiries concerning reproduction outside those terms should be sent to the publishers.

Engineering Materials and Processes ISSN 1619-0181
ISBN 1-85233-665-X Springer-Verlag London Berlin Heidelberg
Springer-Verlag is part of Springer Science+Business Media GmbH
springeronline.com

© Springer-Verlag London Limited 2003
Printed in Great Britain
2nd printing, with corrections 2004

The use of registered names, trademarks, etc. in this publication does not imply, even in the absence of a specific statement, that such names are exempt from the relevant laws and regulations and therefore free for general use.

The publisher makes no representation, express or implied, with regard to the accuracy of the information contained in this book and cannot accept any legal responsibility or liability for any errors or omissions that may be made.

Typesetting: Mac Style Ltd, Scarborough, N. Yorkshire
69/3830-54321 Printed on acid-free paper SPIN 10981375

In celebration of the 85th birthday of my father, Mr. Leslie Wah-Leung Chung

Back row: author's father (Leslie Chung), author's husband (Lan K. Wong), author
Front row: author's mother (Rebecca Chung), author's sister (Maureen Chung)

Preface

Books on composite materials are essentially limited to addressing the fabrication and mechanical properties of these materials, because of the dominance of structural applications (such as aerospace applications) for traditional composite materials. However, non-structural applications are rapidly increasing in importance, due to the need of the electronic, thermal, battery, biomedical and other industries. The scientific concepts that guide the design of functional composites and that of structural composites are quite different, as both performance and cost requirements are different. Therefore, this book is distinct from related books in its emphasis on functional composite materials. The functions addressed include structural, thermal, electrical, electromagnetic, thermoelectric, electro-mechanical, dielectric, magnetic, optical, electrochemical and biomedical functions. The book provides the fundamental concepts behind the ability to provide each function, in addition to covering the fabrication, structure, properties and applications of the relevant composite materials.

Books on composite materials tend to emphasize polymer–matrix composites, although cement–matrix composites are the most widely used structural materials, and metal–matrix, carbon–matrix and ceramic–matrix composites are rising in importance. In contrast, this book covers composite materials with all of the above-mentioned matrices.

The most common approach for books on composite materials is to emphasize mechanics issues, because of the relevance of mechanics to structural applications. A less common approach for books on composite materials involves categorizing composite materials in terms of their matrix and covering the composites in accordance with their matrix material. In contrast, this book takes a new approach, namely the functional approach, i.e., covering composites according to their functions. The functional approach allows readers to appreciate how composites are designed for the needs of various industries. Such appreciation is valuable for students who are preparing themselves for industrial positions (most students are) and for professionals working in various industries. Moreover, the functional approach allows an organized presentation of numerous scientific concepts other than those related to mechanical behavior, thereby enabling coverage of a wide scientific scope.

The book is tutorial in style, but it is up to date and each chapter includes an extensive list of references. Readers need to have taken one course on introductory materials science but do not need to have taken any prior course on composite materials. Therefore, the book is suitable for use as a textbook for upper-level undergraduate students and for graduate students. It is also suitable for use as a reference book for students, engineers, technicians, technology managers and marketing personnel.

Because of the wide scientific scope enabled by the functional approach, the book is expected to be useful to all kinds of engineers (including electrical, thermal, chemical and industrial engineers). In contrast, the conventional approach that emphasizes mechanical behavior limits the readership to materials, mechanical, aerospace and civil engineers.

Deborah D.L. Chung
Buffalo, NY, U.S.A.
April 2002

Contents

1 Applications of composite materials........................ 1
  1.1 Introduction.. 1
  1.2 Structural applications.............................. 3
  1.3 Electronic applications.............................. 6
  1.4 Thermal applications................................ 8
  1.5 Electrochemical applications......................... 9
  1.6 Environmental applications.......................... 11
  1.7 Biomedical applications............................. 12

2 Science of composite materials........................... 15
  2.1 Introduction....................................... 15
  2.2 Polymer–matrix composites........................... 17
    2.2.1 Introduction................................. 17
    2.2.2 Interface engineering......................... 19
    2.2.3 Classification................................ 21
    2.2.4 Fabrication.................................. 22
    2.2.5 Properties................................... 25
  2.3 Cement–matrix composites............................ 25
  2.4 Carbon–matrix composites............................ 28
    2.4.1 Introduction................................. 28
    2.4.2 Fabrication.................................. 28
    2.4.3 Oxidation protection.......................... 33
    2.4.4 Mechanical properties......................... 38
    2.4.5 Thermal conductivity and electrical resistivity..... 39
  2.5 Metal–matrix composites............................. 39
    2.5.1 Introduction................................. 39
    2.5.2 Fabrication.................................. 41
    2.5.3 Wetting of reinforcement by molten metals....... 42
  2.6 Ceramic–matrix composites........................... 46

3 Composite materials for thermal applications.............. 55
  3.1 Introduction....................................... 55
  3.2 Materials of high thermal conductivity................ 55
    3.2.1 Metals, diamond and ceramics.................. 56
    3.2.2 Metal–matrix composites....................... 57
    3.2.3 Carbon–matrix composites..................... 60

  3.2.4 Carbon and graphite 61
  3.2.5 Ceramic–matrix composites 61
  3.2.6 Polymer–matrix composites 62
 3.3 Thermal interface materials. 63
 3.4 Materials for thermal insulation 66
 3.5 Materials for heat retention 67
 3.6 Summary 67

4 Composite materials for electrical applications 73
 4.1 Introduction 73
 4.2 Applications in microelectronics. 73
 4.3 Applications in resistance heating. 77
 4.4 Polymer–matrix composites 78
  4.4.1 Polymer–matrix composites with continuous fillers 78
  4.4.2 Polymer–matrix composites with discontinuous fillers 79
 4.5 Ceramic–matrix composites 84
 4.6 Conclusion 85

5 Composite materials for electromagnetic applications 91
 5.1 Introduction 91
 5.2 Mechanisms behind electromagnetic functions 91
 5.3 Composite materials for electromagnetic functions 93
  5.3.1 Composite materials with discontinuous fillers. 93
  5.3.2 Composite materials with continuous fillers 94
 5.4 Conclusion 95

6 Composite materials for thermoelectric applications. 101
 6.1 Introduction 101
 6.2 Non-structural composites 102
 6.3 Structural composites 103
  6.3.1 Introduction to thermoelectric property tailoring by composite engineering 103
  6.3.2 Tailoring by the choice of fibers 105
  6.3.3 Tailoring by the choice of interlaminar filler 113
 6.4 Conclusion 121

7 Composite materials for dielectric applications 125
 7.1 Background on dielectric behavior. 125
  7.1.1 Dielectric constant 125
  7.1.2 AC loss. 129
  7.1.3 Dielectric strength 133
 7.2 Piezoelectric behavior 133
 7.3 Ferroelectric behavior 140
 7.4 Piezoelectric/ferroelectric composite principles 148

7.5 Pyroelectric behavior.............................. 151
7.6 Introduction to composite materials for dielectric
    applications.. 152
7.7 Composites for electrical insulation 153
    7.7.1 Polymer-matrix composites 153
    7.7.2 Ceramic-matrix composites 154
7.8 Composites for capacitors 154
    7.8.1 Polymer-matrix composites 154
    7.8.2 Ceramic-matrix composites 154
7.9 Composites for piezoelectric, ferroelectric and
    pyroelectric functions 155
    7.9.1 Polymer-matrix composites 155
    7.9.2 Ceramic-matrix composites 156
7.10 Composites for microwave switching and electric field
     grading 156
7.11 Composites for electromagnetic windows.............. 157
7.12 Composites for solid electrolytes 157
7.13 Conclusion 157

8 Composite materials for optical applications 167
8.1 Background on optical behavior...................... 167
    8.1.1 The electromagnetic spectrum 167
    8.1.2 Interaction of electromagnetic radiation with
          materials 168
    8.1.3 Reflection and refraction...................... 171
    8.1.4 Optical fiber 173
    8.1.5 Light sources................................. 178
    8.1.6 Light detection 182
8.2 Composite materials for optical waveguides 184
8.3 Composite materials for optical filters 184
8.4 Composite materials for lasers 185
8.5 Summary... 185

9 Composite materials for magnetic applications 191
9.1 Background on magnetic behavior.................... 191
    9.1.1 Magnetic moment............................. 191
    9.1.2 Ferromagnetic behavior....................... 193
    9.1.3 Paramagnetic behavior 194
    9.1.4 Ferrimagnetic behavior 195
    9.1.5 Antiferromagnetic behavior 198
    9.1.6 Hard and soft magnets........................ 198
    9.1.7 Diamagnetic behavior 199
    9.1.8 Magnetostriction 199
    9.1.9 Magnetoresistance 202
9.2 Metal-matrix composites for magnetic applications 203
9.3 Polymer-matrix composites for magnetic applications .. 204

    9.4 Ceramic–matrix composites for magnetic applications . . 205
    9.5 Multilayers for magnetic applications. 206
    9.6 Magnetic composites for non-destructive evaluation 206
    9.7 Summary 207

10 Composite materials for electrochemical applications 213
    10.1 Background on electrochemical behavior 213
    10.2 Background on batteries 220
    10.3 Background on fuel cells 222
    10.4 Background on electric double-layer capacitors 223
    10.5 Composite materials for electrodes and current
           collectors 224
           10.5.1 Composites for improved electrochemical
                  behavior 224
           10.5.2 Composites for improved conductivity 225
           10.5.3 Composites for improved processability,
                  handleability, chemical stability, and electrolyte
                  absorptivity 225
           10.5.4 Carbon composites 226
    10.6 Composite materials for electrolytes 226
    10.7 Composite materials for multiple functions. 227
    10.8 Summary 228

11 Composite materials for biomedical applications 233
    11.1 Background on biomedical materials and applications . 233
    11.2 Polymer–matrix composites for biomedical
           applications 235
    11.3 Ceramic–matrix composites for biomedical
           applications 237
    11.4 Carbon–matrix composites for biomedical applications 238
    11.5 Metal–matrix composites for biomedical applications . . 238
    11.6 Summary 238

12 Composite materials for vibration damping 245
    12.1 Introduction 245
    12.2 Metals for vibration damping 245
    12.3 Polymers for vibration damping 247
    12.4 Ceramics for vibration damping 248
    12.5 Comparison among representative materials. 248
    12.6 Summary 250

13 Intrinsically smart structural composites 253
    13.1 Introduction 253
    13.2 Cement–matrix composites for smart structures 253
           13.2.1 Background on cement–matrix composites 254
           13.2.2 Cement–matrix composites for strain sensing . . . 254

13.2.3 Cement–matrix composites for damage sensing . 257
13.2.4 Cement–matrix composites for temperature sensing................................. 260
13.2.5 Cement–matrix composites for thermal control . 265
13.2.6 Cement–matrix composites for vibration reduction................................ 267
13.3 Polymer–matrix composites for smart structures...... 268
13.3.1 Background on polymer–matrix composites.... 268
13.3.2 Polymer–matrix composites for strain/stress sensing................................. 272
13.3.3 Polymer–matrix composites for damage sensing................................. 275
13.3.4 Polymer–matrix composites for temperature sensing................................. 277
13.3.5 Polymer–matrix composites for vibration reduction................................ 281
13.4 Conclusion... 282

Index 285

1 Applications of composite materials

1.1 Introduction

Engineering materials constitute the foundation of technology, whether the technology pertains to structural, electronic, thermal, electrochemical, environmental, biomedical or other applications. The history of human civilization evolved from the Stone Age to the Bronze Age, the Iron Age, the Steel Age and to the Space Age (simultaneously the Electronic Age). Each age is marked by the advent of certain materials. The Iron Age brought tools and utensils. The Steel Age brought rails and the Industrial Revolution. The Space Age was brought about by structural materials (e.g., composite materials) that are both strong and lightweight. The Electronic Age was brought about by semiconductors. Materials include metals, polymers, ceramics, semiconductors and composite materials.

Metals (including alloys) consist of atoms and are characterized by metallic bonding (i.e., the valence electrons of each atom being delocalized and shared among all the atoms). Most of the elements in the Periodic Table are metals. Examples of alloys are Cu–Zn (brass), Fe–C (steel) and Sn–Pb (solder). Alloys are classified according to the majority element present. The main classes of alloys are iron-based alloys (for structures), copper-based alloys (for piping, utensils, thermal conduction, electrical conduction, etc.) and aluminum-based alloys (for lightweight structures and for metal–matrix composites). Alloys are almost always in the polycrystalline form.

Ceramics are inorganic compounds such as Al_2O_3 (for spark plugs and for substrates for microelectronics), SiO_2 (for electrical insulation in microelectronics), Fe_3O_4 (ferrite for magnetic memories used in computers), silicates (e.g., clay, cement, glass), SiC (an abrasive), etc. The main classes of ceramics are oxides, carbides, nitrides and silicates. Ceramics are typically partly crystalline and partly amorphous. They consist of ions (often atoms as well) and are characterized by ionic bonding (often covalent bonding as well).

Polymers in the form of *thermoplastics* (e.g., nylon, polyethylene, polyvinyl chloride, rubber) consist of molecules which have covalent bonding within each molecule and van der Waals forces between the molecules. Polymers in the form of *thermosets* (e.g., epoxy, phenolics) consist of a network of covalent bonds. Polymers are amorphous, except for a minority of thermoplastics. Due to the bonding, polymers are typically electrical and thermal insulators. However, conducting polymers can be obtained by doping and conducting polymer–matrix composites can be obtained by the use of conducting fillers.

Semiconductors are characterized by having the highest occupied energy band (the valence band, where the valence electrons reside energetically) full, such that the energy gap between the top of the valence band and the bottom of the empty energy band above (called the conduction band) is small enough for some fraction of the valence electrons to be excited from the valence band to the conduction band by thermal, optical or other forms of energy. Conventional semiconductors, such as silicon, germanium and gallium arsenide (GaAs, a *compound semiconductor*), are covalent network solids. They are usually doped in order to enhance the electrical conductivity. They are typically used in the form of single crystals without dislocations, as grain boundaries and dislocations would degrade the electrical behavior.

Composite materials are multi-phase materials obtained by artificial combination of different materials, so as to attain properties that the individual components by themselves cannot attain. An example is a lightweight structural composite that is obtained by embedding continuous carbon fibers in one or more orientations in a polymer matrix. The fibers provide the strength and stiffness, while the polymer serves as the binder. Another example is *concrete*, which is a structural composite obtained by combining (through mixing) cement (the matrix, i.e., the binder, obtained by a reaction between cement and water known as hydration), sand (fine aggregate), gravel (coarse aggregate) and optionally other ingredients that are known as admixtures. Short fibers and silica fume (a fine SiO_2 particulate) are examples of admixtures. In general, composites are classified according to their matrix material. The main classes of composites are polymer–matrix, cement–matrix, metal–matrix, carbon–matrix and ceramic–matrix composites.

Polymer–matrix and cement–matrix composites are the most common, owing to the low cost of fabrication. Polymer–matrix composites are used for lightweight structures (aircraft, sporting goods, wheelchairs, etc.), in addition to vibration damping, electronic enclosures, asphalt (composite with pitch, a polymer, as the matrix), solder replacement, etc. Cement–matrix composites in the form of concrete (with fine and coarse aggregates), steel-reinforced concrete, *mortar* (with fine aggregate, but no coarse aggregate) or *cement paste* (without any aggregate) are used for civil structures, prefabricated housing, architectural precasts, masonry, landfill cover, thermal insulation, sound absorption, etc. Carbon–matrix composites are important for lightweight structures (e.g., Space Shuttle) and components (e.g., aircraft brakes) that need to withstand high temperatures, but they are relatively expensive owing to the high cost of fabrication. Carbon–matrix composites suffer from their tendency to be oxidized ($C + O_2 \rightarrow CO_2$), thereby becoming vapor. Ceramic–matrix composites are superior to carbon–matrix composites in the oxidation resistance, but they are not as well developed as carbon–matrix composites. Metal–matrix composites with aluminum as the matrix are used for lightweight structures and low-thermal-expansion electronic enclosures, but their applications are limited by the high cost of fabrication and by galvanic corrosion.

Not included in the five categories mentioned above is carbon, which can be in the form of graphite (most common form), diamond and fullerenes (a recently discovered form). They are not ceramics, because they are not compounds.

Graphite (a semi-metal) consists of carbon atom layers stacked in the AB sequence, such that the bonding is covalent (due to sp^2 hybridization) and metallic (two-dimensionally delocalized $2p_z$ electrons) within a layer and is of van der

Waals type between the layers. This bonding makes graphite very anisotropic, so that it is a good lubricant (due to the ease of sliding of the layers with respect to one another). Graphite is also used for pencils because of this property. Moreover, graphite is an electrical and thermal conductor within the layers, but is an insulator in the direction perpendicular to the layers. The electrical conductivity is valuable in the use of graphite for electrochemical electrodes. Graphite is chemically quite inert. However, due to the anisotropy, graphite can undergo a reaction (known as *intercalation*) in which foreign species (called the *intercalate*) are inserted between the carbon layers. Disordered carbon (called *turbostratic carbon*) also has a layered structure but, unlike graphite, it does not have the AB stacking order and the layers are bent. Upon heating, disordered carbon becomes more ordered, as the ordered form (graphite) has the lowest energy. *Graphitization* refers to the ordering process that leads to graphite. Conventional carbon fibers are mostly disordered carbon, such that the carbon layers are arranged preferentially along the fiber axis. A form of graphite called *flexible graphite* is formed by compressing a collection of intercalated graphite flakes that have been exfoliated (i.e., allowed to expand by over 100 times along the direction perpendicular to the layers, typically through heating after intercalation). The exfoliated flakes are held together by mechanical interlocking, as there is no binder. Flexible graphite is typically in the form of sheets, which are resilient in the direction perpendicular to the sheet. The resilience allows flexible graphite to be used as a gasket for fluid sealing.

Diamond is a covalent network solid exhibiting the diamond crystal structure due to sp^3 hybridization (akin to silicon). It is used as an abrasive and as a thermal conductor. The thermal conductivity is the highest among all materials. However, it is an electrical insulator. Owing to the high material cost, diamond is typically used in the form of powder or thin-film coating. Diamond is to be distinguished from *diamond-like carbon* (DLC), which is amorphous carbon having carbon that is sp^3-hybridized. Diamond-like carbon is mechanically weaker than diamond, but it is less expensive than diamond.

Fullerenes are molecules (C_{60}) with covalent bonding within each molecule. Adjacent molecules are held by van der Waals forces. However, fullerenes are not polymers. *Carbon nanotubes* are a derivative of the fullerenes, as they are essentially fullerenes with extra carbon atoms at the equator. The extra atoms cause the fullerene to be longer. For example, 10 extra atoms (i.e., one equatorial band of atoms) exist in the molecule C_{70}. Carbon nanotubes can be single-wall or multi-wall nanotubes, depending on the number of carbon layers in the wall of the nanotube.

1.2 Structural applications

Structural applications refer to applications that require mechanical performance (e.g., strength, stiffness and vibration damping ability) in the material, which may or may not bear the load in the structure. In cases where the material bears the load, the mechanical property requirements are particularly stringent. An example is a building in which steel-reinforced concrete columns bear the load of the structure and unreinforced concrete architectural panels cover the face of the building. Both the columns and the panels serve structural applications and are structural materials, although only the columns bear the load of the structure. Mechanical strength and stiffness are required of the panels, but the requirements are more stringent for the columns.

Structures include buildings, bridges, piers, highways, landfill cover, aircraft, automobiles (body, bumper, shaft, window, engine components, brake, etc.), bicycles, wheelchairs, ships, submarines, machinery, satellites, missiles, tennis rackets, fishing rods, skis, pressure vessels, cargo containers, furniture, pipelines, utility poles, armor, utensils and fasteners.

In addition to mechanical properties, a structural material may be required to have other properties, such as low density (light weight) for fuel saving in the case of aircraft and automobiles, for high speed in the case of race bicycles, and for handleability in the case of wheelchairs and armor. Another property that is often required is corrosion resistance, which is desirable for the durability of all structures, particularly automobiles and bridges. Yet another property that may be required is the ability to withstand high temperatures and/or thermal cycling, as heat may be encountered by the structure during operation, maintenance or repair.

A relatively new trend is for a structural material to be able to serve functions other than the structural function, so that the material becomes *multifunctional* (i.e., killing two or more birds with one stone, thereby saving cost and simplifying design). An example of a non-structural function is the sensing of damage. Such sensing, also called *structural health monitoring*, is valuable for the prevention of hazards. It is particularly important to aging aircraft and bridges. The sensing function can be attained by embedding sensors (such as optical fibers, the damage or strain of which affects the light throughput) in the structure. However, the embedding usually causes degradation of the mechanical properties and the embedded devices are costly and poor in durability compared to the structural material. Another way to attain the sensing function is to detect the change in property (e.g., the electrical resistivity) of the structural material due to damage. In this way, the structural material serves as its own sensor and is said to be *self-sensing*.

Mechanical performance is basic to the selection of a structural material. Desirable properties are high strength, high modulus (stiffness), high ductility, high toughness (energy absorbed in fracture) and high capacity for vibration damping. Strength, modulus and ductility can be measured under tension, compression or flexure at various loading rates, as dictated by the type of loading on the structure. A high compressive strength does not imply a high tensile strength. Brittle materials tend to be stronger under compression than tension due to the microcracks in them. High modulus does not imply high strength, as the modulus describes the elastic deformation behavior whereas the strength describes the fracture behavior. Low toughness does not imply a low capacity for *vibration damping*, as the damping (energy dissipation) may be due to slipping at interfaces in the material, rather than being due to the shear of a viscoelastic phase in the material. Other desirable mechanical properties are fatigue resistance, creep resistance, wear resistance and scratch resistance.

Structural materials are predominantly metal-based, cement-based and polymer-based materials, although they also include carbon-based and ceramic-based materials, which are valuable for high-temperature structures. Among the metal-based structural materials, steel and aluminum alloys are dominant. Steel is advantageous in high-strength applications, whereas aluminum is advantageous in low-density applications. For high-temperature applications, intermetallic compounds (such as NiAl) have emerged, although they suffer from their brittleness. Metal–matrix composites are superior to the corresponding metal–matrix in their high modulus, high creep resistance and low thermal expansion coefficient, but they are expensive because of the processing cost.

Among the cement-based structural materials, concrete is dominant. Although concrete is an old material, improvement in long-term durability is needed, as suggested by the degradation of bridges and highways all over the USA. Improvement pertains to decrease in the *drying shrinkage* (shrinkage of the concrete during curing or hydration), as the shrinkage can cause cracks. It also pertains to decrease in the *fluid permeability*, as water permeating into steel-reinforced concrete can cause corrosion of the reinforcing steel. Moreover, it pertains to improvement in *freeze–thaw durability*, which is the ability of concrete to withstand variations between temperatures below 0°C (freezing of water in concrete) and those above 0°C (thawing of water in concrete).

Among the polymer-based structural materials, fiber-reinforced polymers are dominant, due to their combination of high strength and low density. All polymer-based materials suffer from their inability to withstand high temperatures. This inability can be due to the degradation of the polymer itself or, in the case of a polymer–matrix composite, to the thermal stress resulting from the thermal expansion mismatch between the polymer matrix and the fibers. (The coefficient of thermal expansion is typically much lower for the fibers than for the matrix.)

Most structures involve joints, which may be attained by welding, brazing, soldering, the use of adhesives or fastening. The structural integrity of joints are critical to the integrity of the overall structure.

As structures can degrade or be damaged, repair may be needed. Repair often involves the use of a repair material, which may be the same as or different from the original material. For example, a damaged concrete column may be repaired by removing the damaged portion and patching with a fresh concrete mix. A superior but much more costly way involves the above-mentioned patching, followed by wrapping the column with continuous carbon or glass fibers, using epoxy as the adhesive between the fibers and the column. Due to the tendency for the molecules of a thermoplastic polymer to move upon heating, joining of two thermoplastic parts can be attained by liquid-state or solid-state welding, thereby facilitating repair of a thermoplastic structure. In contrast, the molecules of a thermosetting polymer do not tend to move, and repair of a thermoset structure needs to involve other methods, such as the use of adhesives.

Corrosion resistance is desirable for all structures. Metals, due to their electrical conductivity, are particularly prone to corrosion. In contrast, polymers and ceramics, due to their poor conductivity, are much less prone to corrosion. Techniques of corrosion protection include the use of a *sacrificial anode* (i.e., a material that is more active than the material to be protected, so that it is the party that corrodes) and *cathodic protection* (i.e., the application of a voltage which causes electrons to go into the material to be protected, thereby making the material a cathode). The implementation of the first technique simply involves attaching the sacrificial anode material to the material to be protected. The implementation of the second technique involves applying an electrical contact material on the surface of the material to be protected and passing an electric current through wires embedded in the electrical contact. The electrical contact material must be a good electrical conductor and be able to adhere to the material to be protected, in addition to being wear resistant and scratch resistant.

Vibration damping is desirable for most structures. It is commonly attained by attaching to or embedding in the structure a viscoelastic material, such as rubber. Upon vibration, shear deformation of the viscoelastic material causes energy dissipation. However, due to the low strength and modulus of the viscoelastic material compared to the structural material, the presence of the viscoelastic material (especially if it is embedded) lowers the strength and modulus of the structure. A more ideal way to attain vibration damping is to modify the structural material itself, so that it maintains its high strength and modulus while providing damping. In the case of a composite material being the structural material, modification can involve the addition of a filler (particles or fibers) of a very small size, so that the total filler–matrix interface area is large and slippage at the interface during vibration provides a mechanism of energy dissipation.

1.3 Electronic applications

Electronic applications include electrical, optical and magnetic applications, as the electrical, optical and magnetic properties of materials are largely governed by electrons. There is overlap among these three areas of application.

Electrical applications pertain to computers, electronics, electrical circuitry (e.g., resistors, capacitors and inductors), electronic devices (e.g., diodes and transistors), *optoelectronic devices* (e.g., solar cells, light sensors and light-emitting diodes for conversion between electrical energy and optical energy), *thermoelectric devices* (heaters, coolers and thermocouples for conversion between electrical energy and thermal energy), *piezoelectric devices* (strain sensors and actuators for conversion between electrical energy and mechanical energy), robotics, micromachines (or microelectromechanical systems, MEMS), ferroelectric computer memories, *electrical interconnections* (e.g., solder joints, thick-film conductors and thin-film conductors), *dielectrics* (i.e., electrical insulators in bulk, thick-film and thin-film forms), *substrates* for thin films and thick films, heat sinks, electromagnetic interference (EMI) shielding, cables, connectors, power supplies, electrical energy storage, motors, electrical contacts and brushes (sliding contacts), electrical power transmission, eddy current inspection (the use of a magnetically induced electrical current to indicate flaws in a material), etc.

Optical applications pertain to lasers, light sources, *optical fibers* (materials of low optical absorptivity for communication and sensing), absorbers, reflectors and transmitters of electromagnetic radiation of various wavelengths (for optical filters, low-observable or Stealth aircraft, radomes, transparencies, optical lenses, etc.), photography, photocopying, optical data storage, holography, color control, etc.

Magnetic applications pertain to transformers, magnetic recording, magnetic computer memories, magnetic field sensors, magnetic shielding, magnetically levitated trains, robotics, micromachines, magnetic particle inspection (the use of magnetic particles to indicate the location of flaws in a magnetic material), magnetic energy storage, magnetostriction (strain in a material due to the application of a magnetic field), magnetorheological fluids (for vibration damping that is controlled by a magnetic field), magnetic resonance imaging (MRI, for patient diagnosis in hospitals), mass spectrometry (for chemical analysis), etc.

The large range of applications mentioned above means that all classes of materials are used for electronic applications. Semiconductors are at the heart of electronic and optoelectronic devices. Metals are used for electrical interconnections, EMI shielding, cables, connectors, electrical contacts, electrical power transmission, etc. Polymers are used for dielectrics and cable jackets. Ceramics are used for capacitors, thermoelectric devices, piezoelectric devices, dielectrics and optical fibers.

Microelectronics refers to electronics involving integrated circuits. Because of the availability of high-quality single crystalline semiconductors, the most critical problems that the microelectronic industry faces do not pertain to semiconductors, but are related to *electronic packaging*, which refers to chip carriers, electrical interconnections, dielectrics, heat sinks, etc. Section 3.2 gives more details on electronic packaging applications.

Because of the miniaturization and increasing power of microelectronics, heat dissipation is critical to the performance and reliability of microelectronics. Thus, materials for heat transfer from electronic packages are needed. Ceramics and polymers are both dielectrics, but ceramics are advantageous in their higher thermal conductivity compared to polymers. Materials that are electrically insulating but thermally conducting are needed. Diamond is the best material that exhibits such properties, but it is expensive.

Because of the increasing speed of computers, signal propagation delay needs to be minimized by the use of dielectrics with a low value of relative dielectric constant. (A dielectric with a high value of relative dielectric constant and used to separate two conductor lines acts like a capacitor, thereby causing signal propagation delay.) Polymers tend to be advantageous over ceramics in their low value of relative dielectric constant.

Electronic materials are available in bulk (single crystalline, polycrystalline or, less commonly, amorphous), *thick-film* (typically over 10 µm thick, obtained by applying a paste on a substrate by, say, screen printing, such that the paste contains the relevant material in particle form, together with a binder and a vehicle) or thin-film (typically less than 1500 Å thick, obtained by vacuum evaporation, sputtering, chemical vapor deposition, molecular beam epitaxy or other techniques) forms. Semiconductors are typically available in bulk single-crystalline form (cut into slices called wafers, each of which may be subdivided into "chips"), although bulk polycrystalline and amorphous forms are emerging for solar cells, due to the importance of low cost for solar cells. Conductor lines in microelectronics are mostly in thick-film and thin-film forms.

The dominant material for electrical connections is *solder* (e.g., Sn–Pb alloy). However, the difference in coefficient of thermal expansion (CTE) between the two members that are joined causes the solder to experience thermal fatigue upon thermal cycling, which is encountered during operation. The thermal fatigue can lead to failure of the solder joint. Polymer–matrix composites in paste form and containing electrically conducting fillers are being developed to replace solder. Another problem lies in the lead (poisonous) used in solder to improve the rheology of the liquid solder. Lead-free solders are being developed.

Heat sinks are materials with a high thermal conductivity. They are used to dissipate heat from electronic components. Because they are joined to materials of a low CTE (e.g., a printed circuit board in the form of a continuous-fiber polymer–matrix composite), they need to have a low CTE also. Hence, materials exhibiting both a high thermal conductivity and a low CTE are needed for heat sinks. Copper is a metal with a high thermal conductivity, but its CTE is too high. Therefore, copper is reinforced with continuous carbon fibers, molybdenum particles or other fillers of low CTE.

1.4 Thermal applications

Thermal applications refer to applications that involve heat transfer, whether by conduction, convection or radiation. Heat transfer is needed in heating (of buildings, heating involved in industrial processes such as casting and annealing, cooking, de-icing, etc.) and in cooling (cooling of buildings, refrigeration of food and industrial materials, cooling of electronics, removal of heat generated by chemical reactions such as the hydration of cement, removal of heat generated by friction or abrasion as in a brake and in machining (cutting), removal of heat generated by the impingement of electromagnetic radiation, removal of heat from industrial processes such as welding, etc.).

Conduction refers to the heat flow from points of higher temperature to points of lower temperature in a material. It typically involves metals, due to their high thermal conductivity.

Convection is attained by the movement of a hot fluid. If the fluid is forced to move by a pump or a blower, the convection is known as forced convection. If the fluid moves because of differences in density, the convection is known as natural or free convection. The fluid can be a liquid (e.g., oil) or a gas (e.g., air). It must be able to withstand the heat involved. Fluids are outside the scope of this book.

Radiation, i.e., blackbody radiation, is involved in space heaters. It refers to the continual emission of radiant energy from the body. The energy is in the form of electromagnetic radiation, typically infrared radiation. The dominant wavelength of the emitted radiation decreases with increasing temperature of the body. The higher the temperature, the greater is the rate of emission of radiant energy per unit area of the surface of the body. This rate is proportional to T^4, where T is the absolute temperature. It is also proportional to the emissivity of the body. The emissivity depends on the material of the body. In particular, it increases with increasing roughness of the surface of the body.

Heat transfer can be attained by the use of more than one mechanism. For example, both conduction and forced convection are involved when a fluid is forced to flow through the interconnected pores of a solid, which is a thermal conductor.

Conduction is more tied to material development than convection or radiation. Therefore, materials for thermal conduction are specifically addressed in Chapter 3.

Thermal conduction can involve electrons, ions and/or phonons. Electrons and ions move from a point of higher temperature to a point of lower temperature, thereby transporting heat. Due to the high mass of ions compared to electrons, electrons move much more easily than ions. *Phonons* are lattice vibrational waves, the propagation of which also leads to the transport of heat. Metals conduct by electrons, because they have free electrons. Diamond conducts by phonons, because free electrons are not available and the low atomic weight of carbon intensifies the lattice vibrations. Diamond is the material with the highest thermal conductivity. In contrast, polymers are poor conductors, because free electrons are not available and the weak secondary bonding (van der Waals forces) between the molecules makes it difficult for the phonons to move from one molecule to another. Ceramics tend to be more conductive than polymers, due to the ionic and covalent bonding making it possible for the phonons to propagate. Moreover, ceramics tend to have more mobile electrons or ions than polymer, and the movement of electrons and/or ions contributes to thermal conduction. On the other hand, ceramics tend to be poorer than metals in thermal conductivity, due to the low concentration of free electrons (if any) in ceramics compared to metals.

Materials for thermal applications include thermal conductors, thermal insulators, heat retention materials (i.e., materials with a high heat capacity), thermal interface materials (i.e., materials for improving thermal contacts) and thermoelectric materials (for sensing temperature, heating and cooling). The design of all these materials involves composite science. These materials are addressed in Chapter 3, except for thermoelectric materials, which are addressed in Chapter 6.

1.5 Electrochemical applications

Electrochemical applications involve those pertaining to electrochemical reactions. An *electrochemical reaction* involves an oxidation reaction (such as Fe → Fe^{2+} + 2e$^-$), in which electrons are generated, and a reduction reaction (such as O_2 + $2H_2O$ + 4e$^-$ → 4OH$^-$), in which electrons are consumed. The electrode that releases electrons is called the *anode*; the electrode that receives electrons is called the *cathode*.

When the anode and cathode are electrically connected, say, by using a wire, electrons move from the anode to the cathode. Both the anode and cathode must be electronic conductors. As the electrons move in the wire from the anode to the cathode, ions move in an ionic conductor (called the *electrolyte*) placed between the anode and the cathode, such that cations (positive ions) generated by the oxidation of the anode move in the electrolyte from the anode to the cathode.

Whether an electrode behaves as an anode or a cathode depends on its propensity to be oxidized. The electrode that has the higher propensity for oxidation serves on the anode, while the other electrode serves as the cathode. On the other hand, a voltage can be applied between the anode and the cathode at the location of the wire, such that the positive end of the voltage is at the anode side. The positive end attracts electrons, thus forcing the anode to be oxidized, even when the anode may not be more prone to oxidation than the cathode.

The oxidation reaction is associated with corrosion of the anode. For example, the oxidation reaction Fe → Fe^{2+} + 2e$^-$ causes iron atoms to be corroded away, becoming Fe^{2+} ions which go into the electrolyte. Thus, hindering the oxidation reaction results in corrosion protection.

Electrochemical reactions are relevant not only to corrosion, but also to batteries, fuel cells and industrial processes (such as the reduction of Al_2O_3 to make Al) which make use of electrochemical reactions. The burning of fossil fuels such as coal and gasoline causes pollution to the environment. In contrast, batteries and fuel cells have much less environmental problem.

A *battery* involves an anode and a cathode which are inherently different in their propensity for oxidation. When the anode and cathode are open circuited at the wire, a voltage difference is present between the anode and the cathode, such that the negative end of the voltage is at the anode side. This is because the anode "wants" to release electrons, although the electrons cannot exit because of the open circuit condition. This voltage difference is the output of the battery, which is a source of direct current (DC).

A unit involving an anode and a cathode is called a *galvanic cell*. A battery consists of a number of galvanic cells connected in series, so that the battery voltage is the sum of the voltages of the individual cells.

An example of a battery is the *lead storage battery*, which is used in cars. Lead (Pb) is the anode, while lead dioxide (PbO_2, in the form of a coating on lead) is the cathode. Sulfuric acid (H_2SO_4) is the electrolyte. The oxidation reaction (i.e., anode reaction) is

$$Pb + HSO_4^- \rightarrow PbSO_4 + H^+ + 2e^-$$

The reduction reaction (i.e., cathode reaction) is

$$PbO_2 + HSO_4^- + 3H^+ + 2e^- \rightarrow PbSO_4 + 2H_2O.$$

Discharge is the state of operation of the battery. $PbSO_4$ is a solid reaction product that adheres to the electrodes, thereby hindering further reaction. Therefore, a battery needs to be charged by forcing current through the battery in the opposite direction, thereby breaking down $PbSO_4$, i.e., making the above reactions go in the reverse direction. In a car, the battery is continuously charged by an alternator.

Another example of a battery is the alkaline version of the *dry cell battery*. This battery comprises a zinc anode and an MnO_2 cathode. Because MnO_2 is not an electrical conductor, carbon powder (an electrical conductor) is mixed with MnO_2 powder to form the cathode (a composite). The electrolyte is either KOH or NaOH. The anode reaction is

$$Zn + 2OH^- \rightarrow ZnO + H_2O + 2e^-$$

The cathode reaction is

$$2MnO_2 + H_2O + 2e^- \rightarrow Mn_2O_3 + 2OH^-$$

A *fuel cell* is a galvanic cell in which the reactants are continuously supplied. An example is the hydrogen–oxygen fuel cell. The anode reaction is

$$2H_2 + 4OH^- \rightarrow 4H_2O + 4e^-$$

The cathode reaction is

$$4e^- + O_2 + 2H_2O \rightarrow 4OH^-$$

The overall cell reaction (the anode and cathode reactions added together) is

$$2H_2\ (g) + O_2\ (g) \rightarrow 2H_2O\ (l),$$

which is the formation of water from the reaction of hydrogen and oxygen.

During cell operation, hydrogen gas is fed to a porous carbon plate which contains a catalyst that helps the anode reaction. The carbon is an electrical conductor, which allows electrons generated by the anode reaction to flow. The porous carbon is known as a current collector. Simultaneously oxygen gas is fed to another porous carbon plate which contains a catalyst. The two carbon plates are electrically connected by using a wire, so that electrons generated by the anode reaction at one plate flow through the wire and enter the other carbon plate for consumption in the cathode reaction. As this occurs, the OH⁻ ions generated by the cathode reaction move through the electrolyte (KOH) between the two carbon plates, and are then consumed in the anode reaction at the other carbon plate. The overall cell reaction produces H_2O, which exits the cell at an opening located at the electrolyte between the two carbon plates. The useful output of the cell is the electric current associated with the flow of electrons in the wire from one plate to the other.

Materials required for electrochemical applications include the electrodes, current collector (such as the porous carbon plates of the fuel cell mentioned above), conductive additive (such as carbon powder mixed with MnO_2 powder in the dry cell mentioned above), electrolyte and cell container (which needs to be chemically resistant). An electrolyte can be liquid or solid, as long as it is an ionic conductor. The interface between the electrolyte and an electrode should be intimate and greatly affects cell performance. The ability to recharge a cell is governed by the reversibility of the cell reactions. In practice, the reversibility is not complete, leading to a low charge–discharge cycle life.

Because the ingredients in an electrode or a current collector are usually particles, a binder (e.g., Teflon) is needed in order to provide handleability. The binder affects the mechanical integrity as well as electrolyte absorptivity, electrical conductivity and durability. Thus, the design involves composite science.

1.6 Environmental applications

Environmental applications are those relating to protection of the environment from pollution. Protection can involve removal of a pollutant or reduction in the amount of pollutant generated. Pollutant removal can be attained by extraction through adsorption of the pollutant on the surface of a solid (e.g., activated carbon) with surface porosity. It can also be attained by planting trees, which take in CO_2 gas. Pollutant generation can be reduced by changing the materials and/or processes used in industry, by using biodegradable materials (i.e., materials that can be degraded by nature, so that their disposal is not necessary), by using materials that can be recycled, or by changing the energy source from fossil fuels to batteries, fuel cells, solar cells and/or hydrogen. For this reason, composites with *biodegradable polymer* matrices are attractive. An example of a biodegradable polymer is starch.

Materials have been mainly developed for structural, electronic, thermal or other applications without much consideration given to disposal or recycling problems. It is now recognized that such considerations must be included during the design and development of materials, rather than considering disposal or recycling after the materials have been developed.

Materials for removal of certain molecules or ions in a fluid by adsorption are central to the development of materials for environmental applications. They include carbons, zeolites, aerogels and other porous materials in particulate and fibrous forms. The particles or fibers can be loose, held together by a small amount of binder or, in the case of fibers, held together by stapling, in which mechanical interlocking between fibers occurs through a process involving poking of the fiber agglomerate by a needle. The loose form is least desirable because the loose particles or fibers may be pulled into the fluid stream. Continuous fibers that are woven together do not need a binder or stapling, but they are expensive compared to discontinuous fibers.

Desirable qualities of an adsorption material include large adsorption capacity, pore size large enough for relatively large molecules and ions to lodge, ability to be regenerated or cleaned after use, fluid dynamics for fast movement of the fluid from which the pollutant is to be removed and, in some cases, selective adsorption of certain species.

Activated carbon fibers are superior to activated carbon particles in fluid dynamics, owing to the channels between the fibers. However, they are much more expensive.

Pores on the surface of a material must be accessible from the outside in order to serve as adsorption sites. In general, the pores can be *macropores* (>500 Å), *mesopores* (between 20 and 500 Å), *micropores* (between 8 and 20 Å) or *micro-micropores* (<8 Å). Activated carbons typically have pores that are micropores and micro-micropores.

Materials for removal of particles in a fluid by filtration are typically macroporous materials. They are usually in the form of fibers or particles that are held together by using a small amount of a binder. Fibers are preferred in that they allow the pores to be larger, so that the fluid flow is faster. However, fibers tend to be more expensive than particles. Particularly challenging is the situation in which the fluid is hot. The challenge involves mainly the choice of binder, which should be effective for binding even at a small concentration and should be stable at high temperatures. These porous materials are composites, although they are not designed for mechanical performance. Without the binder, the material can still be used as a filter but, as loose particles or fibers, it cannot be conveniently handled, and the incorporation of the loose particles or fibers in the flowing fluid can occur and is not desirable. A filter is also known as a membrane.

Electronic pollution is an environmental problem that is becoming important. It refers to the electromagnetic waves (particularly radio waves) that are present in the environment as a result of radiation sources such as cellular telephones. Such radiation can interfere with digital electronics such as computers, thereby causing hazards and affecting the operation of society. To alleviate this problem, radiation sources and electronics are shielded by materials that reflect and/ or absorb radiation. For reflection, electrically conducting polymer–matrix composites are important.

1.7 Biomedical applications

Biomedical applications are those involving the diagnosis and treatment of conditions, diseases and disabilities, as well as their prevention. They include implants (e.g., hips, heart valves, skin and teeth), surgical and diagnostic devices, pacemaker (device with electrical leads to the wall of the heart for electrical control of heartbeats), electrodes for collecting or sending electrical or optical signals for diagnosis or treatment, wheelchairs, devices for helping the disabled, exercise equipment, pharmaceutical packaging (for controlled release of drug to the body, or for other purposes) and instrumentation for diagnosis and chemical analysis (such as equipment for analyzing blood and urine). Implants are particularly challenging, as they need to be made of materials that are biocompatible (compatible with fluids such as blood), corrosion resistant, wear resistant and fatigue resistant, and able to maintain these properties over tens of years.

Carbon is a particularly biocompatible material (more so than gold), so carbon–carbon composites are used for implants. Composites with biocompatible polymer matrices are also used for implants. Electrically conducting polymer-matrix composites are used for electrodes for diagnostics. Composites with biodegradable polymer matrices are used for pharmaceuticals.

Materials for bone replacement or bone growth support need to have an elastic modulus similar to that of bone. Tailoring of the modulus can be achieved by composite design, i.e., appropriate choice of reinforcement and its volume fraction.

Review questions

1. What is the difference in chemical bonding between a thermoplastic and a thermoset?
2. Define "composite material".
3. What are the matrix and fillers in concrete?
4. What is the matrix in asphalt?
5. What is the difference between concrete and mortar?
6. What is the reaction that causes carbon to become a vapor in the presence of oxygen and heat?
7. Why are metal–matrix composites not widely used?
8. What is the difference in chemical bonding between a metal and a semiconductor?
9. Graphite is an electrical conductor in the direction parallel to the carbon layers and is a poor conductor in the direction perpendicular to the carbon layers. Explain by considering the chemical bonding.
10. What are the three forms of carbon?
11. A structure can be given the ability to sense by embedding sensors in the structure. Another method is to use a structural material that has the inherent ability to sense. What are the two main disadvantages of the former method compared to the latter method?
12. How is cathodic protection implemented?
13. What are the requirements for the electrical contact material used in cathodic protection?
14. Why is diamond an excellent thermal conductor?
15. Why are polymers poor thermal conductors?

2 Science of composite materials

2.1 Introduction

Composite materials are those containing more than one phase such that the different phases are artificially blended together. They are not multiphase materials in which the different phases are formed naturally by reactions, phase transformations or other phenomena.

A composite material typically consists of one or more fillers (fibrous or particulate) in a certain matrix. A carbon fiber composite is one in which at least one of the fillers is composed of carbon fibers, either short or continuous, unidirectional or multidirectional, woven or non-woven. The matrix is usually a polymer, a metal, a carbon, a ceramic or a combination of different materials. Except for sandwich composites, the matrix is three-dimensionally continuous, whereas the filler can be three-dimensionally discontinuous or continuous. Carbon fiber fillers are usually three-dimensionally discontinuous, unless the fibers are three-dimensionally interconnected by weaving or by the use of a binder such as carbon.

The high strength and modulus of carbon fibers make them useful as a reinforcement for polymers, metals, carbons and ceramics, even though they are brittle. Effective reinforcement requires good bonding between the fibers and the matrix, especially for short fibers. For an ideally unidirectional composite (i.e., one containing continuous fibers all in the same direction) containing fibers of much higher modulus than that of the matrix, the longitudinal tensile strength is quite independent of the fiber–matrix bonding, but the transverse tensile strength and the flexural strength (for bending in longitudinal or transverse directions) increase with increasing fiber–matrix bonding. On the other hand, excessive fiber–matrix bonding can cause a composite with a brittle matrix (e.g., carbon and ceramics) to become more brittle, as the strong fiber–matrix bonding causes cracks to propagate in a straight line in a direction perpendicular to the fiber–matrix interface, without being deflected to propagate along this interface. In the case of a composite with a ductile matrix (e.g., metals and polymers), a crack initiating in the brittle fiber tends to be blunted when it reaches the ductile matrix, even when the fiber–matrix bonding is strong. Therefore, an optimum degree of fiber–matrix bonding (i.e., not too strong and not too weak) is needed for brittle-matrix composites, whereas a high degree of fiber–matrix bonding is preferred for ductile-matrix composites.

The mechanisms of fiber–matrix bonding include chemical bonding, interdiffusion, van der Waals bonding and mechanical interlocking. Chemical bonding gives a relatively large bonding force, provided that the density of chemical bonds across the fiber–matrix interface is sufficiently high and that a brittle reaction product is absent at the fiber–matrix interface. The density of chemical bonds can be increased by chemical treatment of the fibers or by sizings on the fibers. Interdiffusion at the fiber–matrix interface also results in bonding, although its occurrence requires the interface to be rather clean. Mechanical interlocking between the fibers and the matrix is an important contribution to the bonding if the fibers form a three-dimensional network. Otherwise, the fibers should have a rough surface in order for a small degree of mechanical interlocking to take place.

Chemical bonding, interdiffusion and van der Waals bonding require the fibers to be in intimate contact with the matrix. For intimate contact to take place, the matrix or matrix precursor must be able to wet the surfaces of the carbon fibers during infiltration of the matrix or matrix precursor into the carbon fiber preform. Wetting is governed by the surface energies. Chemical treatments and coatings can be applied to the fibers to enhance wetting through their effects on the surface energies. The choice of treatment or coating depends on the matrix. A related method is to add a wetting agent to the matrix or matrix precursor before infiltration. As wettability may vary with temperature, the infiltration temperature can be chosen to enhance wetting. Although wetting is governed by thermodynamics, it is strongly affected by kinetics. Thus, yet another way to enhance wetting is the use of high pressure during infiltration.

The occurrence of a reaction between the fibers and the matrix helps the wetting and bonding between them. However, an excessive reaction degrades the fibers, and the reaction product(s) may be undesirable for the mechanical, thermal or moisture resistance properties of the composite. Therefore, an optimum amount of reaction is preferred.

Carbon fibers are electrically and thermally conductive, in contrast to the nonconducting nature of polymer and ceramic matrices. Therefore, carbon fibers can serve not only as a reinforcement, but also as an additive for enhancing the electrical or thermal conductivity. Furthermore, carbon fibers have nearly zero coefficient of thermal expansion, so they can also serve as an additive for lowering the thermal expansion. The combination of high thermal conductivity and low thermal expansion makes carbon fiber composites useful for heat sinks in electronics and for space structures that require dimensional stability. As the thermal conductivity of carbon fibers increases with the degree of graphitization, applications requiring a high thermal conductivity should use graphitic fibers, such as high-modulus pitch-based fibers and vapor-grown carbon fibers. Carbon fibers are more cathodic than practically any metal, so in a metal–matrix a galvanic couple is formed with the metal as the anode. This causes corrosion of the metal. The corrosion product tends to be unstable in moisture and causes pitting, which aggravates corrosion. To alleviate this problem, carbon fiber metal–matrix composites are often coated.

The carbon fibers in a carbon-matrix composite (called *carbon–carbon composite*) serve to strengthen the composite, as the carbon fibers are much stronger than the carbon matrix owing to the crystallographic texture (preferred crystallographic orientation) in each fiber. Moreover, the carbon fibers serve to toughen the composite, as debonding between the fibers and the matrix provides a mechanism for energy absorption during mechanical deformation. In addition to having attractive mechanical properties, carbon–carbon composites are more thermally conductive than carbon fiber polymer–matrix composites. However, at elevated temperatures (above 320°C), carbon–carbon composites degrade as a result of the oxidation of carbon (especially the carbon matrix), which forms CO_2 gas. To alleviate this problem, carbon–carbon composites are coated.

Carbon fiber ceramic–matrix composites are more oxidation resistant than carbon–carbon composites. The most common form of such composites is carbon fiber reinforced concrete. Although the oxidation of carbon is catalyzed by an alkaline environment and concrete is alkaline, the chemical stability of carbon fibers in concrete is superior to that of competitive fibers, such as polypropylene, glass and steel. Composites containing carbon fibers in more advanced ceramic matrices (such as SiC) are rapidly being developed.

Fiber composites are most commonly fabricated by the impregnation (or infiltration) of the matrix or matrix precursor in the liquid state into the fiber preform, which is most commonly in the form of a woven fabric. In the case of composites in the shape of tubes, the fibers may be impregnated in the form of a continuous bundle (called a tow) from a spool and subsequently the bundles may by wound on a mandrel. Instead of impregnation, the fibers and matrix material may be intermixed in the solid state by commingling reinforcing fibers and matrix fibers, by coating the reinforcing fibers with the matrix material, by sandwiching reinforcing fibers with foils of the matrix material, and in other ways. After impregnation or intermixing, consolidation is carried out, often under heat and pressure.

2.2 Polymer–matrix composites

2.2.1 Introduction

Polymer–matrix composites are much easier to fabricate than metal–matrix, carbon–matrix and ceramic–matrix composites, whether the polymer is a thermoset or a thermoplastic. This is because of the relatively low processing temperatures required for fabricating polymer–matrix composites. For thermosets, such as epoxy, phenolic and furfuryl resin, the processing temperature typically ranges from room temperature to about 200°C; for thermoplastics, such as polyimide (PI), polyethersulfone (PES), polyetheretherketone (PEEK), polyetherimide (PEI) and polyphenyl sulfide (PPS), the processing temperature typically ranges from 300°C to 400°C.

Thermosets (especially epoxy) have long been used as polymer matrices for carbon fiber composites. During curing, usually performed in the presence of heat and pressure, a thermoset resin hardens gradually owing to the completion of polymerization and the associated cross-linking of the polymer molecules. Thermoplastics have recently become important because of their greater ductility and processing speed compared to thermosets, and the recent availability of thermoplastics that can withstand high temperatures. The higher processing speed of thermoplastics is due to the fact that amorphous thermoplastics soften immediately upon heating above the glass transition temperature (T_g) and the softened material can be shaped easily. Subsequent cooling completes the processing. In contrast, the curing of a thermoset resin is a reaction which occurs gradually.

Epoxy is by far the most widely used polymer matrix for carbon fibers. Trade names of epoxy include Epon, Epi-rez and Araldite. Epoxy has an excellent combination of mechanical properties and corrosion resistance, is dimensionally stable, exhibits good adhesion and is relatively inexpensive. Moreover, the low molecular weight of uncured epoxide resins in the liquid state results in exceptionally high molecular mobility during processing. This mobility helps the resin to quickly spread on the surface of carbon fiber, for example.

Epoxy resins are characterized by having two or more epoxide groups per molecule. The chemical structure of an epoxide group is

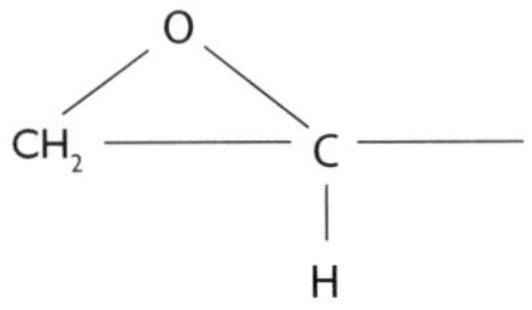

The mers (repeating units) of typical thermoplasts used for carbon fibers are shown below.

PI

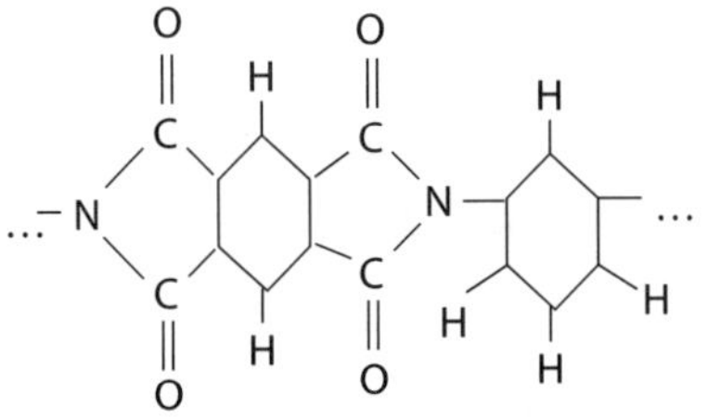

PEEK

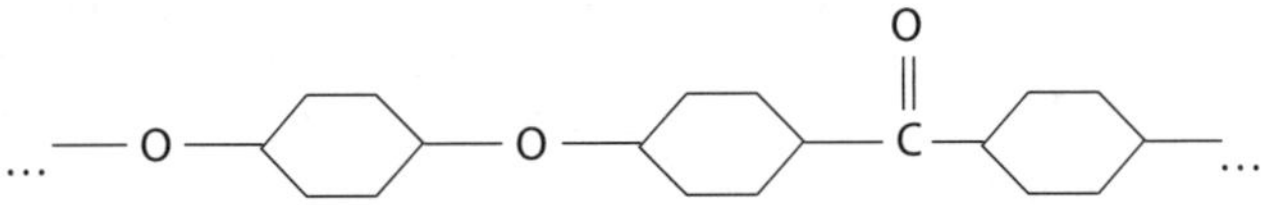

PPS

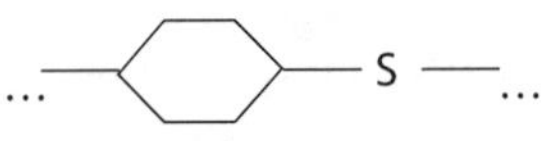

PES

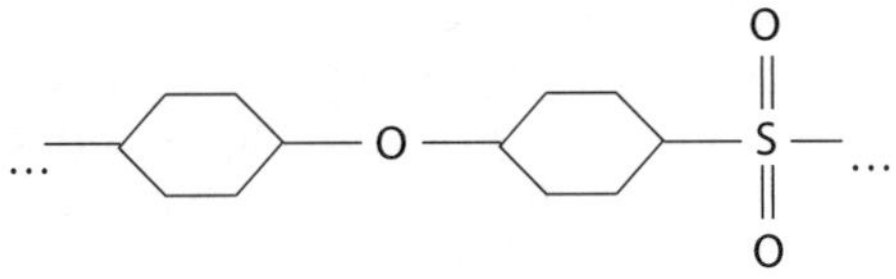

PEI

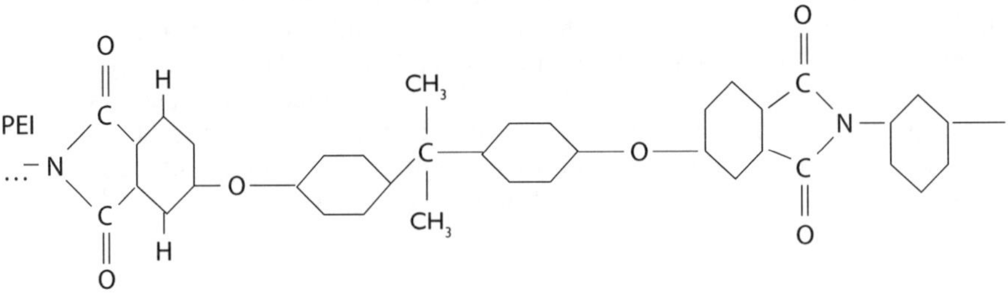

Table 2.1 Properties of thermoplastics for carbon fiber polymer-matrix composites.

	PES	PEEK	PEI	PPS	PI
T_g (°C)	230[a]	170[a]	225[a]	86[a]	256[b]
Decomposition temperature (°C)	550[a]	590[a]	555[a]	527[a]	550[b]
Processing temperature (°C)	350[a]	380[a]	350[a]	316[a]	304[b]
Tensile strength (MPa)	84[c]	70[c]	105[c]	66[c]	138[b]
Modulus of elasticity (GPa)	2.4[c]	3.8[c]	3.0[c]	3.3[c]	3.4[b]
Ductility (% elongation)	30–80[c]	50–150[c]	50–65[c]	2[c]	5[b]
Izod impact (ft lb/in.)	1.6[c]	1.6[c]	1[c]	<0.5[c]	1.5[c]
Density (g/cm³)	1.37[c]	1.31[c]	1.27[c]	1.3[c]	1.37[b]

[a]From Ref. 1.
[b]From Ref. 2.
[c]From Ref. 3.

The properties of the above thermoplastics are listed in Table 2.1 [1–3]. In contrast, epoxies have moduli of elasticity of 2.8–3.4 GPa, ductilities (elongation at break) of 0–6% and a density of 1.25 g/cm³ [3]. Thus, epoxies are much more brittle than PES, PEEK and PEI. In general, the ductility of a semi-crystalline thermoplastic decreases with increasing crystallinity. For example, the ductility of PPS can range from 2% to 20%, depending on the crystallinity [4]. Another major difference between thermoplastics and epoxies lies in the higher processing temperatures of thermoplastics (300–400°C).

2.2.2 Interface engineering

Surface treatment of a reinforcement is valuable for improving the bonding between the reinforcement and the polymer matrix. In the case of the reinforcement being carbon fibers, surface treatments involve oxidation treatments and the use of coupling agents, wetting agents and/or sizings (coatings). Carbon fibers need treatment both for thermosets and thermoplastics. As the processing temperature is usually higher for thermoplastics than thermosets, the treatment must be stable to a higher temperature (300–400°C) when a thermoplastic is used.

Oxidation treatments can be applied by gaseous, solution, electrochemical and plasma methods. Oxidizing plasmas include those involving oxygen [5–8], CO_2 [9] and air [5,10]. The resulting oxygen-containing functional groups on the fiber surface cause improvement in the wettability of the fiber and in the fiber–matrix adhesion. The consequence is enhancement of the *interlaminar shear strength* (ILSS) and flexural strength. Other plasmas (not necessarily oxidizing) that are also effective involve nitrogen [5], acrylonitrile [5] and trimethyl silane [11]. Akin to plasma treatment is ion beam treatment, which involves oxygen or nitrogen ions [12]. Plasma treatments are useful for epoxy as well as thermoplastic matrices [6,13–16]. Oxidation by gaseous methods includes the use of oxygen gas containing ozone [17]. Oxidation by solution methods involves wet oxidation [18,19], such as acid treatments [6,20,21]. Oxidation by electrochemical methods (i.e., anodic treatment) includes the use of ammonium sulfate solutions [22], a diammonium hydrogen phosphate solution containing ammonium rhodanide [23], ammonium bicarbonate solutions [24,25], a phosphoric acid solution [26] and other aqueous electrolytes [27–30]. In general, the various treatments provide chemical modification of the fiber surface, in addition to removal of a loosely adherent surface layer [31–36]. More severe oxidation treatments also serve to roughen the fiber surface, thereby enhancing the mechanical interlocking between the fibers and the matrix [37,38].

Table 2.2 [39] shows the effect of gaseous and solution oxidation treatments on the mechanical properties of high-modulus carbon fibers and their epoxy–matrix composites. The treatments degrade the fiber properties but improve the composite properties. The most effective treatment in Table 2.2 is refluxing in a 10% $NaClO_3$/25% H_2SO_4 mixture for 15 min, as this treatment results in a fiber weight loss of 0.2%, a fiber tensile strength loss of 2%, a composite flexural strength gain of 5% and a composite ILSS gain of 91%. Epoxy-embedded single-fiber tensile testing showed that anodic oxidation of pitch-based carbon fibers in ammonium sulfate solutions increased the interfacial shear strength by 300% [40]. As the modulus of the fiber increases, progressively longer treatment times are required to attain the same improvement in ILSS. Although the treatment increases ILSS, it decreases the impact strength (i.e., impact energy), so the treatment time must be carefully controlled in order to achieve a balance in properties. The choice of treatment time also depends on the particular fiber–resin combination used. For a particular treatment, as the modulus of the fiber increases, the treatment's positive effect on the ILSS and its negative effect on the impact strength both become more severe [39].

Less common methods of modification of the carbon fiber surface involve the use of gamma-ray radiation for surface oxidation [41], and the application of coatings by electrochemical polymerization (e.g., polyphenylene oxide (PPO) as a coating) [42–45], plasma polymerization [43,44,46–49], vapor deposition polymerization [50] and other polymerization techniques [51–54]. A relatively

Table 2.2 Effects of various surface treatments on properties of high-modulus carbon fibers and their epoxy–matrix composites.

Fiber treatment	Fiber properties		Composite properties	
	Wt loss (%)	Tensile strength loss (%)	Flexural strength loss (%)	ILSS gain (%)
400°C in air (30 min)	0	0	0	18
500°C in air (30 min)	0.4	6	12	50
600°C in air (30 min)	4.5	50	Too weak to test	–
60% HNO_3 (15 min)	0.2	0	8	11
5.25% NaOCl (30 min)	0.4	1.5	5	30
10–15% NaOCl (15 min)	0.2	0	8	6
15% $HClO_4$ (15 min)	0.2	0	12	0
5% $KMnO_4$/10% NaOH (15 min)	0.4	0	15	19
5% $KMnO_4$/10% H_2SO_4 (15 min)	6.0(+)	17	13	95
10% H_2O_2/20% H_2SO_4 (15 min)	0.1	5	14	0
42% HNO_3/30% H_2SO_4 (15 min)	0.1	0	4(+)	0
10% $NaClO_3$/15% NaOH (15 min)	0.2	0	12	12
10% $NaClO_3$/25% H_2SO_4 (15 min)	0.2	2	5(+)	91
15% $NaClO_3$/40% H_2SO_4 (15 min)	0.7	4	15	108
10% $Na_2Cr_2O_7$/25% H_2SO_4 (15 min)	0.3	8	15(+)	18
15% $Na_2Cr_2O_7$/40% H_2SO_4 (15 min)	1.7	27	31	18

All liquid treatments at reflux temperature.
From Ref. 39.

simple coating technique is solution dipping [43], as in the case of coating with polyurethane [55]. Tough compliant thermoplastic coatings [56], such as polyurethane [55], silicone [57,58] and polyvinyl alcohol [59], are attractive for enhancing the toughness of the composites. Other thermoplastic coatings, such as polyetherimide (PEI), improve the epoxy matrix properties through diffusion into the epoxy matrix [60]. Still other coatings serve as coupling agents between the fiber and the matrix, due to the presence of reactive functional groups, such as epoxy groups [61] and amine groups [62]. Metal (e.g., nickel and copper) coatings are used not only for enhancing the electrical properties [63,64], but also for improving the adhesion between fiber and the epoxy matrix [65].

2.2.3 Classification

Polymer–matrix composites can be classified according to whether the matrix is a thermoset or a thermoplastic. Thermoset–matrix composites are by tradition far more common, but thermoplastic–matrix composites are under rapid development. The advantages of thermoplastic–matrix composites compared to thermoset–matrix composites include the following:

Lower manufacturing cost

- no cure
- unlimited shelf-life
- reprocessing possible (for repair and recycling)
- fewer health risks due to chemicals during processing
- low moisture content
- thermal shaping possible
- weldability (fusion bonding possible)

Better performance

- high toughness (damage tolerance)
- good hot/wet properties
- high environmental tolerance

The disadvantages of thermoplastic–matrix composites include the following:

- limitations in processing methods
- high processing temperatures
- high viscosities
- prepreg (collection of continuous fibers aligned to form a sheet which has been impregnated with the polymer or polymer precursor) being stiff and dry when solvent is not used (i.e., not drapeable or tacky)
- fiber surface treatments less developed

Fibrous polymer–matrix composites can be classified according to whether the fibers are short or continuous. Continuous fibers have much more effect than short fibers on the composite's mechanical properties, electrical resistivity, thermal conductivity and other properties. However, they give rise to composites that are more anisotropic. Continuous fibers can be in unidirectionally aligned tape or woven fabric form.

2.2.4 Fabrication

Short-fiber or particulate composites are usually fabricated by mixing the fibers or particles with a liquid resin to form a slurry, and then molding to form a composite. The liquid resin is the unpolymerized or partially polymerized matrix material in the case of a thermoset; it is the molten polymer or the polymer dissolved in a solvent in the case of a thermoplastic. The molding methods are those conventionally used for polymers by themselves. For thermoplastics, the methods include *injection molding* (heating above the melting temperature of the thermoplastic and forcing the slurry into a closed die opening by using a screw mechanism), *extrusion* (forcing the slurry through a die opening by using a screw mechanism), *calendering* (pouring the slurry into a set of rollers with a small opening between adjacent rollers to form a thin sheet) and *thermoforming* (heating above the softening temperature of the thermoplastic and forming over a die (using matching dies, a vacuum, or air pressure), or without a die (using movable rollers)). For thermosets, *compression molding* or *matched die molding* (applying a high pressure and temperature to the slurry in a die to harden the thermoset) is commonly used. The casting of the slurry into a mold is not usually suitable because the difference in density between the resin and the fibers causes the fibers to float or sink, unless the viscosity of the resin is carefully adjusted. For forming a composite coating, the fiber–resin or particle–resin slurry can be sprayed instead of molded.

Instead of using a fiber–resin slurry, short fibers in the form of a mat or a continuous spun staple yarn can be impregnated with a resin and shaped using methods commonly used for continuous fiber composites. By using spun staple yarns from the Heltra process, researchers have produced epoxy–matrix composites that retain 97% of the tensile modulus and 70% of the tensile strength of their counterparts containing continuous carbon fiber tows. This is in spite of the discontinuity and slight twist in the fibers of the staple yarns [66].

Yet another method involves using continuous staple yarns in the form of an intimate blend of short carbon fibers and short thermoplastic fibers. The yarns may be woven, if desired. They do not need to be impregnated with a resin to form a composite, as the thermoplastic fibers melt during consolidation under heat and pressure [66].

Continuous fiber composites are commonly fabricated by hand lay-up of unidirectional fiber tapes or woven fabrics and impregnation with a resin. The molding, called *bag molding*, is done by placing the tapes or fabrics in a die and introducing high-pressure gases or a vacuum via a bag to force the individual plies together. Bag molding is widely used to fabricate large composite components for the skins of aircraft.

A method for forming unidirectional fiber composite parts with a constant cross-section (e.g., round, rectangular, pipe, plate, I-shaped) is *pultrusion*, in which fibers are drawn from spools, passed through a polymer resin bath for impregnation, and gathered together to produce a particular shape before entering a heated die.

A method for forming continuous fiber composites of intricate shapes is *resin transfer molding* (RTM), in which a fiber preform (usually prepared by braiding and held under compression in a mold) is impregnated with a resin. The resin is admitted at one end of the mold and is forced by pressure through the mold and preform. Subsequently the resin is cured. This method is limited to resins of low viscosity, such as epoxy. A problem with this process is the formation of surface voids by volatilization of dissolved gases in the resin, partial evaporation of mold releasing agent into the preform or mechanical entrapment of gas bubbles [67].

A method for forming continuous fiber composites in the shape of cylinders or related objects is *filament winding*, which involves wrapping continuous fibers from a spool around a form of mandrel. The fibers can be impregnated with a resin before or after winding. Filament winding is used to make pressure tanks. The winding pattern is a part of the composite design. The temperature of the mandrel, impregnation temperature of the resin, impregnation time, tension of the fibers and pressure of the fiber winding are processing parameters that need to be controlled. For the case of wet winding of carbon fibers to make epoxy–matrix composites, the optimum temperature of the mandrel is 70–80°C, the optimum resin impregnation temperature is 80–85°C, the sufficient impregnation time is 1–2 s, the recommended fiber tension is 8.3–16.6 MPa and the recommended pressure of the winding is 6–8 MPa [68]. In the case of a PEEK–matrix composite, hot air can be used to heat fibers impregnated with the thermoplastic, although infrared heating is more effective; a winding speed of up to 0.5 m/s has been reported [69].

In most of the composite fabrication methods mentioned above, impregnation of the fibers with a resin is involved. In the case of a thermoset, the resin is a liquid that has not been polymerized or is partially polymerized. In the case of a thermoplastic, the resin is either the polymer melt or the polymer dissolved in the solvent. After resin application, solid thermoplastic results from solidification in the case of melt impregnation, and from evaporation in the case of solution impregnation [70]. Both amorphous and semi-crystalline thermoplastics can be melt processed, but only the amorphous resins can normally be dissolved. Because of the high melt viscosities of semi-crystalline thermoplastics (e.g., about 370 Pa.s for PEEK at 370°C, compared to about 0.39 Pa.s for low-viscosity epoxy) due to their long and rigid macromolecular chains, direct melt impregnation of semi-crystalline thermoplastics is difficult [66]. Melt impregnation followed by solidification produces a thermoplastic prepreg that is stiff and lacks tack; solution impregnation usually produces prepregs that are drapeable and tacky, although this character changes as solvent evaporation occurs from the solution. The drapeable and tacky character of thermoplastic prepregs made by solution impregnation is comparable to that of thermoset prepregs. Hence, the main problem with resin impregnation occurs for semi-crystalline thermoplastics.

Instead of thermoplastic impregnation of fibers by using a melt or a solution of the thermoplastic, solid thermoplastic in the form of powder, fibers or slurries can be impregnated [71]. For example, carbon fibers can be immersed in a suspension of a thermoplastic powder in an aqueous liquid medium (which contains at least 20 wt.% of an organic liquid) to impregnate the thermoplastic into the fibers [72].

An alternative to impregnation is the commingling of continuous reinforcing fibers with continuous thermoplastic fibers. This commingling can be on a fabric level, where yarns of different materials are woven together (coweaving); it can be on a yarn level, where yarns of different materials are twisted together; or it can be on a fiber level, where fibers of different materials are intimately mixed within a unidirectional fiber bundle [73]. During processing, such as compression molding or filament winding, the thermoplastic fibers melt, wet the carbon fibers and fuse to form the matrix [74]. However, there is a preferred orientation in the thermoplastic fibers, due to the spinning process used in their production, and this may be a problem. Furthermore, the thermoplastic fibers have a tendency to form drops during heating [75]. In addition, the availability of high-temperature thermoplastic fibers is limited. PEEK is most commonly used for commingling with carbon fibers, but processing must be carried out at a sufficiently high temperature to destroy the previous thermal history of the PEEK matrix [76]. Fiber–matrix adhesion in a commingled system depends on the molding temperature, residence time at the melt temperature and the cooling rate. This is probably due to several complex mechanisms such as matrix adsorption on the fiber surface, matrix degradation leading to chemical bonding and interfacial crystallization [76]. On the other hand, prepregs made from commingled fibers are flexible and drapeable [66], and the use of three-dimensional braiding allows net structural shape formation and enhances the damage tolerance due to lack of delamination [73]. To prevent the ends of the braided preform from unbraiding, the thermoplastic fibers are melted with a soldering gun before cutting, or alternatively, the ends of the braided preform are wrapped with a polyimide tape and cut through the tape. The fiber commingling makes possible a uniform polymer distribution even when the three-dimensional preform is very large, although the heating time during consolidation needs to be longer for dense three-dimensional commingled fiber network braids than for unidirectional prepregs [73].

The shaping of thermoplastic–matrix composite laminates can be performed by thermoforming in the form of *matched-die forming* [77] or *die-less forming* [78]. However, in addition to shaping, deformations in the form of transverse fiber flow (shear flow perpendicular to the fiber axis) and interply slip commonly occur, while intraply slip is less prevalent [77]. Die-less forming uses an adjustable array of universal, computer-controlled rollers to form an initially flat composite material into a long, singly curved part having one arbitrary cross-sectional shape at one end and another arbitrary shape at the other end. Heating and bending of the material are strictly local processes, occurring only within a small active forming zone at any one instant. The initially flat workpiece is translated back and forth along its length in a number of passes. On successive passes, successive portions of the transverse extent of the workpiece pass through the active forming region [78]. Induction heating is used to provide the local heating in die-less forming, because it enables rapid, non-contact, localized and uniform through-thickness heating [79].

The schedule for variation of the temperature and pressure during curing and consolidation of prepregs to form a thermoset–matrix composite must be carefully controlled. *Curing* refers to the polymerization and cross-linking relations that occur upon heating and lead to the polymer, whereas *consolidation* refers to the application of pressure to obtain proper fiber–matrix bonding, low void content, and the final shape of the part. Curing and consolidation are usually performed together as one process. For example, the curing and consolidation of a polyimide–matrix composite involves first heating without pressure at 220°C for 120 min, when melting and imidization occur, and then raising the temperature to 315°C, when the resin initially exhibits melt-flow behavior then solidification. Pressure is applied at the beginning of the 315°C heating stage [80].

The curing of a thermoset–matrix composite requires heat, which is usually obtained by resistance heating, although microwave heating is also possible [81]. An attraction of microwave heating is an increase in the amount of chemical interaction between the carbon fiber surface and the epoxy resin and amine components of the matrix [82].

For thermoplastic–matrix composites, increasing the cooling rate after lamination decreases the crystallinity of the polymer matrix. For cooling rates from 1°F/min to 1000°F/min, a PEEK matrix in the presence of carbon fibers has a crystallinity ranging from 45 to 30 wt.%. Because the fibers act as nucleation sites for polymer crystallization when the polymer melt is sheared, the presence of fibers enhances the polymer crystallinity to a level above that of the neat polymer [83,84]. A greater crystallinity is associated with a higher level of fiber–matrix interaction [85]. Crystallinity can be increased by annealing at up to 310°C; the presence of carbon fibers accelerates the annealing effect [86].

Because of the high processing temperatures (up to 400°C) of high-temperature thermoplastics, traditional tooling materials are not very suitable. Instead of metal tooling materials, carbon fiber polyimide–matrix composites have successfully been used to fabricate parts from prepregs based on polyimide, PEEK, biomaleimides, etc. [87]. The advantages of the composite tooling lies in its low thermal expansion coefficient as well as its thermal stability.

2.2.5 Properties

Carbon fiber polymer–matrix composites have the following attractive properties [88]:

- low density (40% lower than aluminum)
- high strength (as strong as high-strength steels)
- high stiffness (stiffer than titanium, yet much lower in density)
- good fatigue resistance (a virtually unlimited life under fatigue loading)
- good creep resistance
- low friction coefficient and good wear resistance (a 40 wt.% short carbon fiber nylon–matrix composite has a friction coefficient nearly as low as Teflon and unlubricated wear properties approaching those of lubricated steel)
- toughness and damage tolerance (can be designed by using laminate orientation to be tougher and much more damage tolerant than metals)
- chemical resistance (chemical resistance controlled by the polymer matrix)
- corrosion resistance (impervious to corrosion)
- dimensional stability (can be designed for zero coefficient of thermal expansion)
- vibration damping ability (excellent structural damping when compared with metals)
- low electrical resistivity
- high electromagnetic interference (EMI) shielding effectiveness
- high thermal conductivity

2.3 Cement–matrix composites

Cement–matrix composites include concrete, which is one having a fine aggregate (sand), a coarse aggregate (gravel) and optionally other additives (called admixtures). Concrete is the most widely used civil structural material. When the coarse aggregate is absent, the composite is known as a mortar, which is used in masonry (for joining bricks) and for filling cracks. When both coarse and fine aggregates are absent, the material is known as cement paste. Cement paste is rigid after curing, which refers to the hydration reaction involving cement (a silicate) and water to form a rigid gel.

The admixtures can be a fine particulate such as silica (SiO_2) fume for decreasing porosity in the composite. It can be a polymer (used in either a liquid solution form or a solid dispersion form) such as latex, also for decreasing porosity. It can be short fibers (such as carbon fibers, glass fibers, polymer fibers and steel fibers) for increasing toughness and decreasing drying shrinkage (shrinkage during curing – undesirable as it can cause cracks to form). Continuous fibers are seldom used because of their high cost and the impossibility of incorporating continuous fibers in a cement mix. Due to the bidding system for many construction projects, low cost is essential.

Carbon fiber cement–matrix composites are structural materials that are gaining in importance quite rapidly owing to the decrease in carbon fiber cost [89] and the increasing demand of superior structural and functional properties. These composites contain short carbon fibers, typically 5 mm in length. However, due to the weak bond between carbon fiber and the cement matrix, continuous fibers [90–92] are much more effective than short fibers in reinforcing concrete. Surface treatment of carbon fiber (e.g., by heating [93] or by using ozone [94,95], silane [96], SiO_2 particles [97] or hot NaOH solution [98]) is useful for improving the bond between fiber and matrix, thereby improving the properties of the composite. In the case of surface treatment by ozone or silane, the improved bond is due to the enhanced wettability by water. Admixtures such as latex [94,99], methylcellulose [94] and silica fume [100] also help the bond.

The effect of carbon fiber addition on the properties of concrete increases with fiber volume fraction [101], unless the fiber volume fraction is so high that the air void content becomes excessively high [102]. (The air void content increases with fiber content and air voids tend to have a negative effect on many properties, such as compressive strength.) In addition, the workability of the mix decreases with fiber content [101]. Moreover, the cost increases with fiber content. Therefore, a rather low volume fraction of fibers is desirable. A fiber content as low as 0.2 vol.% is effective [103], although fiber contents exceeding 1 vol.% are more common [104,108]. The required fiber content increases with the particle size of the aggregate, as the flexural strength decreases with increasing particle size [109].

Effective use of carbon fibers in concrete requires dispersion of the fibers in the mix. Dispersion is enhanced by using silica fume (a fine particulate) as an admixture [102,110–112]. A typical silica fume content is 15% by weight of cement [102]. The silica fume is typically used along with a small amount (0.4% by weight of cement) of methylcellulose for helping the dispersion of the fibers and the workability of the mix [102]. Latex (typically 15–20% by weight of cement) is much less effective than silica fume for helping fiber dispersion, but it enhances the workability, flexural strength, flexural toughness, impact resistance, frost resistance and acid resistance [102,113,114]. The ease of dispersion increases with decreasing fiber length [112].

The improved structural properties rendered by carbon fiber addition pertain to the increased tensile and flexible strengths, increased tensile ductility and flexural toughness, enhanced impact resistance, reduced drying shrinkage and improved freeze-thaw durability [101-103,105-113,115-126]. The tensile and flexural strengths decrease with increasing specimen size, such that the size effect becomes larger as fiber length increases [127]. The low drying shrinkage is valuable for large structures and for use in repair [128,129] and in joining bricks in a brick structure [130,131]. The functional properties rendered by carbon fiber addition pertain to the strain sensing ability [95,132-145] (for smart structures), temperature sensing ability [146-149], damage sensing ability [132,136,150-152], thermoelectric behavior [147-149], thermal insulation ability [153-155] (to save energy for buildings), electrical conduction ability [156-165] (to facilitate cathodic protection of embedded steel and to provide electrical grounding or connection) and radio wave reflection/absorption ability [166-170] (for EMI shielding, for lateral guidance in automatic highways and for television image transmission).

In relation to the structural properties, carbon fibers compete with glass, polymer and steel fibers [106,115-117,120,124-126,171]. Carbon fibers (isotropic pitch based) [89,171] are advantageous in their superior ability to increase the tensile strength of concrete, even though the tensile strength, modulus and ductility of the isotropic pitch-based carbon fibers are low compared to most other fibers. Carbon fibers are also advantageous in the relative chemical inertness [172]. PAN-based carbon fibers are also used [105,107,111,122], although they are more commonly used as continuous fibers than short fibers. Carbon-coated glass fibers [173,174] and submicron-diameter carbon filaments [164-166] are even less commonly used, although the former is attractive for the low cost of glass fibers and the latter is attractive for its high radio wave reflectivity (which results from the skin effect). C-shaped carbon fibers are more effective for strengthening than round carbon fibers [175], but their relatively large diameter makes them less attractive. Carbon fibers can be used in concrete together with steel fibers, as the addition of short carbon fibers to steel fiber-reinforced mortar increases the fracture toughness of the interfacial zone between steel fiber and the cement matrix [176]. Carbon fibers can also be used in concrete together with steel bars [177,178], or together with carbon fiber-reinforced polymer rods [179].

In relation to most functional properties, carbon fibers are exceptional compared to other fiber types. Carbon fibers are electrically conducting, in contrast to glass and polymer fibers, which are not conducting. Steel fibers are conducting, but their typical diameter (≥ 60 μm) is much larger than the diameter of a typical carbon fiber (15 μm). The combination of electrical conductivity and small diameter makes carbon fibers superior to the other fiber types in the area of strain sensing and electrical conduction. However, carbon fibers are inferior to steel fibers for providing thermoelectric composites, owing to the high electron concentration in steel and the low hole concentration in carbon.

Although carbon fibers are thermally conducting, addition of carbon fibers to concrete lowers the thermal conductivity [153], thus allowing applications related to thermal insulation. This effect of carbon fiber addition is due to the increase in air void content. The electrical conductivity of carbon fibers is higher than that of the cement matrix by about eight orders of magnitude, whereas the thermal conductivity of carbon fibers is higher than that of the cement matrix by only one or two orders of magnitude. As a result, electrical conductivity is increased upon carbon fiber addition in spite of the increase in air void content, but thermal conductivity is decreased upon fiber addition.

2.4 Carbon–matrix composites

2.4.1 Introduction

Carbon fiber carbon–matrix composites, also called carbon–carbon composites, are the most advanced form of carbon, as the carbon fiber reinforcement makes them stronger, tougher and more resistant to thermal shock than conventional graphite. With the low density of carbon, the specific strength (strength/density), specific modulus (modulus/density) and specific thermal conductivity (thermal conductivity/density) of carbon–carbon composites are the highest among composites. Furthermore, the coefficient of thermal expansion is near zero.

The carbon fibers used for carbon–carbon composites are usually continuous and woven. Both two-dimensional and higher-dimensional weaves are used, although the latter has the advantage of an enhanced interlaminar shear strength. The carbon matrix is derived from a pitch, a resin or a carbonaceous gas. Depending on the carbonization/graphitization temperature, the resulting carbon matrix can range from being amorphous to being graphitic. The higher the degree of graphitization of the carbon matrix, the greater the oxidation resistance and the thermal conductivity, but the more brittle the material. As the carbon fibers used can be highly graphitic, it is usually the carbon matrix that limits the oxidation resistance of the composite.

The main disadvantages of carbon–carbon composites lie in the high fabrication cost, poor oxidation resistance, poor interlaminar properties (especially for two-dimensionally woven fibers), difficulty of making joints and insufficient engineering database.

Of the world market in carbon–carbon composites, 79% resides in the United States, 20% in Europe and the former USSR and 1% in Japan. The market is essentially all aerospace, with re-entry thermal protection constituting 37%, rocket nozzles 31% and aircraft brakes 31%. Other applications include furnace heating elements, molten materials transfer, spacecraft and aircraft components and heat exchangers. Future applications include air-breathing engine components, hypersonic vehicle airframe structures, space structures and prosthetic devices.

2.4.2 Fabrication

The fabrication of carbon–carbon composites is carried out by using four main methods, namely (1) *liquid phase impregnation* (LPI), (2) *hot isostatic pressure impregnation carbonization* (HIPIC), (3) *hot pressing* and (4) *chemical vapor infiltration* (CVI).

All of the methods (except, in some cases, CVI) involve first the preparation of a prepreg by either wet winding continuous carbon fibers with pitch or resin (e.g., phenolic) or wetting woven carbon fiber fabrics with pitch or resin. Unidirectional carbon fiber tapes are not as commonly used as woven fabrics, because fabric lay-ups tend to result in more interlocking between the plies. For highly directional carbon–carbon composites, fabrics which have a greater number of fibers in the warp direction than the fill direction may be used. After prepreg preparation and, in the case of fabrics, fabric lay-up, the pitch or resin needs to be pyrolyzed or carbonized by heating at 350–850°C. Due to the shrinkage of the pitch or resin during carbonization (which is accompanied by the evolution of volatiles), additional pitch or resin is impregnated in the case of LPI and HIPIC, and carbonization is carried out under pressure in the case of hot pressing. In LPI carbonization and impregnation are carried out as distinct steps, whereas in HIPIC carbonization and impregnation are performed together as a single step.

The *carbon yield* (or *char yield*) from carbonization is around 50 wt.% for ordinary pitch and 80–88 wt.% for mesophase pitch [180] at atmospheric pressure. Although mesophase pitch tends to be more viscous than ordinary pitch, making impregnation more difficult, mesophase pitch of viscosity below 1 Pa.s at 350°C has been reported [180]. In the case of resins, the carbon yield varies much from one resin to another; for example, it is 57% for phenolic (a thermoset), 79% for polybenzimidazole (PBI, a thermoplastic with T_g = 435°C) [181] and 95% for an aromatic diacetylene oligomer [182]. Significant increases in the carbon yield of pitches can be obtained by the use of high pressure during carbonization; at a pressure in excess of 100 MPa, yields as high as 90% have been observed [183]. The higher the pressure, the more coarse and isotropic will be the resulting microstructure, probably due to the suppression of gas formation and escape during carbonization [183]. Spheres, known as *mesophase*, exhibiting a highly oriented structure similar to liquid crystals and initially around 0.1 µm in diameter, are observed in isotropic liquid pitch above 400°C. Prolonged heating causes the spheres to coalesce, solidify and form larger regions of lamellar order; this favors graphitization upon subsequent heating to ~2500°C. The high pressure during carbonization lowers the temperature at which mesophase forms [183]. At very high pressures (~200 MPa), coalescence of mesophase does not occur. Therefore, an optimum pressure is around 100 MPa [183]. Pressure may or may not be applied during carbonization in LPI but is always applied during carbonization in HIPIC.

In LPI, after carbonization, vacuum impregnation is performed with additional pitch or resin in order to densify the composite. Pressure (e.g., 2 MPa) may be applied to help the impregnation. The carbonization–impregnation cycles are repeated several times (typically three to six) in order to achieve sufficient densification. Both the density and ILSS increase with increasing number of cycles [184]. The first carbonization decreases the density from the value of the green composite, so that subsequent impregnation and recarbonization are necessary [184].

The density levels off after a few cycles of impregnation and recarbonization. This is because the repeated densification cycles cause the mouths of the pores to narrow down, so that it is difficult for the impregnant to enter the pores. As a consequence, impregnant pickup levels off. This problem can be alleviated by intermediate graphitization, wherein the composites are subjected to heat treatment at 2200–3000°C between the carbonization and impregnation steps after the densification cycle when the density levels off. On graphitization, the entrance to the pores opens up due to rearrangement of the crystallites in the matrix. These opened pores then become accessible during further impregnation, thus leading to further density increase [185–187]. The use of mesophase pitch instead of isotropic pitch for impregnation can cut down on the required number of impregnation cycles [188].

In HIPIC, an isostatic inert gas pressure of around 100 MPa is applied to impregnate pitch (rather than resins, which suffer from a low carbon yield) into the pores in the sample while the sample is being carbonized at 650–1000°C. The pressure increases the carbon yield and maintains the more volatile fractions of the pitch in a condensed phase. After this combined step of carbonization and impregnation, graphitization is performed by heating above 2200°C without applied pressure. HIPIC allows the density to reach a higher value than LPI (with or without intermediate graphitization), and a smaller number of cycles is needed for HIPIC to achieve the high density. However, HIPIC is an expensive technique.

One HIPIC process involves vacuum impregnating a dry fiber preform or porous carbon–carbon laminate with molten pitch, placing it inside a metal container (or can) with an excess of pitch surrounding it inside the can. The can is then evacuated and sealed (preferably by using an electron beam weld) and placed within the work zone of a hot isostatic press (HIP) unit. The temperature is then raised at a programmed rate above the melting point of the pitch, but not so high as to result in weight loss due to the onset of carbonization. The pressure is then increased and maintained at around 100 MPa. The pitch initially melts and expands within the can and is forced by isostatic pressure into the pores in the sample. The sealed container acts like a rubber bag, facilitating the transfer of pressure to the workpiece. After that, the temperature is gradually increased towards that required for pitch carbonization (650–1000°C). The pressure not only increases the carbon yield, but also prevents liquid from being forced out of the pores by pyrolysis products. After the HIPIC cycle is complete, the preforms are removed from their container and cleaned up by removing any excess carbonized liquid from the surface [189]. The optimum carbonization pressure for HIPIC is 1000–1500 bar (100–150 MPa). Lower pressures are insufficient to prevent bloating of the composites owing to the evolution of carbonization gases. Higher pressures do not offer any significant improvement, and even seem to be detrimental to the mechanical properties [189].

Another HIPIC process does not use a can, but simply applies an isostatic gas pressure on the surface of molten pitch, which seals the workpiece [190]. HIPIC increases the carbon yield of pitch, especially when the molecular size of the pitch is small. Table 2.3 [190] shows the carbon yield at 0.1 and 10 MPa for pitches of three different molecular weights. The increase in pressure causes the carbon yield to increase dramatically for pitch A (low molecular weight), but only slightly for pitch C (high molecular weight). This is due to the already high carbon yield of pitch C at 0.1 MPa. The improvement in carbon yield due to the use of pressure becomes saturated at a pressure of 10 MPa. The origin of the improvement is attributed to the trapping and decomposition of the evolved hydrocarbon gases under high pressure; the decomposition produces carbon and hydrogen. The increase in pressure causes bulk density to increase, porosity to decrease and flexural strength to increase. Moreover, as the pressure rises, the

Table 2.3 Pitch properties.

Pitch	Molecular weight	Carbon yield (%)	
		0.1 MPa	10 MPa
A	726	45.2	85.9
B	782	54.4	86.4
C	931	84.5	89.8

From Ref. 190.

pores in the carbonized matrices become smaller in size and more spherical in shape [190].

In hot pressing (also called high-temperature consolidation), carbonization is performed at an elevated temperature (1000°C typically, but only 650°C for an aromatic diacetylene oligomer as the matrix precursor [182]) under a uniaxial pressure (2–3 MPa typically, but 38–76 MPa for an aromatic diacetylene oligomer as the matrix precursor [182]) in an inert or reducing atmosphere, or in a vacuum. During hot pressing, graphitization may occur even for thermosetting resins, which are harder to graphitize than pitch [191]. This is known as *stress-graphitization*. Subsequently, further graphitization may be performed by heating at 2200–3000°C without applied pressure. No impregnation is performed after the carbonization. Composites made by hot pressing have flattened pores in the carbon matrix and the part thickness is reduced by about 50%. Excessive pressure (say, 5 MPa) causes the formation of vertical cracks [192].

In CVI (also called CVD, chemical vapor deposition), gas phase impregnation of a hydrocarbon gas (e.g., methane, propylene) into a carbon fiber preform takes place at 700–2000°C, so that pyrolytic carbon produced by the cracking of the gas is deposited in the open pores and surface of the preform. The carbon fiber preform can be in the form of carbon fabric prepregs which have been carbonized and graphitized, or in the form of dry wound carbon fibers. There are three CVI methods, namely the isothermal method, temperature gradient method and pressure gradient method.

In the *isothermal method*, the gas and sample are kept at a uniform temperature. As carbon growth in the pores will cease when they become blocked, there is a tendency for preferential deposition on the exterior surfaces of the sample. This causes the need for multiple infiltration cycles, such that the sample is either skinned by light machining or exposed to high temperatures to reopen the surface pores for more infiltration in subsequent cycles.

In the *temperature gradient method*, an induction furnace is used. The sample is supported by an inductively heated mandrel (a susceptor) so that the inside surface of the sample will be at a higher temperature than the outside surface. The hydrocarbon gas flows along the outside surface of the sample. Due to the temperature gradient, the deposition occurs first at the inside surface of the sample and progresses toward the outside surface, thereby avoiding the crusting problem.

In the *pressure gradient method*, the hydrocarbon gas impinges on the inside surface of the sample, so the gas pressure is higher at the inside surface than the outside surface. The pressure gradient method is not as widely used as the isothermal method or the temperature gradient method.

Both the temperature gradient method and the pressure gradient method are limited to single samples, whereas the isothermal method can handle several samples at once. However, the isothermal method is limited to thin samples because of the crusting problem.

A drawback of CVI is the low rate of deposition resulting from the use of a low gas pressure (1–150 torr), which favors a long mean free path for the reactant and decomposed gases; a long mean free path enhances deposition into the center of the sample. A diluent gas (e.g., He, Ar) is usually used to help the infiltration. For example, a gas mixture containing 3–10 vol.% propylene (C_3H_6) in Ar was used for CVI at 760–800°C [193]. Hydrogen is often used as a carbon surface detergent.

An attraction of CVI is that CVI carbon is harder than char carbon from pitch or resin, so that CVI carbon is particularly desirable for carbon–carbon composites used for brakes and friction products.

Fillers such as carbon black can be added to the resin or pitch prior to carbonization in order to provide bridging between the fibers during subsequent CVI [194].

The methods of CVI and LPI may be combined by first performing CVI on dry wound carbon fibers then performing LPI [195]. The CVI step serves to make the carbon fibers rigid prior to impregnation.

The quality of a carbon–carbon composite depends on the quality of the polymer–matrix composite from which the carbon–carbon composite is made. For example, resin pooling may result in areas of excessive shrinkage cracks after carbonization and graphitization at high temperatures [194].

During carbonization, pitch tends to bloat due to the evolution of gases generated by pyrolysis. This can cause the expulsion of pitch from the carbon fiber preform during carbonization. There are two ways to alleviate this problem. One way is oxidation stabilization, which is oxidation of the pitch at a temperature below the softening point of the pitch, i.e., generally below 300°C [196]. Another way is uniaxial pressing at 500 Pa and between room temperature and 600°C prior to carbonization at 1000°C [197].

The choice of pitch affects the carbon yield, which increases with increasing C/H ratio of the pitch [198]. Mesophase pitch gives a higher carbon yield than isotropic pitch; it can be prepared by removal of the light aromatic materials by solvent extraction with toluene [197].

The combined use of pitch and resin is also possible. Resin (epoxy) can be used to form carbon fiber prepreg sheets, which are then laminated alternately in a die with pitch, which is in the form of a mixture of coal tar pitch, coke powder and carbonaceous bulk mesophase (as binder for the coke). Carbon–carbon composites were thus prepared by hot pressing at 600°C and 49 MPa, followed by heat treatment in N_2 at 1500°C [199].

The choice of carbon fibers plays an important role in affecting the quality of the carbon–carbon composites.

The use of graphitized fibers (fired at 2200–3000°C) with a carbon content in excess of 99% is preferred, because their thermal stability reduces the part warpage during later high-temperature processing to form a carbon–carbon composite [194]. Moreover, graphitized fibers lead to better densified carbon–carbon composites than carbon fibers which have not been graphitized [200]. This is because the adhesion of the fibers with the polymer resin is weaker for graphitized fibers, so that, on carbonization, the matrix can easily shrink away from the fibers, leaving a gap which can be filled in during subsequent impregnation [200]. In contrast, the polar surface groups on carbonized fibers make strong bonds with the resin (phenolic), thus inhibiting the shrinking away of the charred matrix from the fibers and leading to the formation of fine microcracks in the carbon matrix [200].

Circular fibers are preferred to irregularly shaped fibers, as the latter leads to stress concentration points in the matrix around the fiber corners. Microcrack initiation occurs at these points, thus resulting in low strength in the carbon–carbon composite [201].

The microstructure of mesophase pitch-based carbon fibers influences the physical changes that take place during graphitization of the carbon–carbon composite made with these fibers. For medium-modulus carbon fibers having parallel graphite planar sheet-like microstructure, the prestressed carbon matrix shears and orders itself in the fiber direction during graphitization, thereby stretching the fibers. This causes an expansion of the composite in the fiber direction and an increase in the flexural strength of the composite after graphitization. In contrast, for mesophase pitch-based carbon fibers having sheath- and core-type microstructures, the composite does not expand in the fiber direction and the flexural strength of the composite decreases upon graphitization [202].

The weave pattern of the carbon fabric affects densification. The 8H satin weave is preferred over plain weave because of the inhomogeneous matrix distribution around the crossed bundles in the plain weave. Microcracks develop beneath the bundle crossover points. After carbonization, composites made with the plain weave fabric show nearly catastrophic failure with bundle pullout, whereas those made with the satin weave fabric shown shear-type failure with fiber pullout. On densification, the flexural strength of composites made with satin weave fabric increases appreciably, whereas only marginal improvement is obtained in composites made with plain weave fabric [200].

The fiber–matrix bond strength in carbon–carbon composites must be optimal. If the bond strength is too high, the resulting composite may be extremely brittle, exhibiting catastrophic failure and poor strength. If it is too low, the composites fail in pure shear, with poor fiber strength translation [203]. Thus, among (1) non-surface-treated unsized carbon fibers (too low in bond strength), (2) non-surface-treated sized carbon fibers (optimum), (3) surface-treated unsized carbon fibers (too high in bond strength) and (4) surface-treated sized carbon fibers (too high in bond strength), the non-surface-treated sized carbon fibers give carbon–carbon composites of the highest strength. These optimum composites fail in a mixed mode fracture [203]. Similarly, carbon fibers that have been oxygen plasma treated and then argon plasma detreated (optimum) give carbon–carbon composites of higher strength than carbon fibers that have been oxygen plasma treated (but not argon plasma detreated) (too high in bond strength) and than those which have not been treated at all (too low in bond strength) [204]. Moreover, carbon fibers which have been oxidized by nitric acid and then detreated in an argon plasma give composites of higher strength than those which have been oxidized in nitric acid but not detreated [204].

Surface treatment (say, with concentrated nitric acid) of carbon fibers increases the concentration of surface groups, thus strengthening the fiber–matrix bonding. The strengthened fiber–matrix bonding makes it more difficult for the matrix to shrink away from the fiber surface during carbonization, so the fibers get pulled together by the matrix shrinkage and a large composite cross-sectional shrinkage results.

Polyarylacetylene (PAA) is a resin that is typically non-graphitizing, but it is attractive in its high char yield at 88% (compared to a char yield of ~50% for the phenolic resin) and it has easy processability compared to pitch [205].

2.4.3 Oxidation protection

In the absence of oxygen, carbon–carbon composites have excellent high-temperature strength, modulus and creep resistance. For example, the carbon–carbon composites used for the nose cap of the Space Shuttle can withstand 1600°C, whereas more advanced carbon–carbon composites can withstand 2200°C. In contrast, superalloys can withstand only 1200°C and also suffer from having high densities.

Carbon–carbon composites are highly susceptible to oxidation at temperatures above 320°C [206]. The predominant reaction that occurs in air is

$$C + O_2 \rightarrow CO_2 \uparrow.$$

This reaction is associated with a very large negative value of Gibbs free energy change, so it proceeds with a big driving force even at very low O_2 partial pressures [207]. Thus the rate of oxidation is controlled not by the chemical reaction itself, but by transport of the gaseous species to and away from the reaction front [207].

The oxidation of carbon–carbon composites preferentially attacks the fiber–matrix interfaces and weakens the fiber bundles. The unoxidized material fails catastrophically by delamination cracking between plies and at bundle–bundle interfaces within plies. As oxidation progresses, failure becomes a multistep process with less delamination cracking and more cross-bundle cracking. This change of failure mode with oxidation is attributed to more severe attack within bundles than at bundle–bundle interfaces. For a weight loss on oxidation of 10%, the reductions in elastic modulus and flexural strength were 30% and 50%, respectively [208].

Oxidation protection of carbon–carbon composites up to 1700°C involves various combinations of four methods:

1. SiC coatings applied by pack cementation, reaction sintering, silicone resin impregnation/pyrolysis or chemical vapor deposition (CVD) to the outer surface of the composite.
2. Oxidation inhibitors (oxygen getters and glass formers) introduced into the carbon matrix during lay-up and densification cycles to provide additional oxidation protection from within by migrating to the outer surface and sealing cracks and voids during oxidation.
3. Application of a glassy sealant on top of the SiC conversion coating mainly by slurry brush-on, so that the sealants melt, fill voids and stop oxygen diffusion, and, in some cases, act as oxygen getters.
4. Dense SiC or Si_3N_4 overlayers applied by CVD on top of the glassy sealant, if a glassy sealant is used, or otherwise on top of the SiC conversion coating, to control and inhibit the transfer of oxygen to the substrate and to control venting of reaction products to the outside [206,209].

A SiC coating in method 1, known as SiC conversion coating, is graded in composition so that it shades off gradually from pure silicon compounds on the outside surface to pure carbon on the inside [210]. The gradation minimizes spallation resulting from the thermal expansion mismatch between SiC and the carbon–carbon composite. The conversion coating can also be made to be gradated in porosity so that it is denser near the outside surface [206]. The SiC conversion coating is applied by pack cementation, which involves packing the composite in a mixture of silicon carbide and silicon powders, then heating this assembly up to 1600°C. During this process, primarily the following reactions take place [211]:

$$\text{Si}(l) + \text{C} \rightarrow \text{SiC}$$
$$\text{Si}(g) + \text{C} \rightarrow \text{SiC}$$
$$\text{SiO}(g) + 2\text{C} \rightarrow \text{SiC} + \text{CO}(g).$$

The net result is the chemical conversion of the outermost surfaces of the composite to silicon carbide. Typical thicknesses of pack cementation coatings range from 0.3 to 0.7 mm [206]. One disadvantage of this process is that elemental silicon may be trapped in the carbon matrix under the conversion coating. The entrapped silicon tends to vaporize at elevated temperatures and erupt through the coating, leaving pathways for oxygen to migrate to the carbon–carbon substrate [206].

A second method to form a SiC coating is reaction sintering, which involves dipping a carbon–carbon composite into a suspension of silicon powder (average 10 μm size) in an alcohol solution, then sintering at 1600°C for 4 h in argon [212].

A third method to form a SiC coating involves vacuum impregnating and cold isostatic pressing (30,000 psi or 200 MPa) a silicone resin into the matrix of a carbon–carbon composite and subsequent pyrolysis at 1600°C for 2 h in argon [212].

The SiC overlayer in method 4 is more dense than the SiC conversion coating in method 1. It serves as the primary oxygen barrier [209]. It is prepared by CVD, which involves the thermal decomposition/reduction of a volatile silicon compound (e.g., CH_3SiCl, CH_3SiCl_3) to SiC. The reaction is of the form [213]

$$CH_3SiCl_3 \xrightarrow{\text{Heat}/H_2} \text{SiC} + 3\text{HCl}(g).$$

The decomposition occurs in the presence of hydrogen and heat (e.g., 1125°C [214]). If desired, the overlayer can be deposited so that it contains a small percentage of unreacted silicon homogeneously dispersed in the SiC [211]. The excess silicon upon oxidation becomes SiO_2, which has a very low oxygen diffusion coefficient. Such silicon-rich SiC is abbreviated SiSiC. Instead of SiC, Si_3N_4 may be used as the overlayer; Si_3N_4 can also be deposited by CVD.

Silicon-based ceramics (SiC and Si_3N_4), among high-temperature ceramics, have the best thermal expansion compatibility with respect to carbon–carbon composites and exhibit the lowest oxidation rates. Moreover, the thin amorphous SiO_2 scales that grow have low oxygen diffusion coefficients and can be modified with other oxides to control the viscosity [215]. Above 1800°C, these silicon-based ceramic coatings cannot be used because of the reactions at the interface between the SiO_2 scale and the underlying ceramic. Furthermore, the reduction of SiO_2 by carbon produces CO(g) [215].

The glass sealants in method 3 are in the form of glazes comprising usually silicates and borates. If desired, the glaze can be filled with SiC particles [215]. The sealant is particularly important if the SiC conversion coating is porous. Moreover, when microcracks develop in the dense overlayer, the sealant fills the microcracks. Borate (B_2O_3) glazes wet C and SiC quite well, but they are useful up to 1200°C owing to volatilization [216,217]. Moreover, B_2O_3 has poor moisture resistance at ambient temperatures as it undergoes hydrolysis, which results in swelling and crumbling [215]. In addition, B_2O_3 has a tendency to galvanically corrode SiC coatings at high temperatures [218]. However, these problems of B_2O_3 can be alleviated by the use of multicomponent systems, such as $10TiO_2 \cdot 20SiO_2 \cdot 70B_2O_3$ [218]. TiO_2 has a high solubility in B_2O_3; it is used to prevent volatilization of B_2O_3 and increase the viscosity over a wide temperature range. The SiO_2 component acts to increase the moisture resistance at ambient temperatures, reduce B_2O_3 volatility at high temperatures, increase the overall viscosity of the sealant, and prevent galvanic corrosion of the SiC at high temperatures by the B_2O_3 [218].

The inhibitors in method 2 are added to the carbon matrix by incorporation as particulate fillers in the resin or pitch prior to prepregging. They function as oxygen getters and glass formers. These fillers can be in the form of elemental silicon, titanium and boron. Oxidation of these particles within the carbon matrix forms a viscous glass, which serves as a sealant that flows into the microcracks of the SiC coating, covering the normally exposed carbon–carbon surface to prevent oxygen ingress into the carbon–carbon [218]. The mechanism of oxidation inhibition by boron-based inhibitors may involve B_2O_2, a volatile suboxide that condenses to B_2O_3 upon encountering a locally high oxygen partial pressure in coating cracks [219]. Instead of using elemental Ti and Si, a combination of SiC, Ti_5Si_3 and TiB_2 may be used [220]. For a more uniform distribution of the glass sealant, the filler ingredients may be pre-reacted to form alloys such as Si_2TiB_{14} prior to addition to the resin or pitch [220]. Yet another way to obtain the sealant is to use an organoborosilazane polymer solution [221].

The addition of glass-forming additives such as boron, silicon carbide and zirconium boride to the carbon matrix can markedly reduce the reactivity of the composite with air, but the spreading of the glassy phase throughout the composite is slow and substantial fractions of the composite are gasified before the inhibitors become effective. Thus, in the absence of an exterior impermeable coating, the oxidation protection afforded at temperatures above 1000°C by inhibitor particles added to the carbon matrix is strictly limited [222].

The inhibition mechanism of B_2O_3 involves the blockage of active sites (such as the edge carbon atoms) for small inhibitor contents and the formation of a mobile diffusion barrier for oxygen when the B_2O_3 amount is increased [223,224]. The inhibition effect of B_2O_3 is most pronounced at the beginning of oxidation, as shown by the small slope of the weight loss curves near time zero. Thereafter a pseudolinear oxidation regime takes place, as for the untreated composite. The inhibition factor is defined as the ratio of the oxidation rate of the untreated carbon to that of the treated carbon.

For oxidation protection above 1700°C, a four-layer coating scheme is available. This scheme consists of a refractory oxide (e.g., ZrO_2, HfO_2, Y_2O_3, ThO_2) as the outer layer for erosion protection, an SiO_2 glass inner layer as an oxygen barrier and sealant for cracks in the outer coating, followed by another refractory oxide layer for isolation of the SiO_2 from the carbide layer underneath, and finally a refractory carbide layer (e.g., TaC, TiC, HfC, ZrC) to interface with the carbon–carbon substrate and to provide a carbon diffusion barrier for the oxide to prevent carbothermic reduction. The four-layer system is thus refractory oxide/SiO_2 glass/refractory oxide/refractory carbide [183,215]. It should be noted that ZrO_2, HfO_2, Y_2O_3 and ThO_2 have the required thermal stability for long-term use at >2000°C, but they have very high oxygen permeabilities; silica exhibits the lowest oxygen permeability and is the best candidate as an oxygen barrier other than iridium at >1800°C; iridium suffers from a relatively high thermal expansion coefficient, high cost and limited availability [215].

A ternary $HfC-SiC-HfSi_2$ system deposited by CVD has been reported to provide good oxidation protection up to 1900°C [221]. The HfC component is chemically compatible with carbon. Furthermore, HfO_2 forms from HfC by the reaction

$$HfC + \tfrac{3}{2}O_2 \rightarrow HfO_2 + CO$$

and HfO_2 is a very stable oxide at high temperatures. However, HfO_2 undergoes a phase change from monoclinic to tetragonal at 1700°C, with a volume change of 3.4%. To avoid catastrophic failure due to the volume change, HfO_2 is stabilized through the addition of $HfSi_2$. The SiC component acts as a diffusion barrier [225].

Pack cementation is a relatively inexpensive technique for coating carbon–carbon composites in large quantities. The quality of SiC coatings prepared by pack cementation can be improved by first depositing a 10 μm carbon film by CVD on to the surface of the carbon–carbon composite, because the carbon film improves the homogeneity of the carbon–carbon surface and eases the reaction with Si [226]. Similarly, carbon CVD can be used to improve SiC films deposited by reaction sintering or resin impregnation [212]. The carbon CVD involves the pyrolysis of methane in a tube furnace at 1300°C [212].

Pack cementation has been used to form chromium carbide coatings in addition to SiC coatings for oxidation protection of carbon–carbon composites. For chromium carbide coatings, the carbon–carbon composite sample is packed in a mixture of chromium powder, alumina powder and a small quantity of NH_4Cl (an activator) and reacted at 1000°C in argon. The chromium powder produces chromium carbide by reaction with the carbon–carbon composite. At the same time, HCl dissociated from NH_4Cl reacts with the chromium powder to form chromous halide liquid, which reacts with the carbon–carbon composite to form chromium carbide. The latter kind of chromium carbide permeates the openings in the former kind of chromium carbide. Upon oxidation of the chromium carbide coating, a dense layer of Cr_2O_3 is formed and serves to prevent oxidation of the carbon–carbon composite [227].

Another form of oxidation protection can be provided by treatments of carbon–carbon composites by various acids [228] and bromine [229].

The fundamental approaches for oxidation protection of carbons can be categorized into four groups [230]: (1) prevention of catalysis, (2) retardation of gas access to the carbon, (3) inhibition of carbon–gas reactions, and (4) improvement in the carbon crystallite structure. Approach 2 is the dominant one applied to carbon–carbon composites, as it provides the greatest degree of oxidation protection. However, the other approaches need to be exploited as well. In particular, approach 4 means that pitch and CVI carbon are preferred to resins as precursors for carbon–carbon matrices, as pitch and CVI carbon are more graphitizable than resins [231]. Nevertheless, the stress applied to the matrix by the adjacent fibers during carbonization causes alignment of the matrix molecules near the fibers [232]. Furthermore, the microstructure of the carbon fibers affects strongly the microstructure of the carbon matrix, even when the fiber fraction is only about 50 wt.%, and the microstructure of the carbon matrix affects the amount of accessible porosity, thereby affecting the oxidation behavior [233].

The application of coatings on carbon–carbon composites can deteriorate the room temperature mechanical properties of carbon–carbon composites. For example, after application of a 0.25–0.50 mm thick SiC conversion coating, the flexural strength decreases by 29% [206]. On the other hand, oxidation of a carbon–carbon composite to a burn-off of 20% causes the flexural strength to decrease by 64% [234].

2.4.4 Mechanical properties

The mechanical properties of carbon–carbon composites are much superior to those of conventional graphite. Three-dimensional carbon–carbon composites are particularly attractive. Their preform structure can be tailored in three directions. The three-dimensional integrated preform structure results in superior damage tolerance and minimum delamination crack growth under interlaminar shearing compared with two-dimensional laminate carbon–carbon composites. Unlike conventional materials, the crack in three-dimensional carbon–carbon composites diffuses in a tortuous manner, probably tracking pre-existing voids or microcracks. The failure of three-dimensional composites involves a series of stable crack propagation steps across the matrix and yarn bundles, followed by unstable crack propagation. The dominant damage mechanisms are bundle breakage and matrix cracking [235].

Heat treatment temperature has a significant effect on the mechanical properties of carbon–carbon composites. Composites carbonized at 1000°C upon subsequent graphitization at 2700°C show a 54% increase in the flexural strength, a 40% decrease in the interlaminar shear strength and a 93% increase in the flexural modulus [236]. This suggests that the fiber–matrix interaction is different before and after graphitization. The increase in flexural modulus is probably due to further graphitization of the carbon fibers under the influence of the carbon matrix around them, even though the fibers have been graphitized prior to this [236]. The choice of graphitization temperature affects the toughness of the composite. For a pitch-based matrix, the optimum graphitization temperature is 2400°C, where the microstructure is sufficiently ordered to accommodate some slip from shear forces but is disordered enough to prevent long-range slip [237]. A similar graphitization temperature may be used for a polymer-based matrix [238].

The tensile and flexural properties of carbon–carbon composites are fiber dominated, whereas the compression behavior is mainly affected by density and matrix morphologies. The tensile moduli are sometimes higher than the values calculated according to the fiber content because of the contribution of the sheath matrix morphology (stress-graphitization of the matrix). Tensile strength levels are lower than the calculated values owing to the residual stresses resulting from thermal processing. A high glass-like carbon fraction in the matrix is associated with enhanced strength and modulus, both in tension and compression [239].

The effect of surface treatment of the carbon fibers is significant on the mechanical properties of the resulting carbon–carbon composites. Surface-treated fibers having strong bonding with the polymer exhibit high flexural strength in polymer composites, but result in carbonized composites of poor flexural strength. For graphitized composites, the flexural strength increases monotonically with increasing treatment time. Graphitization causes composites with surface-treated fibers to increase in flexural strength and interlaminar shear strength and those with untreated fibers to decrease in flexural strength and interlaminar shear strength. Hence, the fiber–matrix bonding is very poor in graphitized composites containing untreated fibers and is stronger in graphitized composites containing treated fibers [236].

After carbonization, a residual stress remains at the interface between the fibers and the matrix. This is because, during carbonization, fiber–matrix interaction causes the crystallite basal planes to be aligned parallel to the fiber axis [232]; the resulting Poisson's effect elongates the fibers along the fiber axis and compresses them in the radial direction. This effect is indicated during carbonization by the transverse shrinkage and longitudinal expansion of the composite. Its transverse shrinkage is larger than the shrinkage of the resin alone (which is the matrix material), while its longitudinal expansion is larger than the expansion of the fibers alone [191]. The residual stress can cause warpage [240].

The high-temperature resistance of carbon–carbon composites containing boron or zirconium diboride glass-forming oxidation inhibitors can be impaired by the reactions between the inhibitors and the carbonaceous components of the composite. These reactions, which probably form carbides, affect both fibers and matrix. They result in almost complete crystallization of the composite components. This crystallization transforms the microstructure of the composite, weakening it and producing brittle failure behavior. For boron, the reaction occurs at temperatures between 2320°C and 2330°C; for zirconium diboride, it occurs at temperatures between 2330°C and 2350°C [241].

For two-dimensional carbon–carbon composites containing plain weave fabric reinforcements under tension, the mode of failure of the fiber bundles depends on their curvature. Fiber bundles with small curvatures fail due to tensile stress or to a combination of tensile and bending stresses. Fiber bundles with large curvatures fail due to shear stresses at the point where the local fiber direction is most inclined to the applied load [242].

The carbon matrix significantly enhances the carbon fibers' resistance to creep deformation owing to the ability of the matrix to distribute loads more evenly and to impose a plastic flow-inhibiting, triaxial stress state in the fibers [243]. The thermally activated process for creep is controlled by vacancy formation and motion.

2.4.5 Thermal conductivity and electrical resistivity

Carbon–carbon composites with high thermal conductivity are important in first wall components for nuclear fusion reactors, hypersonic aircraft, missiles and spacecraft, thermal radiator panels and electronic heat sinks.

The thermal conductivity of carbon–carbon composites at <1000°C increases with the heat treatment temperature, particularly above 2800°C [244], as more graphitic carbon is associated with a higher thermal conductivity. Parallel to the fiber axis, the thermal conductivity is high and the electrical resistivity is low. Perpendicular to the laminate, the thermal conductivity is low and the electrical resistivity is high. Graphitization increases the thermal conductivity and decreases electrical resistivity. After a 3000°C heat treatment, the thermal conductivity is 500 W m^{-1} K^{-1} at 300°C [245], compared to a value of 225 W m^{-1} K^{-1} for Al and a value of 363 W m^{-1} K^{-1} for Cu at the same temperature. The low density of carbon makes the specific thermal conductivity of carbon–carbon composites outstandingly high compared to other materials. The use of porous carbon–carbon composites with even lower densities [246] may further increase the specific thermal conductivity.

2.5 Metal–matrix composites

2.5.1 Introduction

Metal–matrix composites are gaining importance because the reinforcement serves to reduce the coefficient of thermal expansion (CTE) and increase the strength and modulus. If a relatively graphitic kind of carbon fiber is used, the thermal conductivity can be enhanced also. The combination of low CTE and high thermal conductivity makes them very attractive for electronic packaging (e.g., heat sinks). Besides good thermal properties, their low density makes them particularly desirable for aerospace electronics and orbiting space structures; orbiters are thermally cycled by moving through the earth's shadow.

Compared to the metal itself, a carbon fiber metal–matrix composite is characterized by a higher strength-to-density ratio (i.e., specific strength), a higher modulus-to-density ratio (i.e., specific modulus), better fatigue resistance, better high-temperature mechanical properties (a higher strength and a lower creep rate), a lower CTE and better wear resistance.

Compared to carbon fiber polymer–matrix composites, a carbon fiber metal–matrix composite is characterized by higher temperature resistance, higher fire resistance, higher transverse strength and modulus, lack of moisture absorption, higher thermal conductivity, lower electrical resistivity, better radiation resistance and absence of outgassing.

On the other hand, a metal–matrix composite has the following disadvantages compared to the metal itself and the corresponding polymer–matrix composite: higher fabrication cost and limited service experience.

Carbon fibers used for metal–matrix composites are mostly in the form of continuous fibers, but short fibers are also used. The matrices used include aluminum, magnesium, copper, nickel, tin alloys, silver–copper and lead alloys. Aluminum is by far the most widely used matrix metal because of its low density, low melting temperature (which makes composite fabrication and joining relatively convenient), low cost and good machinability. Magnesium is comparably low in melting temperature, but its density is even lower than that of aluminum. Applications include structures (aluminum, magnesium), electronic heat sinks and substrates (aluminum, copper), soldering and bearings (tin alloys), brazing (silver–copper) and high-temperature applications (nickel).

The fabrication of metal–matrix composites often involves the use of an intermediate, called a preform, in the form of sheets, wires, cylinders or near-net shapes. The preform contains the reinforcement usually held together by a binder, which can be a polymer (e.g., acrylic, styrene), a ceramic (e.g., silica, aluminum metaphosphate [247]) or the matrix metal itself. For example, continuous fibers are wound around a drum and bound with a resin, and subsequently the wound fiber cylinder is cut off the drum and stretched out to form a sheet. During subsequent composite fabrication, the organic binder evaporates. As another example, short fibers are combined with a ceramic or polymeric binder and a liquid carrier (usually water) to form a slurry; this is then filtered under pressure or wet pressed to form a wet "cake", which is subsequently dried to form a preform. In the case of using the matrix metal as the binder, a continuous fiber bundle is immersed in the molten matrix metal so as to be infiltrated with it, thus forming a wire preform; alternatively, fibers placed on a matrix metal foil are covered and fixed in place with a sprayed matrix metal, thus forming a sprayed preform.

A binder is not always needed, although it helps the fibers to stay uniformly distributed during subsequent composite fabrication. Excessive ceramic binder amounts should be avoided, as they can make the resulting metal–matrix composite more brittle. For ceramic binders, a typical amount ranges from 1 to 5 wt.% of the preform [247]. In the case of woven fabrics as reinforcement a binder is less important, as the weaving itself serves to hold the fibers together in a uniform fashion.

2.5.2 Fabrication

The most popular method for the fabrication of metal–matrix composites is the *infiltration* of a preform by a liquid metal under pressure. The low viscosity of liquid metals compared to resins or glasses makes infiltration very appropriate for metal–matrix composites. Nevertheless, pressure is required because of the difficulty for the liquid metal to wet the reinforcement. The pressure can be provided by a gas (e.g., argon) or a piston, as illustrated in Figure 2.1. When a piston is used, the process can be quite fast and is known as squeeze casting.

A second method for fabricating metal–matrix composites is *diffusion bonding*. In this method, a stack of alternating layers of fibers and metal foils is hot pressed (at, say, 24 MPa for 20 min) to cause bonding in the solid state. This method is not very suitable for fiber cloths or continuous fiber bundles because of the difficulty for the metal to flow to the space between the fibers during diffusion bonding. In contrast, the infiltration method involves melting the metal, so metal flow is relatively easy, making infiltration a more suitable method for fiber cloths or continuous fiber bundles. A variation of diffusion bonding involves the hot pressing of metal-coated fibers without the use of metal foils. In this case, the metal coating provides the metal for the metal–matrix composite. In general, the diffusion bonding method is complicated by the fact that the surface of the metal foil or metal coating tends to be oxidized and the oxide makes the bonding more difficult. Hence, a vacuum is usually required for diffusion bonding.

A third method of fabricating metal–matrix composites involves *hot pressing* above the solidus of the matrix metal. This method requires lower pressures than diffusion bonding, but the higher temperature of the pressing tends to cause reinforcement degradation, resulting from the interfacial reaction between the reinforcement and the matrix metal. A way to alleviate this problem is to insert a metal sheet of a lower solidus than the matrix metal alternately between wire preform layers, then to hot press at a temperature between the two solidus temperatures [248].

A combination of the second and third methods involves first heating fibers laid up with matrix metal sheets between them in a vacuum in a sealed metal

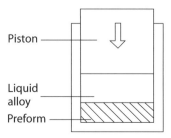

Figure 2.1 Schematic view of squeeze casting, i.e., direct infiltartion of a preform, using a piston to provide pressure.

container above the liquidus of the matrix metal, then immediately hot pressing the container at a temperature below the solidus of the matrix metal [248].

A fourth method of fabricating metal–matrix composites involves *plasma spraying* of the metal on to continuous fibers. As this process usually results in a composite of high porosity, subsequent consolidation (e.g., by hot isostatic pressing) is usually necessary. Compared to the other methods, plasma spraying has the advantage of being able to produce continuous composite parts, although the subsequent consolidation step may limit their size.

Slurry casting, a fifth method, is complicated by the tendency for the carbon fibers (low in density) to float on the metal melt. To overcome this problem, which causes non-uniformity in the fiber distribution, compocasting is necessary. Compocasting (rheocasting) involves vigorously agitating a semi-solid alloy so that the primary phase is non-dendritic, thereby giving a fiber–alloy slurry with thixotropic properties.

The infiltration method can be used to produce near-net shape composites, so that subsequent shaping is not necessary. As infiltration using a gas pressure involves a smaller rate of pressure increase compared to using a piston, infiltration is more suitable than squeeze casting for near-net shape processing. If shaping is necessary, it can be achieved by plastic forming (e.g., extrusion, swaging, forging and rolling) in the case of short fibers. Plastic forming tends to reduce porosity and give preferential orientation to short fibers. These effects result in improved mechanical properties. For continuous fiber metal–matrix composites, shaping cannot be achieved by plastic forming and cutting is necessary.

2.5.3 Wetting of reinforcement by molten metals

The difficulty for molten metals to wet the surface of carbon or ceramic reinforcements complicates the fabrication of metal–matrix composites. This difficulty is particularly severe for high-modulus carbon fibers (e.g., Amoco's Thornel P-100) which have graphite planes mostly aligned parallel to the fiber surface. The edges of the graphite planes are more reactive with the molten metals than the graphite planes themselves, so low-modulus carbon fibers are more reactive and thus are wetted more easily by the molten metals. Although this reaction between the fibers and the metal helps the wetting, it produces a brittle carbide and degrades the strength of the fibers.

In order to enhance the wetting, the reinforcement is coated with a metal or a ceramic. The metals used as coatings are formed by plating and include Ni, Cu and Ag; they generally result in composites of much lower strength than those predicted by the rule of mixtures (ROM). In the case of nickel-coated and copper-coated carbon fibers in an aluminum matrix, metal aluminides (Al_3M) form and embrittle the composites. In the case of nickel-coated carbon fibers in a magnesium matrix, nickel reacts with magnesium to form Ni–Mg compounds and a low-melting (508°C) eutectic [249]. On the other hand, copper-coated fibers are suitable for copper [250], tin or other metals as the matrix. A metal coating that is particularly successful involves sodium, which wets carbon fibers and coats them with a protective intermetallic compound by reaction with one or more other molten metals (e.g., tin). This is called the sodium process [251,252]. A related process immerses the fibers in liquid NaK [253]. However, these processes involving sodium suffer from sodium contamination of the fibers, probably due to the intercalation of sodium into graphite [254]. Nevertheless, aluminum-matrix composites containing unidirectional carbon fibers treated by the sodium process exhibit tensile strengths close to those calculated by using the ROM, indicating that the fibers are not degraded by the sodium process [254].

Examples of ceramics used as coatings on carbon fibers are TiC, SiC, B_4C, TiB_2, TiN, K_2ZrF_6 and ZrO_2. Methods used to deposit the ceramics include (1) reaction of the carbon fibers with a molten metal alloy, called the liquid metal transfer agent (LMTA) technique, (2) chemical vapor deposition (CVD) and (3) solution coating.

The LMTA technique involves immersing the fibers in a melt of copper or tin (called a liquid metal transfer agent, which must not react with carbon) in which a refractory element (e.g., W, Cr, Ti) is dissolved, and subsequent removal of the transfer agent from the fiber surface by immersion in liquid aluminum (suitable for fabricating an aluminum-matrix composite). For example, for forming a TiC coating, the alloy can be Cu-10% Ti at 1050°C or Sn-1% Ti at 900-1055°C. In particular, by immersing the fibers in Sn-1% Ti at 900-1055°C for 0.25-10 min, a 0.1 μm layer of TiC is formed on the fibers, although they are also surrounded by the tin alloy. Subsequent immersion for 1 min in liquid aluminum causes the tin alloy to dissolve in the liquid aluminum [255]. The consequence is a wire preform suitable for fabricating aluminum-matrix composites. Other than titanium carbide, tungsten carbide and chromium carbide have been formed on carbon fibers by the LMTA technique.

The CVD technique has been used for forming coatings of TiB_2, TiC, SiC, B_4C and TiN. The B_4C coating is formed by reactive CVD on carbon fibers, using a BCl_3/H_2 mixture as the reactant [256]. The TiB_2 deposition uses $TiCl_4$ and BCl_3 gases, which are reduced by Zn vapor. The TiB_2 coating is particularly attractive because of the exceptionally good wetting between TiB_2 and molten aluminum. During composite fabrication, the TiB_2 coating is displaced and dissolved in the matrix, while an oxide (γ-Al_2O_3 for a pure aluminum matrix, $MgAl_2O_4$ spinel for a 6061 aluminum matrix) is formed between the fiber and the matrix. The oxygen for the oxide formation comes from the sizing on the fibers; the sizing is not completely removed from the fibers before processing [257]. The oxide layer serves as a diffusion barrier to aluminum, but allows diffusion of carbon, thereby limiting Al_4C_3 growth to the oxide-matrix interface [258]. Moreover, the oxide provides bonding between the fiber and the matrix. Because of the reaction at the interface between the coating and the fiber, the fiber strength is degraded after coating. To alleviate this problem, a layer of pyrolytic carbon is deposited between the fiber and the ceramic layer [259]. The CVD process involves high temperatures, e.g., 1200°C for SiC deposition using CH_3SiCl_4 [260]; this high temperature degrades the carbon fibers. Another problem of the CVD process is the difficulty of obtaining a uniform coating around the circumference of each fiber. Moreover, it is expensive and causes the need to scrub and dispose most of the corrosive starting material, as most of the starting material does not react at all. The most serious problem with the TiB_2 coating is that it is not air stable; it cannot be exposed to air before immersion in the molten metal or wetting will not take place. This problem limits the shape of materials that can be fabricated, especially since the wire preforms are not very flexible [257].

A high compliance (or a low modulus) is preferred for the coating in order to increase the interface strength. An increase in the interface (or interphase) strength results in an increase in the transverse strength. The modulus of SiC coatings can be varied by controlling the plasma voltage in plasma-assisted chemical vapor deposition (PACVD). Modulus values in the range of 19–285 GPa have been obtained in PACVD SiC, compared to a value of 448 GPa for CVD SiC. Unidirectional carbon fiber (Thornel P-55) aluminum–matrix composites in which the fibers are coated with SiC exhibit an interface strength and transverse strength which increase with decreasing modulus of the SiC coating [261–263].

The most attractive coating technique developed to date is the *solution coating method*. In the case of an organometallic solution, fibers are passed through a toluene solution containing an organometallic compound, followed by hydrolysis or pyrolysis of the organometallic compounds to form the coating. Thus, the fibers are passed sequentially through a furnace in which the sizing on the fibers is vaporized, followed by an ultrasonic bath containing an organometallic solution. The coated fibers are then passed through a chamber containing flowing steam in which the organometallic compound on the fiber surface is hydrolyzed to oxide and finally through an argon atmosphere drying furnace in which any excess solvent or water is vaporized and any unhydrolyzed organometallic is pyrolyzed [257]. In contrast to the TiB_2 coatings, the SiO_2 coatings formed by organometallic solution coating are air stable.

The organometallic compounds used are alkoxides, in which metal atoms are bound to hydrocarbon groups by oxygen atoms. The general formula is $M(OR)_x$, where R is any hydrocarbon group (e.g., methyl, ethyl, propyl) and x is the oxidation state of the metal atom M. When exposed to water vapor, these alkoxides hydrolyze, i.e., [257]:

$$M(OR)_x + \tfrac{x}{2}H_2O \rightarrow MO_{x/2} + xROH.$$

For example, the alkoxide tetraethoxysilane (also called tetraethylorthosilicate) is hydrolyzed by water as follows [257]:

$$Si(OC_2H_5)_4 + 2H_2O \rightarrow SiO_2 + 4C_2H_5OH.$$

Alkoxides can also be pyrolyzed to yield oxides, e.g. [257]:

$$Si(OC_2H_5)_4 \rightarrow SiO_2 + 2C_2H_5OH + 2C_2H_4.$$

Most alkoxides can be dissolved in toluene. By controlling the solution concentration and the time and temperature of immersion, it is possible to control the uniformity and thickness of the resulting oxide coating. The thickness of the oxide coatings on the fibers varies from 700 to 1500 Å. The oxide is amorphous and contains carbon, which originates in the carbon fiber.

Liquid magnesium wets SiO_2-coated low-modulus carbon fibers (e.g., T-300) and infiltrates the fiber bundles, owing to reactions between the molten magnesium and the SiO_2 coating. The reactions include the following [257].

$2Mg + SiO_2 \rightarrow 2MgO + Si$	$\Delta G°_{670°C} = -76$ kcal
$MgO + SiO_2 \rightarrow MgSiO_3$	$\Delta G°_{670°C} = -23$ kcal
$2Mg + 3SiO_2 \rightarrow 2MgSiO_3 + Si$	$\Delta G°_{670°C} = -122$ kcal
$2MgO + SiO_2 \rightarrow Mg_2SiO_4$	$\Delta G°_{670°C} = -28$ kcal
$2Mg + 2SiO_2 \rightarrow Mg_2SiO_4 + Si$	$\Delta G°_{670°C} = -104$ kcal

The interfacial layer between the fiber and the Mg matrix contains MgO and magnesium silicates. However, immersion of SiO_2-coated high-modulus fibers (e.g., P-100) in liquid magnesium causes the oxide coating to separate from the fibers, due to the poor adherence of the oxide coating to the high-modulus fibers. This problem can be solved by first depositing a thin amorphous carbon coating on the fibers by passing the fiber bundles through a toluene solution of petroleum pitch, followed by evaporation of the solvent and pyrolysis of the pitch.

The most effective air-stable coating for carbon fibers used in aluminum-matrix composites is a mixed boron–silicon oxide applied from organometallic solutions [257].

Instead of SiO_2, TiO_2 can be deposited on carbon fibers by the organometallic solution method. For TiO_2, the alkoxide can be titanium isopropoxide [264].

SiC coatings can be formed by using polycarbosilane (dissolved in toluene) as the precursor, which is pyrolyzed to SiC. These coatings are wet by molten copper containing a small amount of titanium, due to a reaction between SiC and Ti to form TiC [257].

Instead of using an organometallic solution, another kind of solution coating method uses an aqueous solution of a salt. For example, the salt potassium zirconium hexafluoride (K_2ZrF_6) or potassium titanium hexafluoride (K_2TiF_6) is used to deposit microcrystals of K_2ZrF_6 or K_2TiF_6 on the fiber surface [257,265,266]. These fluoride coatings are stable in air. The following reactions supposedly take place [265] between K_2ZrF_6 and the aluminum matrix:

$$3K_2ZrF_6 + 4Al \rightarrow 6KF + 4AlF_3 + 3Zr \qquad (2.1)$$

$$3Zr + 9Al \rightarrow 3Al_3Zr \qquad (2.2)$$

$$Zr + O_2 \rightarrow ZrO_2. \qquad (2.3)$$

In the case of an Al–12 wt.% Si alloy (rather than pure Al) as the matrix, the following reaction may also occur [266]:

$$Zr + 2Si \rightarrow ZrSi_2. \qquad (2.4)$$

The fluorides KF and AlF_3 are thought to dissolve the thin layer of Al_2O_3 which is on the liquid aluminum surface, thus helping the liquid aluminum to wet the carbon fibers. Furthermore, the reactions (2.1) and (2.2) are strongly exothermic and may therefore cause a local temperature increase near the fiber–matrix interface. The increased temperature probably gives rise to a liquid phase at the fiber–matrix interface [265].

Although the K_2ZrF_6 treatment causes the contact angle between carbon and liquid aluminum at 700–800°C to decrease from 160°C to 60–75°C [265], it causes degradation of the fiber tensile strength during aluminum infiltration [267].

Another example of a salt solution coating method involves the use of the salt zirconium oxychloride ($ZrOCl_2$) [268]. Dip coating the carbon fibers in the salt solution and subsequently heating at 330°C causes the formation of a ZrO_2 coating of less than 1 μm in thickness. The ZrO_2 coating serves to improve fiber–matrix wetting and reduce the fiber–matrix reaction in aluminum-matrix composites.

Instead of treating the carbon fibers, wetting of the carbon fibers by molten metals can be improved by addition of alloying elements into the molten metals. For aluminum as the matrix, effective alloying elements include Mg, Cu and Fe [254].

2.6 Ceramic–matrix composites

Although cement is a ceramic material, ceramic–matrix composites usually refer to those with silicon carbide, silicon nitride, alumina, mullite, glasses and other ceramic matrices that are not cement.

Ceramic–matrix fiber composites are gaining increasing attention because the good oxidation resistance of the ceramic matrix (compared to a carbon matrix) makes the composites attractive for high-temperature applications (e.g., aerospace and engine components). The fibers serve mainly to increase toughness and strength (tensile and flexural) of the composite due to their tendency to be partially pulled out during the deformation. This pullout absorbs energy, thereby toughening the composite. Although the fiber pullout has its advantages, bonding between the fibers and the matrix must still be sufficiently good in order for the fibers to strengthen the composite effectively. Therefore, the control of the bonding between the fibers and the matrix is important for the development of these composites.

In the case of reinforcement being carbon fibers, the reinforcement has a second function, which is to increase the thermal conductivity of the composite, as the ceramic is mostly thermally insulating whereas carbon fibers are thermally conductive. In electronic, aerospace and engine components, the enhanced thermal conductivity is attractive for heat dissipation.

A third function of the reinforcement is to decrease the drying shrinkage in the case of ceramic matrices prepared by using slurries or slips [269]. In general, the drying shrinkage decreases with increasing solid content in the slurry. Fibers are more effective than particles for decreasing the drying shrinkage. This function is attractive for the dimensional control of parts made from the composites.

Fiber-reinforced glasses are useful for space structural applications, such as mirror-back structures and supports, booms and antenna structures. In low earth orbit, these structures experience a temperature range from −100 to 80°C, so they need an improved thermal conductivity and a reduced coefficient of thermal expansion. In addition, increased toughness, strength and modulus are desirable. Due to the environment degradation resistance of carbon fiber-reinforced glasses, they are also potentially useful for gas turbine engine components. Additional attractions are low friction, high wear resistance and low density.

The glass matrices used for fiber-reinforced glasses include borosilicate glasses (e.g., Pyrex), aluminosilicate glasses, soda-lime glass and fused quartz. Moreover, a lithia aluminosilicate glass–ceramic [270] and a $CaO-MgO-Al_2O_3-SiO_2$ glass–ceramic [271] have been used.

The fibers can be short or continuous. In both cases, the composites can be formed by viscous glass consolidation, i.e., either hot pressing a mixture of fibers and glass powder [272,273], or winding glass-impregnated continuous fiber under tension above the annealing temperature of the glass [274].

Short (≤3 mm) fiber borosilicate glass composites are prepared by hot pressing, in vacuum or argon, using an isopropyl alcohol slurry of fibers and Pyrex powder (<50 μm particle size) at 700–1000°C (depending on the fiber content, which ranges from 10 to 40 vol.%) and 6.9 MPa [272]. Fibers longer than 3 mm are more difficult to disperse in the slurry. The resulting composite has fibers that are two-dimensionally random, as some alignment occurs during pressing.

Continuous-fiber composites are made by allowing fibers from a spool to pass through a glass powder slurry (containing water and a water-soluble acrylic binder), winding the slurry-impregnated fibers onto the sides of a hexagonal prism (mandrel or take-up drum), cutting up the flat unidirectional tapes from the mandrel, stacking up the pieces (plies) in the correct orientation, burning out the stack to remove the binder and hot pressing the stack at a temperature above the working temperature of the glass. This process is known as *slurry infiltration* or *viscous glass consolidation*. During hot pressing the glass must flow into the space between adjacent fibers. Since glass does not wet carbon, a sufficient pressure is necessary [273].

Viscous glass consolidation can also be carried out without hot pressing, but by winding under tension (e.g., about 15 ksi or 100 MPa) a glass powder-impregnated continuous-fiber tow onto a collection mandrel at a temperature above the annealing temperature of the glass [274].

Other than viscous glass consolidation, another method for forming short- or continuous-fiber glass–matrix composite is the *sol–gel method*, i.e., infiltration by a sol and subsequent sintering [275], using metal alkoxides and/or metal salts as precursors and using an acid or base catalyst to promote hydrolysis and polymerization [276]. Composite fabrication consists of two steps: (1) the preparation of fiber–gel prepregs and (2) thermal treatment and densification by hot pressing [277]. In the first step, the sol-impregnated prepregs are allowed to gel, dried at room temperature for a day, dried at 50°C for a day, then heat-treated at 300–400°C for 3 h. In the second step, hot pressing is performed in a nitrogen atmosphere with a pressure of 10 MPa [277]. For the case of a borosilicate glass, the hot pressing temperature is between 900°C and 1200°C [276]. The sol–gel method is advantageous in that homogenization of components during impregnation can readily be achieved and sintering temperatures are substantially lowered because of the smaller particle size compared to the slurry infiltration method. The easier impregnation or infiltration decreases the tendency for the particles to damage the fibers. However, the sol–gel method has the disadvantage of excessive shrinkage during subsequent heat treatment [278].

The slurry infiltration and sol–gel methods can be combined by using particle-filled sols. The particle filling helps to reduce shrinkage [278].

Yet another method is melt infiltration, which requires heating the glass at a temperature much above the softening temperature in order for the glass to infiltrate the fiber preform. This high temperature may lead to chemical reaction between the fibers and the matrix [278].

The reaction between SiO_2 and carbon fibers is of the form

$$3C + SiO_2 \rightarrow SiC + 2CO.$$

To reduce the extent of this reaction, SiC-coated carbon fibers are used for SiO_2-matrix composites.

Fibers such as carbon fibers have the following effects on the glass: (1) increasing the toughness (and, in some cases, the strength as well), (2) decreasing the coefficient of thermal expansion and (3) increasing thermal conductivity.

The increase in toughness is present in glass with short and randomly oriented fibers, as well as glass with continuous fibers. However, the strength is usually decreased by fibers in the former and is usually increased by fibers in the latter. Even for continuous fibers the strengthening is limited by widespread matrix cracking (both transverse and longitudinal) [279].

The high-temperature strength of carbon fiber-reinforced glasses in air is limited by the oxidation of the carbon fibers. This oxidation cannot be prevented by the glass matrix [280]. In an inert atmosphere, the high-temperature strength is limited by softening of the glass matrix.

Review questions

1. Describe a process for making a carbon–carbon composite from carbon fibers and a pitch.
2. Describe a process for making a carbon fiber thermoplastic–matrix composite part with a specific shape.
3. Give the reasons for the preference for using a *small* volume fraction of short carbon fibers in concrete. That is, why is a large volume fraction not desirable?
4. What are the effects of short carbon fiber addition on the important properties of concrete?
5. Describe a method for enhancing the oxidation resistance of carbon–carbon composites.
6. What are the attractive properties of a metal–matrix composite compared to polymer–matrix composite?
7. Consider a carbon–matrix composite containing continuous carbon fibers.
    (a) What are the two main attractive properties of a carbon–matrix composite?
    (b) Carbon–matrix composites suffer from poor oxidation resistance. Why?
8. An adhesive can be a thermoset or a thermoplastic. What is the main advantage of a thermoplastic compared to a thermoset?
9. Give the main advantage of an SiC-matrix composite compared to a carbon–matrix composite.
10. What are the improvements in properties (structural and functional) rendered by the addition of short carbon fibers to concrete?
11. What are the main advantages, in relation to the properties, of a carbon fiber thermoplastic–matrix composite compared to a carbon fiber thermoset–matrix composite?
12. Why is the time for making a thermoplastic–matrix composite short compared to that for making a thermoset–matrix composite?
13. Give an example of a thermoset and an example of a thermoplastic.

References

[1] W.-T. Whang and W.-L. Liu, *SAMPE Q.*, 1990, **22**(1), 3–9.
[2] D.C. Sherman, C.-Y. Chen and J.L. Cercena, *Proc. Int. SAMPE Symp. and Exhib., 33, Materials: Pathway to the Future*, edited by G. Carrillo, E.D. Newell, W.D. Brown, and P. Phelan, 1988, pp. 538–539.
[3] D.R. Askeland, *The Science and Engineering of Materials*, 2nd ed., PWS-Kent, 1989, pp. 538–539.
[4] S.D. Mills, D.M. Lee, A.Y. Lou, D.F. Register and M.L. Stone, *Proc. 20th Int. SAMPE Tech. Conf.*, 1988, pp. 263–270.
[5] G. Bogoeva-Gaceva, E. Maeder, L. Hauessler and A. Dekanski, *Composites – Part A: Applied Science & Manufacturing*, 1997, **28**(5), 445–452.
[6] J. Jang and H. Yang, *J. Mater. Sci.*, 2000, **35**(9), 2297–2303.
[7] G. Wu, C.-H. Hung and J.-C. Lu, *International SAMPE Symposium & Exhibition*, 1999, **44**(I), 1090–1097.
[8] H. Zhuang and J.P. Wightman, *J. Adhesion*, 1997, **62**(1–4), 213–245.

[9] R.E. Allred and W.C. Schimpf, *J. Adhesion Sci. & Tech.*, 1994, **8**(4), pt. 2, 383-394.
[10] G.J. Farrow, K.E. Atkinson, N. Fluck and C. Jones, *Surface & Interface Analysis*, 1995, **23**(5), 313-318.
[11] E.A. Friis, B. Kumar, F.W. Cooke and H.K. Yasuda, *Proc. 1996 5th World Biomaterials Congress*, Vol. 1, Soc. for Biomaterials, St. Louis Park, MN, 1996, pp. 913.
[12] S.-S. Lin and P.W. Yip, *Proc. of the 1993 Fall Meeting*, Interface Control of Electrical, Chemical, and Mechanical Properties Materials Research Society Symposium Proceedings, Vol. 318, Materials Research Society, Pittsburgh, PA, 1994, pp. 381-386.
[13] N. Dilsiz, *J. Adhesion Sci. & Tech.* 2000, **14**(7), 975-987.
[14] M.C. Paiva, C.A. Bernardo and M. Nardin, *Carbon*, 2000, **38**(9), 1323-1337.
[15] G. Akovali and N. Dilsiz, *Polym. Eng. & Sci.*, 1996, **36**(8), 1081-1086.
[16] N. Chand, E. Schulz and G. Hinrichsen, *J. Mater. Sci. Letters*, 1996, **15**(15), 1374-1375.
[17] T. Yoshikawa and A. Kojima, *Quarterly Report of RTRI* (Railway Technical Research Institute) (Japan), 1991, **32**(3), 190-199.
[18] P.W.M. Peters and H. Albertsen, *J. Mater. Sci.*, 1993, **28**(4), 1059-1066.
[19] H. Albertsen, J. Ivens, P. Peters, M. Wevers and I. Verpoest, *Composites Sci. & Tech.*, 1995, **54**(2), 133-145.
[20] L. Ibarra, A. Macias and E. Palma, *J. Applied Polym. Sci.*, 1996, **61**(13), 2447-2454.
[21] D. Cho, *J. Mater. Sci. Letters*, 1996, **15**(20), 1786-1788.
[22] T.R. King, D.F. Adams and D.A. Buttry, *Composites*, 1991, **22**(5), 380-387.
[23] E. Fitzer, N. Popovska and H.-P. Rensch, *J. Adhesion*, 1991, **36**(2-3), 139-149.
[24] M.R. Alexander and F.R. Jones, *Surface & Interface Analysis*, 1994, **22**(1), 230-235.
[25] A. Fukunaga and S. Ueda, *Composites Sci. & Tech.*, 2000, **60**(2), 249-254.
[26] S.-J. Park and M.-H. Kim, *J. Mater. Sci.*, 2000, **35**(8), 1901-1905.
[27] C. Jones, *Surface & Interface Analysis*, 1993, **20**(5), 357-367.
[28] C.A. Baillie and M.G. Bader, *J. Mater. Sci.*, 1994, **29**(14), 3822-3836.
[29] J.A. Hrivnak and R.L. McCullough, *J. Thermoplastic Composite Mater.*, 1996, **9**(4), 304-315.
[30] I.A. Rashkovan and Y.G. Korabel'nikov, *Composites Sci. & Tech.*, 1997, **57**(8), 1017-1022.
[31] G. Bogoeva-Gaceva, D. Burevski, A. Dekanski and A. Janevski, *J. Mater. Sci.*, 1995, **30**(13), 3543-3546.
[32] D.W. Dwight, *Proc. 53rd Annual Technical Conf. - ANTEC*, Vol. 2, Soc. Plastics Eng., Brookfield, CT, 1995, pp. 2744-2747.
[33] N. Tsujioka, Z. Maekawa, H. Hamada and M. Hojo, *Zairyo/J. Soc. Mater. Sci.* (Japan), 1997, **46**(2), 163-169.
[34] L. Ibarra and D. Panos, *Polym. Int.*, 1997, **43**(3), 251-259.
[35] L. Ibarra and D. Panos, *J. Applied Polym. Sci.*, 1998, **67**(10), 1819-1826.
[36] J.M.M. de Kok and T. Peijs, *Composites - Part A: Applied Science & Manufacturing*, 1999, **30**(7), 917-932 (1999).
[37] P.W. Yip and S.S. Lin, *Mater. Res. Soc. Symp. Proc.*, Vol. 170, Interfaces Compos., edited by C.G. Pantano and E.J.H. Chen, 1990, pp. 339-344.
[38] T.C. Chang and B.Z. Jang, *Mater. Res. Soc. Symp. Proc.*, Vol. 170 Interfaces Compos., edited by C.G. Pantano and E.J.H. Chen, 1990, pp. 321-326.
[39] W.W. Wright, *Compos. Polym.*, 1990, **3**(4), 231-257.
[40] T.R. King, D.F. Adams and D.A. Buttry, *Composites*, 1991, **22**(5), 380-387.
[41] J. Tian, Q. Wang, S. Yang and Q. Xue, *Applied Mathematics & Mechanics* (English Edition), 1998, **19**(10), 1-5.
[42] J.A. King, D.A. Buttry and D.F. Adams, *Polym. Composites*, 1993, **14**(4), 301-307.
[43] B. Harris, O.G. Braddell, D.P. Almond, C. Lefebvre and J. Verbist, *J. Mater. Sci.*, 1993, **28**(12), 3353-3366.
[44] B. Harris, O.G. Braddell, C. Lefebvre and J. Verbist, *Plastics Rubber & Composites*, 1992, **18**(4), 221-240.
[45] F. Zhang and L. Hu, *Fuhe Cailiao Xuebao/Acta Materiae Compositae Sinica*, 1997, **14**(2), 12-16.
[46] A.P. Kettle, F.R. Jones, M.R. Alexander, R.D. Short, M. Stollenwerk, J. Zabold, W. Michaeli, W. Wu, E. Jacobs and I. Verpoest, *Composites - Part A: Applied Science & Manufacturing*, 1998, **29**(3), 241-250.
[47] N. Dilsiz and G. Akovali, *Composite Interfaces*, 1996, **3**(5-6), 401-410.
[48] N. Dilsiz, E. Ebert, W. Weisweiler and G. Akovali, *J. Colloid & Interface Sci.*, 1995, **170**(1), 241-248.
[49] R. Li, L. Ye and Y.-W. Mai, *Composites - Part A: Applied Science & Manufacturing*, 1997, **28**(1), 73-86.
[50] J.P. Armistead and A.W. Snow, *ASTM Special Technical Publication*, 1996, (1290), 168-181.
[51] T. Duvis, C.D. Papaspyrides and T. Skourlis, *Composites Sci. & Tech.*, 1993, **48**(1-4), 127-133.
[52] T. Skourlis, T. Duvis and C.D. Papaspyrides, *Composites Sci. & Tech.*, 1993, **48**(1-4), 119-125.
[53] A. Zheng, X. Wu and S. Li, *Huadong Huagong Xueyuan Xuebao/J. East China Inst. of Chem. Tech.*, 1994, **20**(4), 485-491.

[54] P.C. Varelidis, R.L. McCullough and C.D. Papaspyrides, *Composites Sci. & Tech.*, 1999, **59**(12), 1813-1823.
[55] M. Tanoglu, G.R. Palmese, S.H. McKnight and J.W. Gillespie Jr, *Proc. 1998 56th Annual Tech. Conf. - ANTEC*, Vol. 2, Soc. Plastic Eng., Brookfield, CT, 1998, pp. 2346-2350.
[56] V. Giurgiutiu, K.L. Reifsnider, R.D. Kriz, B.K. Ahn and J.J. Lesko, *Proc. 36th AIAA/ASME/ASCE/AHS/ASC Structures, Structural Dynamics and Mater. Conf. and AIAA/ASME Adaptive Structures Forum*, Vol. 1, AIAA, New York, 1995, pp. 453-469.
[57] M. Labronici and H. Ishida, *Composite Interfaces*, 1998, **5**(3), 257-275.
[58] M. Labronici and H. Ishida, *Composite Interfaces*, 1998, **5**(2), 87-116.
[59] J.-K. Kim and Y.-W. Mai, *Proc. Int. Conf. on Adv. Composite Mater.*, Minerals, Metals & Materials Soc. (TMS), Warrendale, PA, 1993, pp. 69-77.
[60] S. Shin and J. Jang, *J. Mater. Sci.*, 2000, **35**(8), 2047-2054.
[61] J. Gulyas, S. Rosenberger, E. Foldes and B. Pukanszky, *Polym. Composites*, 2000, **21**(3), 387-395.
[62] J.A. King, D.A. Buttry and D.F. Adams, *Polym. Compos.*, 1993, **14**(4), 292-300.
[63] G. Lu, X. Li and H. Jiang, *Compos. Sci. & Tech.*, 1996, **56**(2), 193-2000.
[64] F.A. Hussain and A.M. Zihlif, *J. Thermoplastic Compos. Mater.*, 1993, **6**(2), 120-129.
[65] E. Hage Jr, S.F. Costa and L.A. Pessan, *J. Adhesion Sci. & Tech.*, 1997, **11**(12), 1491-1499.
[66] P.J. Ives and D.J. Williams, *Proc. Int. SAMPE Symp. and Exhib.*, 33, Materials: Pathway to the Future, edited by G. Carrillo, E.D. Newell, W.D. Brown and P. Phelan, 1988, pp. 385-396.
[67] W.R. Stabler, G.B. Tatterson, R.L. Sadler and A.H.M. El-Shiekh, *SAMPE Q.*, 1992, **23**(2), 38.
[68] K.C. Ik and K.U. Gon, *Proc. 7th Int. Conf. on Composite Materials*, Guangzhou, China, Nov. 1989, Vol. 2, edited by Y. Wu, Z.-L. Gu, and R. Wu, International Academic Publishers, Beijing, PR China, and Pergamon, Oxford, UK and New York, 1989, pp. 101-109.
[69] O. Dickman, K. Lindersson and L. Svensson, *Plastics and Rubber Processing and Applications*, 1990, **13**, 9-14.
[70] S.D. Copeland, J.C. Seferis and M. Carrega, *J. Appl. Polym. Sci.*, 1992, **44**(1), 41-53.
[71] T. Hartness, *Proc. Int. SAMPE Symp. and Exhib.*, 33, Materials: Pathway to the Future, edited by G. Carrillo, E.D. Newell, W.D. Brown and P. Phelan, 1988, pp. 1458-1471.
[72] H. Kosuda, Y. Nagata and Y. Endoh, US Patent 4,897,286 (1990).
[73] F.K. Ko, J.-N. Chu and C.T. Hua, *J. Appl. Polym. Sci., Appl. Polym. Symp.*, 1991, **47**, 501-519.
[74] F. Ko, P. Fang and H. Chu, *Proc. Int. SAMPE Symp. and Exhib.*, 33, Materials: Pathway to the Future, edited by G. Carrillo, E.D. Newell, W.D. Brown and P. Phelan, 1988, pp. 899-911.
[75] R. Weiss, *Cryogenics*, 1991, **31**(4), 319-322.
[76] T. Vu-Khanh and J. Denault, *Proc. American Society for Composites, 6th Tech. Conf.*, Technomic, Lancaster, 1991, pp. 473-482.
[77] J.D. Muzzy, X. Wu and J.S. Colton, *Polym. Compos.*, 1990, **11**(5), 280-285.
[78] A.K. Miller, M. Gur and A. Peled, *Materials & Manufacturing Processes*, 1990, **5**(2), 273-300.
[79] A.K. Miller, C. Chang, A. Payne, M. Gur, E. Menzel and A. Peled, *SAMPE J.*, 1990, **26**(4), 37-54.
[80] A. Farouk and T.H. Kwon, *Polym. Compos.*, 1990, **11**(6), 379-386.
[81] J. Wei, M.C. Hawley, J. Jow and J.D. DeLong, *SAMPE J.*, 1991, **27**(1), 33-39.
[82] K.J. Hook, R.K. Agrawal and L.T. Drzal, *J. Adhes.*, 1990, **32**(2-3), 157-170.
[83] S.P. Grossman and M.F. Amateau, *Proc. Int. SAMPE Symp. and Exhib.*, 33, Materials: Pathway to the Future, edited by G. Carrillo, E.D. Newell, W.D. Brown and P. Phelan, 1988, pp. 681-692.
[84] J.M. Barton, J.R. Lloyd, A.A. Goodwin and J.N. Hay, *Br. Polymer J.*, 1990, **23**, 101-109.
[85] S. Saiello, J. Kenny and L. Nicolais, *J. Mater. Sci.*, 1990, **25**, 3493-3496.
[86] C.-C.M. Ma, S.-W. Yur, C.-L. Ong and M.-F. Sheu, *Proc. Int. SAMPE Symp. and Exhib.*, 34, Tomorrow's Materials: Today, edited by G.A. Zakrzewski, D. Mazenko, S.T. Peters and C.D. Dean, 1989, pp. 350-361.
[87] *Sprechsaal*, 1990, **123**(4), 403-408.
[88] *Hercules Composite Structures*, Hercules Inc.
[89] J.W. Newman, *Int. SAMPE Symp. Exhib.*, Vol. 32, SAMPE, Covina, CA, 1987, pp. 938-944.
[90] S. Furukawa, Y. Tsuji and S. Otani, *Proc. 30th Japan Congress on Materials Research*, Soc. of Materials Science, Kyoto, Jpn., 1987, pp. 149-152.
[91] K. Saito, N. Kawamura and Y. Kogo, *Advanced Materials: The Big Payoff*, National SAMPE Technical Conf., Vol. 21, Covina, CA, 1989, pp. 796-802.
[92] S. Wen and D.D.L. Chung, *Cen. Concr. Res.*, 1999, **29**(3), 445-449.
[93] T. Sugama, L.E. Kukacka, N. Carciello and D. Stathopoulos, *Cem. Concr. Res.*, 1989, **19**(3), 355-365.
[94] X. Fu, W. Lu and D.D.L. Chung, *Cem. Concr. Res.*, 1996, **26**(7), 1007-1012.
[95] X. Fu, W. Lu and D.D.L. Chung, *Carbon*, 1998, **36**(9), 1337-1345.
[96] Y. Xu and D.D.L. Chung, *Cem. Concr. Res.*, 1999, **29**(5), 773-776.

[97] T. Yamada, K. Yamada, R. Hayashi and T. Herai, *Int. SAMPE Symp. Exhib.*, Vol. 36, SAMPE, Covina, CA, pt. 1, 1991, pp. 362–371.
[98] T. Sugama, L.E. Kukacka, N. Carciello and B. Galen, *Cem. Concr. Res.*, 1988, **18**(2), 290–300.
[99] B.K. Larson, L.T. Drzal and P. Sorousian, *Composites*, 1990, **21**(3), 205–215.
[100] A. Katz, V.C. Li and A. Kazmer, *J. Materials Civil Eng.*, 1995, **7**(2), 125–128.
[101] S.B. Park and B.I. Lee, *Cem. Concr. Composites*, 1993, **15**(3), 153–163.
[102] P. Chen, X. Fu and D.D.L. Chung, *ACI Mater. J.* 1997, **94**(2), 147–155.
[103] P. Chen and D.D.L. Chung, *Composites*, 1993, **24**(1), 33–52.
[104] A.M. Brandt and L. Kucharska, *Materials for the New Millennium*, Proc. Mater. Eng. Conf., Vol. 1, ASCE, New York, 1996, pp. 271–280.
[105] H.A. Toutanji, T. El-Korchi, R.N. Katz and G.L. Leatherman, *Cem. Concr. Res.*, 1993, **23**(3), 618–626.
[106] N. Banthia and J. Sheng, *Cem. Concr. Composites*, 1996, **18**(4), 251–269 (1996).
[107] H.A. Toutanji, T. El-Korchi and R.N. Katz, *Com. Con. Composites*, 1994, **16**(1), 15–21.
[108] S. Akihama, T. Suenaga and T. Banno, *Int. J. Cem. Composites & Lightweight Concrete*, 1984, **6**(3), 159–168.
[109] M. Kamakura, K. Shirakawa, K. Nakagawa, K. Ohta and S. Kashihara, *Sumitomo Metals*, (1983).
[110] A. Katz and A. Bentur, *Cem. Concr. Res.*, 1994, **24**(2), 214–220.
[111] Y. Ohama and M. Amano, *Proc. 27th Japan Congress on Materials Research*, Soc. Mater. Sci., Kyoto, Japan, 1983, pp. 187–191.
[112] Y. Ohama, M. Amano and M. Endo, *Concrete Int.: Design & Construction*, 1985, **7**(3), 58–62.
[113] K. Zayat and Z. Bayasi, *ACI Mater. J.*, 1996, **93**(2), 178–181.
[114] P. Soroushian, F. Aouadi and M. Nagi, *ACI Mater. J.*, 1991, **88**(1), 11–18.
[115] B. Mobasher and C.Y. Li, *ACI Mater. J.*, 1996, **93**(3), 284–292.
[116] N. Banthia, A. Moncef, K. Chokri and J. Sheng, *Can. J. Civil Eng.*, 1994, **21**(6), 999–1011.
[117] B. Mobasher and C.Y. Li, *Infrastructure: New Materials and Methods of Repair*, Proc. Mater. Eng. Conf., n 804, ASCE, New York, 1994, pp. 551–558.
[118] P. Soroushian, M. Nagi and J. Hsu, *ACI Mater. J.*, 1992, **89**(3), 267–276.
[119] P. Soroushian, *Construction Specifier*, 1990, **43**(12), 102–108.
[120] A.K. Lal, *Batiment Int./Building Research & Practice*, 1990, **18**(3), 153–161.
[121] S.B. Park, B.I. Lee and Y.S. Lim, *Cem. Concr. Res.*, 1991, **21**(4), 589–600.
[122] S.B. Park and B.I. Lee, *High Temperatures – High Pressures*, 1990, **22**(6), 663–670.
[123] P. Soroushian, Mohamad Nagi and A. Okwuegbu, *ACI Mater. J.*, 1992, **89**(5), 491–494.
[124] M. Pigeon, M. Azzabi and R. Pleau, *Cem. Concr. Res.*, 1996, **26**(8), 1163–1170.
[125] N. Banthia, K. Chokri, Y. Ohama and S. Mindess, *Adv. Cem. Based Mater.*, 1994, **1**(3), 131–141.
[126] N. Banthia, C. Yan and K. Sakai, *Com. Concr. Composites*, 1998, **20**(5), 393–404.
[127] T. Urano, K. Murakami, Y. Mitsui and H. Sakai, *Composites – Part A: Applied Science & Manufacturing*, 1996, **27**(3), 183–187.
[128] A. Ali and R. Ambalavanan, *Indian Concrete J.*, **72**(12), 669–675.
[129] P. Chen, X. Fu and D.D.L. Chung, *Cem. Concr. Res.*, 1995, **25**(3), 491–496.
[130] M. Zhu and D.D.L. Chung, *Cem. Concr. Res.*, 1997, **27**(12), 1829–1839.
[131] M. Zhu, R.C. Wetherhold and D.D.L. Chung, *Cem. Concr. Res.*, 1997, **27**(3), 437–451.
[132] P. Chen and D.D.L. Chung, *Smart Mater. Struct.*, 1993, **2**, 22–30.
[133] P. Chen and D.D.L. Chung, *Composites – Part B*, 1996, **27B**, 11–23.
[134] P. Chen and D.D.L. Chung, *J. Am. Ceram. Soc.*, 1995, **78**(3), 816–818.
[135] D.D.L. Chung, *Smart Mater. Struct.*, 1995, **4**, 59–61.
[136] P. Chen and D.D.L. Chung, *ACI Mater. J.*, 1996, **93**(4), 341–350.
[137] X. Fu and D.D.L. Chung, *Cem. Concr. Res.*, 1996, **26**(1), 15–20.
[138] X. Fu, E. Ma, D.D.L. Chung and W.A. Anderson, *Cem. Concr. Res.*, 1997, **27**(6), 845–852.
[139] X. Fu and D.D.L. Chung, *Cem. Concr. Res.*, 1997, **27**(9), 1313–1318.
[140] X. Fu, W. Lu and D.D.L. Chung, *Cem. Concr. Res.*, 1998, **28**(2), 183–187.
[141] Z. Shi and D.D.L. Chung, *Cem. Concr. Res.*, 1999, **29**(3), 435–439.
[142] Q. Mao, B. Zhao, D. Sheng and Z. Li, *J. Wuhan U. Tech.*, Mater. Sci. Ed., 1996, **11**(3), 41–45.
[143] Q. Mao, B. Zhao, D. Shen and Z. Li, *Fuhe Cailiao Xuebao/Acta Materiae Compositae Sinica*, 1996, **13**(4), 8–11.
[144] M. Sun, Q. Mao and Z. Li, *J. Wuhan U. Tech.*, Mater. Sci. Ed., 1998, **13**(4), 58–61.
[145] B. Zhao, Z. Li and D. Wu, *J. Wuhan Univ. Tech.*, Mater. Sci. Ed., 1995, **10**(4), 52–56.
[146] S. Wen and D.D.L. Chung, *Cem. Concr. Res.*, 1999, **29**(6), 961–965.
[147] M. Sun, Z. Li, Q. Mao and D. Shen, *Cem. Concr. Res.*, 1998, **28**(4), 549–554.
[148] M. Sun, Z. Li, Q. Mao and D. Shen, *Cem. Concr. Res.*, 1998, **28**(12), 1707–1712.
[149] S. Wen and D.D.L. Chung, *Cem. Concr. Res.*, 1999, **29**(12), 1989–1993.

[150] D. Bontea, D.D.L. Chung and G.C. Lee, *Cem. Concr. Res.*, 2000, **30**(4), 651–659.
[151] S. Wen and D.D.L. Chung, *Cem. Concr. Res.*, 2000, **30**(12), 1979–1982.
[152] J. Lee and G. Batson, *Materials for the New Millennium*, Proc. 4th Mater. Eng. Conf., Vol. 2 ASCE, New York, 1996, pp. 887–896.
[153] X. Fu and D.D.L. Chung, *ACI Mater. J.*, 1999, **96**(4), 455–461.
[154] Y. Xu and D.D.L. Chung, *Cem. Concr. Res.*, 1999, **29**(7), 1117–1121.
[155] Y. Shinozaki, *Adv. Mater.: Looking Ahead to the 21st Century*, Proc. 22nd National SAMPE Tech. Conf. Vol. 22 SAMPE, Covina, CA, 1990, pp. 986–997.
[156] X. Fu and D.D.L. Chung, *Cem. Concr. Res.*, 1995, **25**(4), 689–694.
[157] J. Hou and D.D.L. Chung, *Cem. Concr. Res.*, 1997, **27**(5), 649–656.
[158] G.G. Clemena, *Materials Performance*, 1988, **27**(3), 19–25.
[159] R.J. Brousseau and G.B. Pye, *ACI Mater. J.*, 1997, **94**(4), 306–310.
[160] P. Chen and D.D.L. Chung, *Smart Mater. Struct.*, 1993, **2**, 181–188.
[161] P. Chen and D.D.L. Chung, *J. Electron. Mater.*, 1995, **24**(1), 47–51.
[162] X. Wang, Y. Wang and Z. Jin, *Fuhe Cailiao Xuebao/Acta Materiae Compositae Sinica*, 1998, **15**(3), 75–80.
[163] N. Banthia, S. Djeridane and M. Pigeon, *Cem. Concr. Res.*, 1992, **22**(5), 804–814.
[164] P. Xie, P. Gu and J.J. Beaudoin, *J. Mater. Sci.*, 1996, **31**(15), 4093–4097.
[165] Z. Shui, J. Li, F. Huang and D. Yang, *J. Wuhan Univ. Tech.*, Mater. Sci. Ed., 1995, 10(4), 37–41.
[166] X. Fu and D.D.L. Chung, *Cem. Concr. Res.*, 1998, **28**(6), 795–801.
[167] X. Fu and D.D.L. Chung, *Carbon*, 1998, **36**(4), 459–462.
[168] X. Fu and D.D.L. Chung, *Cem. Concr. Res.*, 1996, **26**(10), 1467–1472; 1997, **27**(2), 314.
[169] T. Fujiwara and H. Ujie, *Tohoku Kogyo Daigaku Kiyo, 1: Rikogakuhen.*, 1987, (7), 179–188.
[170] Y. Shimizu, A. Nishikata, N. Maruyama and A. Sugiyama, *Terebijon Gakkaishi/J. Inst. Television Engineers of Japan*, 1986, **40**(8), 780–785.
[171] P. Chen and D.D.L. Chung, *ACI Mater. J.*, 1996, **93**(2), 129–133.
[172] T. Uomoto and F. Katsuki, *Doboku Gakkai Rombun-Hokokushu/Proc. Japan Soc. Civil Engineers*, 1994–1995, (490), pt. 5-23, 167–174.
[173] C.M. Huang, D. Zhu, C.X. Dong, W.M. Kriven, R. Loh and J. Huang, *Ceramic Eng. Sci. Proc.*, 1996, **17**(4), 258–265.
[174] C.M. Huang, D. Zhu, X. Cong, W.M. Kriven, R.R. Loh and J. Huang, *J. Am. Ceramic Soc.*, 1997, **80**(9), 2326–2332.
[175] T-J. Kim and C-K. Park, *Cem. Concr. Res.* 1998, **28**(7), 955–960.
[176] S. Igarashi and M. Kawamura, *Doboku Gakkai Rombun-Hokokushu/Proc. Japan Soc. Civil Eng.*, 1994, (502), pt 5-25, 83–92.
[177] M.Z. Bayasi and J. Zeng, *ACI Structural J.*, 1997, **94**(4), 442–446.
[178] G. Campione, S. Mindess and G. Zingone, *ACI Mater. J.*, 1999, **96**(1), 27–34.
[179] T. Yamada, K. Yamada and K. Kubomura, *J. Composite Mater.* 1995, **29**(2), 179–194.
[180] R. Fujiura, T. Kojima, K. Kanno, I. Mochida and Y. Korai, *Carbon*, 1993, **31**(1), 97–102.
[181] R.B. Sandor, *Proc. Int. SAMPE Tech. Conf., 22, Advanced Materials: Looking Ahead to the 21st Century*, L.D. Michelove, R.P. Caruso, P. Adams and W.H. Fossey Jr, 1990, pp. 647–657; *SAMPE Q.*, 1991, **22**(3), 23–28.
[182] J. Economy, H. Jung and T. Gogeva, *Carbon*, 1992, **30**(1), 81–85.
[183] G. Savage, *Met. Mater. (Inst. Met.)*, 1988, **4**(9), 544–548.
[184] B. Rhee, S. Ryu, E. Fitzer and W. Fritz, *High Temp. - High Pressures*, 1987, **19**(6), 677–686.
[185] J. Chlopek and S. Blzewicz, *Carbon*, 1991, **29**(2), 127–131.
[186] O.P. Bahl, L.M. Manocha, G. Bhatia, T.L. Dhami and R.K. Aggarwal, *J. Sci. Ind. Res.*, 1991, **50**(7), 533–538.
[187] H.A. Aglan, *Int. SAMPE Symp. Exhib.*, 1991, **36**(2), 2237–2248.
[188] A.J. Hosty, B. Rand and F.R. Jones, *Inst. Phys. Conf. Ser.*, Vol. 111 New Materials and Their Applications 1990, IOP, Bristol, UK and Philadelphia, 1990, pp. 521–530.
[189] G. Gray and G.M. Savage, *Materials at High Temperatures*, 1991, **9**(2), 102–109.
[190] T. Hosomura and H. Okamoto, *Mater. Sci. Eng.*, 1991, **A143**(1–2), 223–229.
[191] S. Kimura, Y. Tanabe and E. Yasuada, *Proc. 4th Japan–US Conf. Compos. Mater.*, Technomic, Lancaster, PA, 1989, pp. 867–874.
[192] E. Yasuda, Y. Tanabe and K. Taniguchi, *Rep. Res. Lab. Eng. Mater., Tokyo Inst. Technol.*, 1988, **13**, 113–119.
[193] S. Marinkovic and S. Dimitrijevic, *Carbon*, 1985, **23**(6), 691–699.
[194] J.W. Davidson, *Proc. Metal and Ceramic Matrix Composite Processing Conf.*, Vol. II, US Dept. of Defense Information Analysis Centers, 1984, pp. 181–185.
[195] V. Markovic, *Fuel*, 1987, **66**(11), 1512–1515.

[196] P.M. Sheaffer and J.L. White, US Patent 4,986,943 (1991).
[197] A.J. Hosty, B. Rand and F.R. Jones, *Inst. Phys. Conf. Ser.*, Vol. 111, New Materials and their Applications 1990, IOP, Bristol, UK and Philadelphia, 1990, pp. 531-530.
[198] I. Charit, H. Harel, S. Fischer and G. Marom, *Thermochim. Acta*, 1983, **62**, 237-248.
[199] T. Chang and A. Okura, *Trans. Iron Steel Inst. Jpn.*, 1987, **27**(3), 229-237.
[200] L.M. Manocha and O.P. Bahl, *Carbon*, 1988, **26**(1), 13-21.
[201] L.M. Manocha, O.P. Bahl and Y.K. Singh, *Tanso*, 1989, **140**, 255-260.
[202] L.M. Manocha, O.P. Bahl and Y.K. Singh, *Carbon*, 1991, **29**(3), 351-360.
[203] L.M. Manocha, *Composites (Guildford, UK)*, 1988, **19**(4), 311-319.
[204] W. Kowbel and C.H. Shan, *Carbon*, 1990, **28**(2-3), 287-299.
[205] R.J. Zaldivar, R.W. Kobayashi and G.S. Rellick, *Carbon*, 1991, **29**(8), 1145-1153.
[206] R.E. Yeager and S.C. Shaw, *Proc. Metal and Ceramic Matrix Composite Processing Conf.*, Vol. II, US Dept. of Defense Information Analysis Centers, 1984, pp. 145-180.
[207] K.S. Goto, K.H. Han and G.R. St Pierre, *Trans. Iron Steel Inst. Jpn.*, 1986, **26**(7), 597-603.
[208] P. Crocker and B. McEnaney, *Carbon*, 1991, **29**(7), 881-885.
[209] J.E. Sheehan, *Carbon*, 1989, **27**(5), 709-715.
[210] H.V. Johnson, US Patent 1,948,382.
[211] L.M. Manocha, O.P. Bahl, and Y.K. Singh, *Carbon*, 1989, **27**(3), 381-387.
[212] T.-M. Wu, W.-C. Wei and S. Hsu, *Carbon*, 1991, **29**(8), 1257-1265.
[213] R.C. Dickinson, *Mater. Res. Soc. Symp. Proc.*, **125** (Materials Stability and Environmental Degradation), A. Barkatt, E.D. Verink Jr and L.R. Smith, 1988, pp. 3-11.
[214] F.J. Buchanan and J.A. Little, *Surf. Coat. Technol.*, 1991, **46**(2), 217-226.
[215] J.R. Strife and J.E. Sheehan, *Am. Ceram. Soc. Bull.*, 1988, **67**(2), 369-374.
[216] D.W. McKee, *Carbon*, 1987, **25**(4), 551-557.
[217] D.W. McKee, *Carbon*, 1986, **24**(6), 659-665.
[218] P.E. Gray, US Patent 4,894,286 (1990).
[219] T.D. Nixon and J.D. Cawley, *J. Am. Ceram. Soc.*, 1992, **75**(3), 703-708.
[220] P.E. Gray, US Patent 4,937,101 (1990).
[221] L.M. Niebylski, US Patent 4,910,173 (1990).
[222] D.W. McKee, *Carbon*, 1988, **26**(5), 659-665.
[223] P. Ehrburger, *Carbon Fibers Filaments and Composites*, J.L. Figueiredo, C.A. Bernardo, R.T.K. Baker and K.J. Huttinger, Kluwer Academic, Dordrecht, 1990, pp. 327-336.
[224] P. Ehrburger, P. Baranne and J. Lahaye, *Carbon*, 1986, **24**(4), 495-499.
[225] B. Bavarian, V. Arrieta and M. Zamanzadeh, *Proc. Int. SAMPE Symp. and Exhib.*, 35, Advanced Materials: Challenge Next Decade, G. Janicki, V. Bailey and H. Schjelderup, 1990, pp. 1348-1362.
[226] T.-M. Wu, W.-C. Wei and S. Hsu, *Carbon*, 1991, **29**(8), 1257-1265.
[227] K.H. Han, H. Ono, K.S. Goto and G.R. Pierre, *J. Electrochem. Soc.*, 1987, **134**(4), 1003-1009.
[228] E.J. Hippo, N. Murdie and W. Kowbel, *Carbon*, 1989, **27**(3), 331-336.
[229] C.T. Ho and d.D.L. Chung, *Carbon*, 1990, **28**(6), 815-824.
[230] E.J. Hippo, N. Murdie and A. Hyjazie, *Carbon*, 1989, **27**(5), 815-824.
[231] R.A. Meyer and S.R. Gyetvay, *ACS Symp., Ser.*, Vol. 303, Petroleum-Derived Carbons, American Chemical society, Washington, DC, 1986, pp. 380-394.
[232] L.H. Peebles Jr, R.A. Meyer and J. Jortner, *Proc. 2nd Int. Conf. Compos. Interfaces, Interfaces Polym., Ceram., Met. Matrix Compos.*, H. Ishida, Elsevier, New York, 1988, pp. 1-16.
[233] L.E. Jones, P.A. Thrower and P.L. Walker Jr, *Carbon*, 1986, **24**(1), 51-59.
[234] J.X. Zhao, R.C. Bradt and P.L. Walker Jr, *Carbon*, 1985, **23**(1), 9-13.
[235] H. Aglan, *J. Mater. Sci. Lett.*, 1992, **11**(4), 241-243.
[236] L.M. Manocha, O.P. Bahl and Y.K. Singh, *Proc. Int. Conf. Interfacial Phenomena in Composite Materials '89*, F.R. Jones, Butterworth, 1989, pp. 310-315.
[237] W. Huettner, *Carbon Fibers Filaments and Composites*, J.L. Figueiredo, C.A. Bernardo, R.T.K. Baker and K.J. Huttinger, Kluwer Academic, Dordrecht, 1990, pp. 275-300.
[238] R.B. Sandor, *SAMPE Q.*, 1991, **22**(3), 23-28.
[239] H. Weisshaus, S. Kenig and A. Siegmann, *Carbon*, 1991, **29**(8), 1203-1220.
[240] L.A. Feldman, *J. Mater. Sci. Lett.*, 1986, **5**, 1266-1268.
[241] S. Ragan and G.T. Emmerson, *Carbon*, 1992, **30**(3) 339-344.
[242] P.B. Pollock, *Carbon*, 1990, **28**(5), 717-732.
[243] G. Sines, Z. Yang and B.D. Vickers, *J. Am. Ceram. Soc.*, 1989, **72**(1), 54-59.
[244] R.B. Dinwiddie, T.D. Burchell and C.F. Baker, *Ext. Abstr. Program-Bienn. Conf. Carbon*, 1991, **20**, 642-643.
[245] J.W. Sapp Jr, D.A. Bowers, R.B. Dinwiddie and T.D. Burchell, *Ext. Abstr. Program - Bienn. Conf. Carbon*, 1991, **20**, 644-645.

[246] X. Shui and D.D.L. Chung, *Ext. Abstr. Program - Bienn. Conf. Carbon*, 1991, **20**, 376-377.
[247] J.-M. Chiou and D.D.L. Chung, *J. Mater. Sci.*, 1993, **28**, 1435-1446; *J. Mater. Sci.*, 1993, **28**, 1447-1470; *J. Mater. Sci.*, 1993, **28**, 1471-1487.
[248] A. Sakamoto, C. Fujiwara and T. Tsuzuku, *Proc. Jpn. Congr. Mater. Res.*, 1990, **33**, 73-79.
[249] I.W. Hall, *Mettallography*, 1987, **20**(2), 237-246.
[250] D.A. Foster, *Proc. Int. SAMPE Symp. and Exhib.*, 34, Tomorrow's Materials Today, G.A. Zakrzewski, D. Mazenko, S.T. Peters and C.D. Dean, 1989, pp. 1401-1410.
[251] M.F. Amateau, *J. Compos. Mater.*, 1976, **10**, 279.
[252] D.M. Goddard, *J. Mater. Sci.*, 1978, **13**(9), 1841-1848.
[253] A.P. Levitt and H.E. Band, US Patent 4,157,409 (1979).
[254] *Adhesion and Bonding in Composites*, R. Yosomiya, K. Morimoto, A. Nakajima, Y. Ikada and T. Suzuki, Marcel Dekker, New York, 1990, pp. 235-256. (Chapter on Interfacial Modifications and Bonding of Fiber-Reinforced Metal Composite Material.)
[255] D.D. Himbeault, R.A. Varin and K. Piekarski, *Proc. Int. Symp. Process. Ceram. Met. Matrix Compos.*, H. Monstaghaci, Pergamon, New York, 1989, pp. 312-323.
[256] H. Vincent, C. Vincent, J.P. Scharff, H. Mourichoux and J. Bouix, *Carbon*, 1992, **30**(3), 495-505.
[257] H. Katzman, *Proc. Metal and Ceramic Matrix Composite Processing Conf.*, Vol. I, US Dept. of Defense Information Analysis Centers, 1984, pp. 115-140; *J. Mater. Sci.*, 1987, **22**, 144-148; *Mater. Manufacturing Processes*, 1990, **5**(1), 1-15.
[258] L.D. Brown and H.L. Marcus, *Proc. Metal and Ceramic Matrix Composite Processing Conf.*, Vol. II, US Dept. of Defense Information Analysis Centers, 1984, pp. 91-113.
[259] G. Leonhardt, E. Kieselstein, H. Podlesak, E. Than and A. Hofmann, *Mater. Sci. Eng.*, 1991, **A135**, 157-160.
[260] K. Honjo and A. Shindo, *Proc. 1st Compos. Interfaces Int. Conf.*, H. Ishida and J.L. Koenig, North-Holland, New York, 1986, pp, 101-107.
[261] J.A. Cornie, A.S. Argon and V. Gupta, *MRS Bull.*, 1991, **16**(4), 32-38.
[262] H. Landis, Ph.D. dissertation, MIT, 1988.
[263] A.S. Argon, V. Gupta, K.S. Landis and J.A. Cornie, *J. Mater. Sci.*, 1989, **24**, 1207-1218.
[264] J.P. Clement and H.J. Rack, *Proc. Am. Soc. Compos. Symp. High Temp. Compos.*, Technomic, Lancaster, PA, 1989, pp. 11-20.
[265] S. Schamm, J.P. Rocher and R. Naslain, *Proc. 3rd Eur. Conf. Compos. Mater., Dev. Sci. Technol. Compos. Mater.*, A.R. Bunsell, P. Lamicq and A. Massiah, Elsevier, London, 1989, pp. 157-163.
[266] S.N. Patankar, V. Gopinathan and P. Ramakrishnan, *Scripta Metall.*, 1990, **24**, 2197-2202.
[267] S.N. Patankar, V. Gopinathan and P. Ramakrishnan, *J. Mater. Sci. Lett.*, 1990, **9**, 912-913.
[268] R.V. Subramanian and E.A. Nyberg, *J. Mater. Res.*, 1992, **7**(3), 677-688.
[269] P.-W. Chen and D.D.L. Chung, *Composites*, 1993, **24**(1), 33-52.
[270] D.P.H. Hasselman, *Therm. Conduct.*, 1988, **19**, 383-402.
[271] H.S. Kim, R.D. Rawlings and P.S. Rogers, *Br. Ceram. Proc.*, 1989, **42**, 59-68.
[272] R.A.J. Sambell, D.H. Bowen and D.C. Phillips, *J. Mater. Sci.*, 1972, **7**, 663-675.
[273] W.K. Tredway, K.M. Prewo and C.G. Pantano, *Carbon*, 1989, **27**(5), 717-727.
[274] R.A. Allaire, US Patent 4,976,761 (1990).
[275] R.A.J. Sambell, D.C. Phillips and D.H. Bowen, *Carbon Fibres, Their Composites and Applications, Proc. of Int. Conf.*, Plastics Institute, 1974, pp. 105-113.
[276] V. Gunay, P.F. James, F.R. Jones and J.E. Bailey, *Br. Ceram. Proc.*, 1989, **45**, 229-240.
[277] V. Gunay, P.F. James, F.R. Jones and J.E. Bailey, *Inst. Phys. Conf. Ser.*, Vol. 111, New Materials and their Applications 1990, IOP, Bristol, UK and Philadelphia, 1990, pp. 217-226.
[278] B. Rand and R.J. Zeng, *Carbon Fibers Filaments and Composites*, J.L. Figueiredo, C.A. Bernardo, R.T.K. Baker and K.J. Hittinger, Kluwer Academic, Dordrecht, 1990, pp. 367-398.
[279] F.A. Habib, R.G. Cooke and B. Harris, *Br. Ceram. Trans. J.*, 1990, **89**, 115-124.
[280] K.M. Prewo and J.A. Batt, *J. Mater. Sci.*, 1988, **23**, 523.

3 Composite materials for thermal applications

3.1 Introduction

The transfer of heat by conduction is involved in the use of a heat sink to dissipate heat from an electronic package, the heating of an object on a hot plate, the operation of a heat exchanger, the melting of ice on an airport runway by resistance heating, the heating of a cooking pan on an electric range, and in numerous industrial processes that involve heating or cooling. Effective transfer of heat by conduction requires materials (such as a heat sink material) of high thermal conductivity. In addition, it requires a good thermal contact between the two surfaces (such as the surface of a heat sink and the surface of a printed circuit board) across which heat transfer occurs. Without good thermal contacts, the use of expensive thermal conducting materials for the components is not cost effective. The attainment of a good thermal contact requires a thermal interface material, such as a thermal grease, which must be thin (small in thickness) between the mating surfaces, must conform to the topography of the mating surface and preferably should have a high thermal conductivity as well. This chapter addresses materials for thermal conduction, including materials of high thermal conductivity and thermal interface materials. In addition, this chapter addresses materials for thermal insulation and heat retention, which are important in buildings for the purpose of energy saving. Thermoelectric applications can be considered a subset of thermal applications, but they are addressed in Chapter 6 rather than this chapter.

3.2 Materials of high thermal conductivity

Materials of high thermal conductivity are needed for the conduction of heat for the purpose of heating or cooling. One of the most critical needs is in the electronics industry. Because of the miniaturization and increasing power of microelectronics, heat dissipation is key to the reliability, performance and further miniaturization of microelectronics. The heat dissipation problem is so severe that even expensive thermal conductors such as diamond, metal–matrix composites and carbon–matrix composites are being used in high-end microelectronics. Because of the low coefficient of thermal expansion (CTE) of semiconductor chips and their substrates, heat sinks also need to have a low CTE. Thus, the requirement for the thermal conductor material is not just high thermal conductivity, but low CTE as well. For example, copper is a good thermal conductor but its CTE is high. Therefore, copper–matrix composites containing low-CTE fillers such as carbon fibers or molybdenum particles are used. For lightweight electronics, such as laptop computers and avionics, an additional requirement for the thermal conductor material is low density. As aluminum and carbon are light compared to copper, aluminum, carbon and their composites are used for this purpose. Compared to aluminum, carbon has the additional advantage of being corrosion resistant.

3.2.1 Metals, diamond and ceramics

Table 3.1 gives the thermal conductivity of various metals. Copper is most commonly used when materials of high thermal conductivity are required. However, copper suffers from a high value of CTE. A low CTE is needed when the adjoining component has a low CTE. When the CTEs of the two adjoining materials are sufficiently different and the temperature is varied, thermal stress occurs and may even cause warpage. This is the case when copper is used as heat sink for a printed wiring board, which is a continuous fiber polymer–matrix composite that has a lower CTE than copper. Molybdenum and tungsten are metals that have low CTE, but their thermal conductivity is poor compared to copper. The alloy Invar (64Fe–36Ni) is outstandingly low in CTE among metals, but it is very poor in thermal conductivity. Diamond is most attractive, as it has very high thermal conductivity and low CTE, but it is expensive. Aluminum is not as conductive as copper, but it has a low density, which is attractive for aircraft electronics and applications (e.g., laptop computers) which require low weight [1,2]. Aluminum nitride is not as conductive as copper, but it is attractive in its low CTE. Diamond and most ceramic materials are very different from metals in their electrical insulation ability. In contrast, metals are conducting both thermally and electrically. For applications which require thermal conductivity and electrical insulation, diamond and appropriate ceramic materials can be used, but metals cannot.

Table 3.1 Thermal properties and density of various materials

Material	Thermal conductivity (W/m.K)	Coefficient of thermal expansion ($10^{-6}\,°C^{-1}$)	Density (g/cm^3)
Aluminum	247	23	2.7
Gold	315	14	19.32
Copper	398	17	8.9
Lead	30	39	11
Molybdenum	142	4.9	10.22
Tungsten	155	4.5	19.3
Invar	10	1.6	8.05
Kovar	17	5.1	8.36
Diamond	2000	0.9	3.51
Beryllium oxide	260	6	3
Aluminum nitride	320	4.5	3.3
Silicon carbide	270	3.7	3.3

3.2.2 Metal–matrix composites

One way to lower the CTE of a metal is to form a metal–matrix composite [3] (Section 2.5) by using a low CTE filler. Ceramic particles such as AlN and SiC are used for this purpose, because of their combination of high thermal conductivity and low CTE. As the filler usually has lower CTE and lower thermal conductivity than the metal matrix, the higher the filler volume fraction in the composite, the lower the CTE and the lower is the thermal conductivity.

Metal–matrix composites with discontinuous fillers (commonly particles) are attractive for their processability into various shapes. However, layered composites in the form of a matrix–filler–matrix sandwich are useful for planar components. Discontinuous fillers are most commonly ceramic particles. The filler sheets are most commonly low-CTE metal alloy sheets (e.g., Invar or 64Fe-36Ni, and Kovar or 54Fe-29Ni-17Co). Aluminum and copper are common metal matrices owing to their high conductivity.

Aluminum–matrix composites

Aluminum is the most dominant matrix for metal–matrix composites for both structural and electronic applications. This is because of the low cost of aluminum and the low melting point of aluminum (660°C) facilitating composite fabrication by methods that involve the melting of the metal.

Liquid-phase methods for the fabrication of metal–matrix composites include liquid metal infiltration, which usually involves using pressure (from a piston or compressed gas) to push the molten metal into the pores of a porous preform comprising the filler (commonly particles that are not sintered) and a small amount of a binder [4–6]. Pressureless infiltration is less common but is possible [7,8]. The binder prevents the filler particles from moving during infiltration, and also provides sufficient compressive strength to the preform, so that the preform will not be deformed during infiltration. This method thus provides near-net-shape fabrication, i.e., the shape and size of the composite product are the same as those of the preform. Since machining of the composite is far more difficult than that of the preform, near-net-shape fabrication is desirable.

In addition to near-net-shape fabrication capability, liquid metal fabrication is advantageous in being able to provide composites with high filler volume fractions (up to 70%). A high filler volume fraction is necessary in order to attain a low enough CTE ($<10 \times 10^{-6}$/°C) in the composite, even if the filler is a low-CTE ceramic (e.g., SiC), since the aluminum matrix has a relatively high CTE [9,10]. However, to attain a high volume fraction using liquid metal infiltration, the binder used must be in a small amount (so as not to clog the pores in the preform) and still be effective. Hence, the binder technology [11–13] is critical.

The ductility of a composite decreases as the filler volume fraction increases, so that a composite with a low enough CTE is quite brittle. Although the brittleness is not acceptable for structural applications, it is acceptable for electronic applications.

Another liquid-phase technique is stir casting [1], which involves stirring the filler in the molten metal and then casting. This method suffers from non-uniform distribution of filler in the composite due to the difference in density between filler and molten metal and the consequent tendency for the filler to either float or sink in the molten metal prior to solidification. Stir casting also suffers from not being capable of producing composites with a high filler volume fraction.

Yet another liquid-phase technique is plasma spraying [14], which involves spraying a mixture of molten metal and filler onto a substrate. This method suffers from the relatively high porosity of the resulting composite and the consequent need for densification by hot isostatic pressing or other methods, which tend to be expensive.

A solid-phase technique is *powder metallurgy*, which involves mixing the matrix metal powder and filler and subsequent sintering under heat and pressure [14]. This method is relatively difficult for the aluminum matrix because aluminum has a protective oxide on it and the oxide layer on the surface of each aluminum particle hinders sintering. Furthermore, this method is usually limited to low volume fractions of the filler.

The most common filler used is silicon carbide (SiC) particles, owing to the low cost and low CTE of SiC [15]. However, SiC suffers from its reactivity with aluminum. The reaction is

$$3SiC + 4Al \rightarrow 3Si + Al_4C_3.$$

It becomes more severe as the composite is heated. The aluminum carbide is a brittle reaction product which lines the filler-matrix interface of the composite, thus weakening the interface. Silicon, the other reaction product, dissolves in the aluminum matrix, lowering the melting temperature of the matrix, and causing non-uniformity in the phase distribution and mechanical property distribution [16]. Furthermore, the reaction consumes a part of the SiC filler [17].

A way to diminish this reaction is to use an Al–Si alloy matrix, since the silicon in the alloy matrix promotes the opposite reaction and thus inhibits this reaction. However, the Al–Si matrix is less ductile than the Al matrix, thus causing the mechanical properties of the Al–Si matrix composite to be very poor compared to those of the corresponding Al-matrix composite. Thus, the use of an Al–Si alloy matrix is not a solution to the problem.

An effective solution is to replace Si–C by aluminum nitride (AlN) particles, which do not react with aluminum, thus resulting in superior mechanical properties in the composite [18]. The fact that AlN tends to have a higher thermal conductivity than SiC helps the thermal conductivity of the composite. Since the cost of the composite fabrication process dominates the cost of producing the composites, the higher material cost of AlN compared to SiC does not matter, especially for electronic packaging. Aluminum oxide (Al_2O_3) also does not react with aluminum, but it is low in thermal conductivity and tends to suffer from particle agglomeration [18].

Other than ceramics such as SiC and AlN, another filler used in aluminum-matrix composites is carbon in the form of fibers of diameter around 10 μm [19–23] and, less commonly, filaments of diameter less than 1 μm [24]. Carbon also suffers from reactivity with aluminum to form aluminum carbide. However, fibers are more effective than particles for reducing the CTE of the composite. Carbon fibers can even be continuous in length. Moreover, carbon, especially when graphitized, is much more thermally conductive than ceramics. In fact, carbon fibers that are sufficiently graphitic are even more thermally conductive than the metal matrix, so that the thermal conductivity of the composite increases with increasing fiber volume fraction. However, these fibers are expensive. The mesophase-pitch-based carbon fiber K-1100 from Amoco Performance Products (Alpharetta, GA) exhibits longitudinal thermal conductivity of 1000 W m^{-1} K^{-1} [25,26].

Both carbon and SiC suffer from forming a galvanic couple with aluminum, which is the anode – the component in the composite that is corroded. The corrosion becomes more severe in the presence of heat and/or moisture.

The thermal conductivity of aluminum-matrix composites depends on the filler and its volume fraction, the alloy matrix heat treatment condition, as well as the filler-matrix interface [18,27].

To increase the thermal conductivity of SiC aluminum-matrix composite, a diamond film can be deposited on the composite [28]. The thermal conductivity of single-crystal diamond is 2000–2500 W m^{-1} K^{-1}, although a diamond film is not single crystalline.

Copper-matrix composites

Because copper is heavy anyway, the filler does not have to be lightweight. Thus, low CTE but heavy metals such as tungsten [29,30], molybdenum [31,32] and Invar [33–35] are used as fillers. These metals (except Invar) have the advantage that they are quite conductive thermally and are available in particle and sheet forms, so that they are suitable for particulate as well as layered [36,37] composites. Yet another advantage of the metal fillers is the better wettability of the molten matrix metal with metal fillers than with ceramic fillers, in case the composite is fabricated by a liquid-phase method.

An advantage of copper over aluminum is its non-reactivity with carbon, so carbon is a highly suitable filler for copper. Additional advantages are that carbon is lightweight and carbon fibers are available in a continuous form. Furthermore, copper is a rather noble metal, as shown by its position in the Electromotive Series, so it does not suffer from the corrosion which plagues aluminum. Carbon used as a filler in copper is in the form of fibers of diameter around 10 μm [22,38–45]. As carbon fibers that are sufficiently graphitic are even more thermally conductive than copper, the thermal conductivity of a copper-matrix composite can exceed that of copper.

Less common fillers for copper are ceramics such as silicon carbide, titanium diboride (TiB$_2$) and alumina [46–48].

The melting point of copper is much higher than that of aluminum, so the fabrication of copper-matrix composites is commonly done by powder metallurgy, although liquid metal infiltration is also used [22,49,50]. In the case of liquid metal infiltration, the metal matrix is often a copper alloy (e.g., Cu–Ag) chosen for the reduced melting temperature and good castability [50].

Powder metallurgy conventionally involves mixing the metal matrix powder and the filler, and subsequent pressing and then sintering under either heat or both heat and pressure. The problem with this method is that it is limited to low volume fractions of the filler. In order to attain high volume fractions, a less conventional method of powder metallurgy is recommended. This method involves coating the matrix metal on the filler units, followed by pressing and sintering [32,46,51,52]. The mixing of matrix metal powder with the coated filler is not necessary, although it can be done to decrease the filler volume fraction in the composite. The metal coating on the filler forces the distribution of matrix metal to be uniform even when the metal volume fraction is low (i.e., when the filler volume fraction is high). On the other hand, with the conventional method, the matrix metal distribution is not uniform when the filler volume fraction is high, thus causing porosity and the presence of filler agglomerates, in each of which the filler units directly touch one another; this microstructure results in low thermal conductivity and poor mechanical properties.

Continuous carbon fiber copper–matrix composites can be made by coating the fibers with copper and then diffusion bonding (i.e., sintering) [38,40, 44,53–55]. This method is akin to the above-mentioned less conventional method of powder metallurgy.

Less common fillers used in copper include diamond powder [50,56], aluminosilicate fibers [57] and Ni–Ti alloy rod [58]. The Ni–Ti alloy is attractive for its negative CTE of $-21 \times 10^{-6}/°C$.

The coating of a carbon fiber copper–matrix composite with a diamond film has been done to enhance thermal conductivity [43].

Beryllium–matrix composites

Beryllium oxide (BeO) has a high thermal conductivity (Table 3.1). Beryllium–matrix BeO–platelet composites with 20–60 vol.% BeO exhibit low density (2.30 g/cm^3 at 40 vol.% BeO, compared to 2.9 g/cm^3 for Al/SiC at 40 vol.% SiC), high thermal conductivity (232 W m^{-1} K^{-1} at 40 vol.% BeO, compared to 130 W m^{-1} K^{-1} for Al/SiC at 40 vol.% SiC), low CTE (7.5 $\times$ 10^{-6}/°C at 40 vol.% BeO, compared to 12.1 $\times$ 10^{-6}/°C at 40 vol.% SiC), and high modulus (317 GPa at 40 vol.% BeO, compared to 134 GPa for Al/SiC at 40 vol.% SiC) [59,60].

3.2.3 Carbon–matrix composites

Carbon is an attractive matrix for composites for thermal conduction because of its thermal conductivity (although not as high as those of metals) and low CTE (lower than those of metals). Furthermore, carbon is corrosion resistant (more corrosion resistant than metals) and lightweight (lighter than metals). Yet another advantage of the carbon matrix is its compatibility with carbon fibers, in contrast to the common reactivity between a metal matrix and its fillers. Hence, carbon fibers are the dominant filler for carbon–matrix composites. Composites with both filler and matrix being carbon are called carbon–carbon composites (Section 2.4) [61]. Their primary applications in relation to thermal conduction are heat sinks [62], thermal planes [63] and substrates [64]. There is considerable competition between carbon–carbon composites and metal–matrix composites for the same applications.

The main drawback of carbon–matrix composites is their high cost of fabrication, which typically involves making a pitch-matrix or resin-matrix composite and subsequent carbonization (by heating at 1000–1500°C in an inert atmosphere) of the pitch or resin to form a carbon–matrix composite. After carbonization, the porosity is substantial in the carbon matrix, so pitch or resin is impregnated into the composite and then carbonization is carried out again. Quite a few impregnation–carbonization cycles are needed in order to reduce the porosity to an acceptable level, thus resulting in the high cost of fabrication. Graphitization (by heating at 2000–3000°C in an inert atmosphere) may follow carbonization (typically in the last cycle) in order to increase the thermal conductivity, which increases with the degree of graphitization. However, graphitization is an expensive step. Some or all of the impregnation-carbonization cycles may be replaced by chemical vapor infiltration (CVI), in which a carbonaceous gas infiltrates the composite and decomposes to form carbon.

Carbon–carbon composites have been made by using conventional carbon fibers of diameter around 10 μm [62,63,65], as well as carbon filaments grown catalytically from carbonaceous gases and of diameter less than 1 μm [24]. By using graphitized carbon fibers, thermally conductivities exceeding that of copper can be reached.

To increase the thermal conductivity, carbon–carbon composites have been impregnated with copper [66,67] and have been coated with a diamond film [68].

3.2.4 Carbon and graphite

An all-carbon material (called *ThermalGraph*, a tradename of Amoco Performance Products, Alpharetta, GA), made by consolidating oriented precursor carbon fibers without a binder and subsequent carbonization and optional graphitization, exhibits thermal conductivity ranging from 390 to 750 W m^{-1} K^{-1} in the fiber direction of the material.

Another material is pyrolytic graphite (called TPG) encased in a structural shell [69]. The graphite (highly textured with the c-axes of the grains preferentially perpendicular to the plane of the graphite), has an in-plane thermal conductivity of 1700 W m^{-1} K^{-1} (four times that of copper), but it is mechanically weak owing to the tendency to shear in the plane of the graphite. The structural shell serves to strengthen by hindering shear.

Pitch-derived carbon foams with thermal conductivity up to 150 W m^{-1} K^{-1} after graphitization are attractive for their high specific thermal conductivity (thermal conductivity divided by density) [70].

3.2.5 Ceramic–matrix composites

Background on ceramic–matrix composites is given in Section 2.6.

The SiC matrix is attractive owing to its high CTE compared to the carbon matrix, although it is not as thermally conductive as carbon. The CTE of carbon–carbon composites is too low (0.25 × 10^{-6}/°C), thus resulting in reduced fatigue life in chip-on-board (COB) applications with silica chips (CTE = 2.6 × 10^{-6}/°C). The SiC-matrix carbon fiber composite is made from a carbon–carbon composite by converting the matrix from carbon to SiC [65]. To improve the thermal conductivity of the SiC-matrix composite, coatings in the form of chemical vapor deposited AlN or Si have been used. The SiC-matrix metal (Al or Al–Si) composite, as made by a liquid-exchange process, also exhibits relatively high thermal conductivity [71].

Borosilicate glass matrix is attractive owing to its low dielectric constant (4.1 at 1 MHz for B_2O_3–SiO_2–Al_2O_3–Na_2O glass), compared to 8.9 for AlN, 9.4 for alumina

(90%), 42 for SiC, 6.8 for BeO, 7.1 for cubic boron nitride, 5.6 for diamond and 5.0 for glass–ceramic. A low value of dielectric constant is desirable for electronic packaging applications. On the other hand, glass has a low thermal conductivity. Hence, fillers with relatively high thermal conductivity are used with the glass matrix. An example is continuous SiC fibers, the glass–matrix composites of which are made by tape casting, followed by sintering [72]. Another example is aluminum nitride with interconnected pores (about 28 vol.%), the composites of which are obtained by glass infiltration to a depth of about 100 μm [72–74].

3.2.6 Polymer–matrix composites

Polymer–matrix composites (Section 2.2) with continuous or discontinuous fillers are used for thermal management. Composites with continuous fillers (fibers, whether woven or not) are used as substrates, heat sinks and enclosures. Composites with discontinuous fillers (particle or fibers) are used for die (semiconductor chip) attachment, electrically/thermally conducting adhesives, encapsulations and thermal interface materials. Composites with discontinuous fillers can be in a paste form during processing, thus allowing application by printing (screen printing or jet printing) and injection molding. Composites with continuous fillers cannot undergo paste processing, but the continuous fillers provide lower thermal expansion and higher conductivity than discontinuous fillers.

Composites can have thermoplastic or thermosetting matrices. Thermoplastic matrices have the advantage that a connection can be reworked by heating for the purpose of repair, whereas thermosetting matrices do not allow reworking. On the other hand, controlled-order thermosets are attractive for their thermal stability and dielectric properties [75]. Polymers exhibiting low dielectric constant, low dissipation factor, low coefficient of thermal expansion and compliance are preferred [76].

Composites can be electrically conducting or electrically insulating; the electrical conductivity is provided by a conductive filler. The composites can be both electrically and thermally conducting, as attained by the use of metal or graphite fillers; they can be electrically insulating but thermally conducting, as attained by the use of diamond, aluminum nitride, boron nitride or alumina fillers [77–80]. An electrically conducting composite can be isotropically conducting [81,82] or anisotropically conducting [83]. A z-axis conductor is an example of an anisotropic conductor; it is a film which is electrically conducting only in the z-axis, i.e., in the direction perpendicular to the plane of the film. As one z-axis film can replace a whole array of solder joints, z-axis films are valuable for solder replacement, processing cost reduction and repairability improvement.

Epoxy-matrix composites with continuous glass fibers and made by lamination are most commonly used for printed wiring boards, owing to the electrically insulating property of glass fibers and the good adhesive behavior and established industrial usage of epoxy. Aramid (Kevlar) fibers can be used instead of glass fibers to provide lower dielectric constant [84]. Alumina (Al_2O_3) fibers can be used for increasing the thermal conductivity [85]. By selecting the fiber orientation and loading in the composite, the dielectric constant can be decreased and thermal conductivity can be increased [86]. By impregnating the yarns or fabrics with a silica-based sol and subsequent firing, thermal expansion can be reduced [87].

Matrices other than epoxy can be used. Examples are polyimide and cyanate ester [88].

For heat sinks and enclosures, conducting fibers are used, since the conducting fibers enhance thermal conductivity and the ability to shield electromagnetic interference (EMI). EMI shielding is particularly important for enclosures [89]. Carbon fibers are most commonly used for these applications, owing to their conductivity, low thermal expansion and wide availability as a structural reinforcement. For high thermal conductivity, carbon fibers made from mesophase pitch [90–96] or copper-plated carbon fibers are preferred [97–99].

3.3 Thermal interface materials

Thermal interface materials are needed to improve thermal contacts. By placing a thermal interface material at the interface between two components across which heat must flow, the thermal contact between the two components is improved. A primary market for thermal interface materials is in the electronics industry, as heat dissipation is critical to the performance, reliability and further miniaturization of microelectronics. For example, a thermal interface material is used to improve the thermal contact between a heat sink and a printed circuit board, or between a heat sink and a chip carrier.

A thermal interface material can be a thermal fluid, a thermal grease (paste), a resilient thermal conductor or solder which is applied in the molten state. A thermal fluid, thermal grease or molten solder is spread on the mating surfaces. A resilient thermal conductor is sandwiched by the mating surfaces and held in place by pressure. Thermal fluids are most commonly mineral oil. Thermal greases (pastes) are most commonly conducting particle (usually metal or metal oxide) filled silicone. Resilient thermal conductors are most commonly conducting particle filled elastomers. Out of these four types of thermal interface materials, thermal greases (based on polymers, particularly silicone) and solder are by far most commonly used. Resilient thermal conductors are not as well developed as thermal fluids or greases.

As the materials to be interfaced are good thermal conductors (such as copper), the effectiveness of a thermal interface material is enhanced by high thermal conductivity and low thickness of the interface material and low thermal contact resistance between the interface material and each mating surface. As the mating surfaces are not perfectly smooth, the interface material must be able to flow or deform, so as to conform to the topography of the mating surfaces. If the interface material is a fluid, grease or paste, it should have a high fluidity (workability) so as to conform and to have a small thickness after mating. On the other hand, the thermal conductivity of the grease or paste increases with increasing filler content and this is accompanied by decrease in the workability. Without a filler, as in the case of an oil, thermal conductivity is poor. A thermal interface material in the form of a resilient thermal conductor sheet (e.g., a felt consisting of conducting fibers clung together without a binder, a resilient polymer–matrix composite containing a thermally conducting filler, and a form of graphite known as flexible graphite) usually cannot be as thin or as conformable as one in the form of a fluid, grease or paste, so its effectiveness requires a very high thermal conductivity within it.

Solder is commonly used as a thermal interface material for enhancing the thermal contact between two surfaces. This is because solder can melt at rather low temperatures and the molten solder can flow and spread itself thinly on the adjoining surfaces, thus resulting in high thermal contact conductance at the interface between solder and each of the adjoining surfaces. Furthermore, solder in the metallic solid state is a good thermal conductor. In spite of the high thermal conductivity of solder, the thickness of the solder greatly affects the effectiveness of the solder as a thermal interface material; a small thickness is desirable [100]. Moreover, the tendency for solder to react with copper to form intermetallic compounds [101] reduces the thermal contact conductance of the solder–copper interface.

Thermal pastes are predominantly based on polymers, particularly silicone [102–105], although thermal pastes based on sodium silicate have been reported to be superior in providing high thermal contact conductance [106]. The superiority of sodium-silicate-based pastes over silicone-based pastes is primarily due to the low viscosity of sodium silicate compared to silicone, and the importance of high fluidity in the paste so that the paste can conform to the topography of the surfaces with which it interfaces.

A particularly attractive thermal paste is based on polyethylene glycol (PEG, a polymer) of a low molecular weight (400 amu) [107]. These pastes are superior to silicone-based pastes and are as good as sodium-silicate-based pastes because of the low viscosity of PEG and the contribution of lithium ions (a dopant) in the paste to thermal conduction. Compared to sodium-silicate-based pastes, PEG-based pastes are advantageous in their long-term compliance, in contrast to the long-term rigidity of sodium silicate. Compliance is attractive for decreasing the thermal stress, which can cause thermal fatigue.

Table 3.2 gives the thermal contact conductance for different thermal interface materials sandwiched by copper. Included in the comparison are results obtained with the same testing method on silicone-based paste, sodium-silicate-based pastes and solder [106,107]. PEG (i.e., A) gives a much higher thermal contact conductance (11.0×10^4 W m^{-2} °C^{-1}) than silicone (3.08×10^4 W m^{-2} °C^{-1}) owing to its relatively low viscosity, but the conductance is lower than that given by sodium silicate (14.1×10^4 W m^{-2} °C^{-1}), in spite of its low viscosity, owing to the molecular nature of PEG. The addition of the Li salt (1.5 wt.%) to PEG (i.e., to obtain C) raises the conductance from 11.0×10^4 to 12.3×10^4 W m^{-2} °C^{-1}, even though the viscosity is increased. The further addition of water and DMF (i.e., F) raises the conductance to 16.0×10^4 W m^{-2} °C^{-1} and decreases the viscosity. Thus, the addition of water and DMF is very influential, as water and DMF help the dissociation of the lithium salt. The still further addition of BN particles (18.0 vol.%) (i.e., F_2) raises the conductance to 18.9×10^4 W m^{-2} °C^{-1}. The positive effect of BN is also shown by comparing the results of C and D (which are without water or DMF) and by comparing the results of A and B (which are without Li$^+$). In the absence of the lithium salt, water and DMF also help, although not greatly, as shown by comparing A and J. The viscosity increases with lithium salt content, as shown by comparing J, E, F, G, H and I. Comparison of E, F, G, H and I shows that the optimum lithium salt content for the highest conductance is 1.5 wt.%. That an intermediate lithium salt content gives the highest conductance is probably because of the enhancement of the thermal conductivity by Li$^+$ ions and the increase in viscosity caused by lithium salt addition. Both high conductivity and low viscosity (good conformability) are desirable for a high contact

Composite materials for thermal applications

Table 3.2 Thermal contact conductance for various thermal interface materials between copper disks at 0.46 Ma contact pressure [105,106].

Thermal interface material		Interface material thickness (μm) (± 10)	Thermal contact conductance (10^4 W/m^{-2} °C-1)	Viscosity (cps) (± 0.3)
Description	Designation			
PEG	A	<25	11.0 ± 0.3	127[b]
PEG + BN (18 vol.%)	B	25	12.3 ± 0.3	/
PEG + Li salt (1.5 wt.%)	C	25	12.3 ± 0.3	143[b]
PEG + Li salt (1.5 wt.%) + BN (18 vol.%)	D	25	13.4 ± 0.4	/
PEG + water + DMF	J	<25	12.5 ± 0.2	75.6[b]
J + Li salt (0.75 wt.%)	E	<25	11.4 ± 0.3	79.7[b]
J + Li salt (1.5 wt.%)	F	<25	16.0 ± 0.5	85.6[b]
J + Li salt (3.0 wt.%)	G	<25	11.6 ± 0.2	99.0[b]
J + Li salt (4.5 wt.%)	H	<25	9.52 ± 0.25	117[b]
J + Li salt (6.0 wt.%)	I	<25	7.98 ± 0.16	120[b]
F + BN (16.0 vol.%)	F1	25	18.5 ± 0.8	/
F + BN (18.0 vol.%)	F2	25	18.9 ± 0.8	/
F + BN (19.5 vol.%)	F3	25	15.3 ± 0.2	/
F + BN (21.5 vol.%)	F4	25	14.0 ± 0.5	/
G + BN (16.0 vol.%)	G1	25	17.0 ± 0.5	/
G + BN (18.0 vol.%)	G2	25	17.3 ± 0.6	/
G + BN (19.5 vol.%)	G3	25	14.9 ± 0.6	/
G + BN (21.5 vol.%)	G4	25	13.4 ± 0.4	/
H + BN (18.0 vol.%)	H1	25	13.9 ± 0.4	/
Solder	/	25	20.8 ± 0.6[a]	/
Sodium silicate + BN (16.0 vol.%)	/	25	18.2 ± 0.7	/
Sodium silicate + BN (17.3 vol.%)	/	25	15.5 ± 0.4	/
Sodium silicate	/	<25	14.1 ± 0.5	206[c]
Silicone/BN	/	25	10.9 ± 1.5	/
Silicone	/	<25	3.08 ± 0.03	8800[d]
None	/	/	0.681 ± 0.010	/

[a] At zero contact pressure (not 0.46 MPa).
[b] Measured using the Ubbelohde method.
[c] Measured using the Ostwald method.
[d] Value provided by the manufacturer.

conductance. Comparison of F_1, F_2, F_3 and F_4 shows that the optimum BN content is 18.0 vol.%, as also indicated by comparing G_1, G_2, G_3 and G_4. Among all the PEG-based pastes, the highest conductance is given by F_2, as it has the optimum lithium salt content as well as the optimum BN content. An optimum in the BN content also occurs for BN-filled sodium-silicate-based pastes [106]. It is due to the increase in both the thermal conductivity and the viscosity of the paste as the BN content increases. The best PEG-based paste (i.e., F_2) is similar to the best sodium-silicate-based paste in conductance. Both are better than BN-filled silicone, but both are slightly inferior to solder. Although solder gives the highest conductance, it suffers from the need for heating during soldering. In contrast, heating is not needed in the use of PEG-based pastes, silicone-based pastes or silicate-based pastes.

Phase-change materials, which are solid at room temperature but melt at temperatures around the service temperature, are attractive as thermal interface materials. This is because the phase change (melting) provides a mechanism of heat absorption and the molten state is associated with high fluidity. An example of a phase-change material is paraffin wax, which melts at 48°C. The addition of boron nitride particles to the wax makes the material effective as a thermal interface material; the thermal contact conductance (between copper, as in Table 3.2) reaches 17×10^4 W m^{-2} °C^{-1} at 22°C and 0.3 MPa [108].

Flexible graphite (a flexible foil that is all graphite) is effective as a thermal interface material if the thickness is low (0.13 mm), the density is low (1.1 g/cm^3) and the contact pressure is high (11.1 MPa) [109]. The effectiveness is much lower than that of solder and thermal pastes, but flexible graphite is advantageous in its thermal stability.

3.4 Materials for thermal insulation

Materials for thermal insulation are characterized by a low thermal conductivity, which is most commonly attained by the use of air (a thermal insulator), as in the case of polymer foams and glass fiber felts. Composite materials for thermal insulation are mainly polymer–matrix and cement–matrix composites. Either type of composite typically contains a foamy or porous phase, which can be the matrix or the filler. However, such a phase is detrimental to the mechanical properties, which are important for structural composites.

The use of a structural material which is itself a thermal insulator is in contrast to the combined use of a structural material (which is not a thermal insulator) and a thermal insulator (which is not a structural material). The former is attractive due to the space saving and the durability of a structural material. However, the development of a structural material which is also a thermal insulator is scientifically challenging. An avenue is to use the interfaces rather than pores in a structural composite (such as the filler–matrix interface, with the filler being either particles or fibers) to provide thermal barriers [110]. In the case of a cement–matrix composite, an avenue involves the addition of a polymer (e.g., latex particles) admixture, since the thermal conductivity of a polymer is lower than that

Table 3.3 Thermal conductivity and specific heat of cement-matrix composites in the form of cement pastes [107]

Cement paste	Thermal conductivity (Wm^{-1} K^{-1}) (± 0.03)	Specific heat (J g^{-1} K^{-1}) (± 0.001)
Plain	0.52	0.703
+ latex (20% by weight of cement)	0.38	0.712
+ latex (25% by weight of cement)	0.32	0.723
+ latex (30% by weight of cement)	0.28	0.736
+ methylcellulose (0.4% by weight of cement)	0.42	0.732
+ methylcellulose (0.6% by weight of cement)	0.38	0.737
+ methylcellulose (0.8% by weight of cement)	0.32	0.742
+ silica fume	0.36	0.765
+ silica fume + methylcellulose[a]	0.33	0.771
+ methylcellulose[a] + fibers[b] (0.5% by weight of cement)	0.44	0.761
+ methylcellulose[a] + fibers[b] (1.0% by weight of cement)	0.34	0.792
+ silica fume + methylcellulose[a] + fibers[b] (0.5% by weight of cement)	0.28	0.789

[a]0.4% by weight of cement.
[b]Carbon fibers.

of cement [110]. Table 3.3 lists the thermal conductivity of various cement pastes, which involve the use of silica fume (fine particles), latex (a polymer), methylcellulose (molecules) and short carbon fibers as admixtures.

3.5 Materials for heat retention

Heat retention is important for buildings for the purpose of energy saving and temperature stability. This function requires materials with a high specific heat, which is the heat required to raise the temperature of a gram of the material by 1°C.

Polymers tend to have higher values of specific heat than metals, cement or carbon, owing to the ability of the polymer molecules to move with respect to one another and for segments of a polymer molecule to move. The movement provides a mechanism for absorbing heat. In the case of a cement–matrix composite, the addition of a polymer as an admixture thus causes the specific heat to increase [110].

Instead of using a phase that has a high specific heat, one can use the interfaces in a composite to raise the specific heat. This is because the slippage (however slight) at the interfaces during thermal vibrations provides a mechanism for absorbing heat. The interfaces can be provided by particles or fibers in the composite.

Table 3.3 lists the specific heat of various cement pastes.

3.6 Summary

Materials for thermal applications include those for thermal conduction, thermal insulation and heat retention. In particular, materials for thermal conduction include materials exhibiting high thermal conductivity, as well as thermal interface materials. The former includes metals, diamond, carbon, graphite, ceramics, metal–matrix composites, carbon–matrix composites and ceramic–matrix composites. The latter includes polymer-based pastes, silicate-based pastes, phase-change materials, solder and flexible graphite. A primary area of application is in the microelectronics industry.

Review questions

1. What is the main advantage of the coated filler method versus the admixture method in powder metallurgy?
2. Give the two main advantages of copper compared to aluminum as a material for thermal management.
3. Describe a method of fabricating an aluminum–matrix composite containing 60 vol.% SiC particles.
4. An aluminum–matrix SiC composite tends to degrade upon heating. Why?
5. Describe a method of fabricating a copper–matrix composite containing 70 vol.% molybdenum particles.

6. What are the two main criteria that govern the effectiveness of a thermal interface material?
7. Describe a method of obtaining concrete for thermal insulation without compromising on the mechanical properties.
8. Describe the advantageous properties of an aluminum–matrix silicon carbide particle composite compared to aluminum.
9. What are the main advantage and the main disadvantage of aluminum–matrix composites compared to copper–matrix composites as heat sink materials?
10. What is the advantage of using copper-coated molybdenum particles rather than a mixture of copper particles and molybdenum particles for making a copper–matrix molybdenum particle composite by powder metallurgy?
11. Stir casting tends to produce metal–matrix composites with a non-uniform distribution of the reinforcing particles. Why?
12. What is the main disadvantage of an Al/SiC composite compared to an Al/AlN composite?
13. Why is liquid metal infiltration less commonly used for making copper–matrix composites than for making aluminum–matrix composites?
14. Why is an excessive amount, exceeding the optimum amount, of boron nitride particles in a polyethylene-glycol-based thermal paste not desirable for the thermal paste performance, in spite of the thermal conductivity of boron nitride?
15. Give an example of a metal alloy that has an exceptionally low value of the thermal expansion coefficient.

References

[1] W.M. Peck, *American Society of Mechanical Engineers, Heat Transfer Division, (Publication) HTD*, Vol. 329, No. 7, 1996, ASME, New York, pp. 245–253.
[2] A.L. Geiger and M. Jackson, *Adv. Mater. Proc.*, 1989, **136**(1), 6 pp.
[3] P.K. Rohatgi, *Defence Science J.*, 1993, **43**(4), 323–349.
[4] S. Lai and D.D.L. Chung, *J. Mater. Sci.*, 1994, **29**, 3128–3150.
[5] X. Yu, G. Zhang and R. Wu, *Cailiao Gongcheng/J. Mater. Eng.*, June 1994, No. 6, 9–12.
[6] B.E. Novich and R.W. Adams, Proc. *1995 Int. Electronics Packaging Conf.*, Int. Electronics Packaging Society, Wheaton, IL, 1995, pp. 220–227.
[7] X.F. Yang and X.M. Xi, *J. Mater. Res.*, 1995, **10**(10), 2415–2417.
[8] J.A. Hornor and G.E. Hannon, *6th Int. Electronics Packaging Conf.*, 1992, pp. 295–307.
[9] Y.-L. Shen, A. Needleman and S. Suresh, *Met. Mater. Trans.*, 1994, **25A**(4), 839–850.
[10] Y.-L. Shen, *Mater. Sci. Eng. A*, Sept. 1998, (2), 269–275.
[11] J. Chiou and D.D.L. Chung, *J. Mater. Sci.*, 1993, **28**, 1435–1446.
[12] J. Chiou and D.D.L. Chung, *J. Mater. Sci.*, 1993, **28**, 1447–1470.
[13] J. Chiou and D.D.L. Chung, *J. Mater. Sci.*, 1993, **28**, 1471–1487.
[14] M.E. Smagorinski, P.G. Tsantrizos, S. Grenier, A. Cavasin, T. Brzezinski and G. Kim, *Mater. Sci. Eng. A*, Mar. 1998, (1), 86–90.
[15] K. Schmidt, C. Zweben and R. Arsenault, *ASTM Special Technical Publication*, Philadelphia, PA, 1990, No. 1080, pp. 155–164.
[16] S. Lai and D.D.L. Chung, *J. Mater. Sci.*, 1994, **29**, 2998–3016.
[17] S. Lai and D.D.L. Chung, *J. Mater. Chem.*, 1996, **6**, 469–477.
[18] S. Lai and D.D.L. Chung, *J. Mater. Sci.*, 1994, **29**, 6181–6198.
[19] S. Wagner, M. Hook and E. Perkoski, *Proc. Electricon '93*, Electronics Manufacturing Productivity Facility, 1993, p. 11/1–11/14.
[20] O.J. Ilegbusi, *(Paper) ASME*, no. 93-WA/EEP-28, 1993, pp. 1–7.
[21] M.A. Lambert and L.S. Fletcher, HTD Vol. 292, *Heat Transfer in Electronic Systems*, ASME, New York, 1994, pp. 115–122.
[22] M.A. Lambert and L.S. Fletcher, *J. Heat Transfer – Trans. ASME*, 1996, **118**(2), 478–480.

[23] M.A. Kinna, *6th Int. SAMPE Electronics Conf.*, 1992, pp., 547–555.
[24] J.-M. Ting, M.L. Lake and D.R. Duffy, *J. Mater. Res.*, 1995, **10**(6), 1478–1484.
[25] T.F. Fleming, C.D. Levan and W.C. Riley, *Proc. Technical Conf., Int. Electronics Packaging Conf.*, Wheaton, Ill, Int. Electronics Packaging Society, 1995, pp. 493–503.
[26] T.F. Fleming and W.C. Riley, *Proc. SPIE - The Int. Soc. for Optical Eng.*, Soc. Photo-Optical Instrumentation Engineers, Bellingham, WA, 1993, Vol. 1997, pp. 136–147.
[27] H. Wang and S.H.J. Lo, *J. Mater. Sci. Lett.*, 1996, **15**(5), 369–371.
[28] R.E. Morgan, S.L. Ehlers and J. Sosniak, *6th Int. SAMPE Electronics Conf.*, 1992, pp. 320–333.
[29] S. Yoo, M.S. Krupashankara, T.S. Sudarshan and R.J. Dowding, *Mater. Sci. Tech.*, 1998, **14**(2), 170–174.
[30] Anonymous, *Metal Powder Report*, 1997, no. 4, 28–31.
[31] T.W. Kirk, S.G. Caldwell and J.J. Oakes, *Particulate Materials and Processes, Advances in Powder Metallurgy*, Publ. By Metal Powder Industries Federation, Princeton, NJ, Vol. 9, 1992, pp. 115–122.
[32] P. Yih and D.D.L. Chung, *J. Electron. Mater.*, 1995, **24**(7), 841–851.
[33] S. Jha, *1995 Proc. 45th Electronic Components & Technology Conf.*, IEEE, New York, 1995, pp. 542–547.
[34] R. Chanchani and P.M. Hall, *IEEE Trans. Components, Hybrids, and Manufacturing Technology*, 1990, **13**(4), 743–750.
[35] C. Woolger, *Materials World*, 1996, **4**(6), 332–333.
[36] R. Chanchani and P.M. Hall, *IEEE Trans. Components, Hybrids, and Manufacturing Technology*, 1990, **13**(4), 743–750.
[37] J.R. Hanson and J.L. Hauser, *Electronic Packaging & Production*, 1986, **26**(11), 48–51.
[38] J. Korab, G. Korb and P. Sebo, *Proc. IEEE/CPMT Int. Electronics Manufacturing Technology (IEMT) Symp.*, 1998, IEEE, Piscataway, NJ, pp. 104–108.
[39] J. Korab, G. Korb, P. Stefanik and H.P. Degischer, *Composites*, Part A, 1999, **30**(8), 1023–1026.
[40] G. Korb, J. Koráb and G. Groboth, *Composites*, Part A, 1998, **29A**, 1563–1567.
[41] M.A. Kinna, *6th Int. SAMPE Electronics Conf.*, 1992, pp. 547–555.
[42] M.A. Lambert and L.S. Fletcher, HTD Vol. 292, *Heat Transfer in Electronic Systems*, ASME, New York, 1994, pp. 115–122.
[43] C.H. Stoessel, C. Pan, J.C. Withers, D. Wallace and R.O. Loutfy, *Mater. Res. Soc. Symp. Proc.*, Vol. 390 (Electronic Packaging Materials Science VII), 1995, pp. 147–152.
[44] Y. LePetitcorps, J.M. Poueylaud, L. Albingre, B. Berdeu, P. Lobstein and J.F. Silvain, *Key Engineering Materials*, 1997, **127–131**, 327–334.
[45] K. Prakasan, S. Palaniappan and S. Seshan, *Composites*, Part A, 1997, **28**(12), 1019–1022.
[46] P. Yih and D.D.L. Chung, *Int. J. Powder Met.*, 1995, **31**(4), 335–340.
[47] M. Ruhle, *Key Engineering Materials*, 1996, **116–117**, 1–40.
[48] M. Ruhle, *J. European Ceramic Soc.*, 1996, **16**(3), 353–365.
[49] K. Prakasan, S. Palaniappan and S. Seshan, *Composites*, Part A, 1997, **28**(12), 1019–1022.
[50] J.A. Kerns, N.J. Colella, D. Makowiecki and H.L. Davidson, *Int. J. Microcircuits and Electronic Packaging*, 1996, **19**(3), 206–211.
[51] P. Yih and D.D.L. Chung, *J. Mater. Sci.*, 1997, **32**(7), 1703–1709.
[52] P. Yih and D.D.L. Chung, *J. Mater. Sci.*, 1997, **32**(11), 2873–2894.
[53] C.H. Stoessel, J.C. Withers, C. Pan, D. Wallace and R.O. Loutfy, *Surf. Coatings Tech.*, 1995, **76–77**, 640–644.
[54] Y.Z. Wan, Y.L. Wang, G.J. Li, H.L. Luo and G.X. Cheng, *J. Mater. Sci. Lett.*, 1997, 16, 1561–1563.
[55] D.A. Foster, *34th Int. SAMPE Symp. Exhib.*, Book 2 (of 2), SAMPE, Covina, CA, 1989, pp. 1401–1410.
[56] H.L. Davidson, N.J. Colella, J.A. Kerns and D. Makowiecki, *1995 Proc. 45th Electronic Components & Technology Conf.*, IEEE, New York, 1995, pp. 538–541.
[57] K. Prakasan, S. Palaniappan and S. Seshan, *Composites*, Part A, 1997, **28**(12), 1019–1022.
[58] H. Mavoori and S. Jin, *JOM*, 1998, **50**(6), 70–72.
[59] T. Parsonage, *7th Int. SAMPE Electronics Conf.*, 1994, pp. 280–295.
[60] T.B. Parsonage, *National Electronic Packaging and Production Conf. - Proc. Technical Program (West and East)*, 1996, Reed Exhibition Companies, Norwalk, CT, pp. 325–334.
[61] K.M. Kearns, *Proc. Int. SAMPE Symp. Exhib.*, SAMPE, Covina, CA, 1998, **43**(2), 1362–1369.
[62] W.H. Pfeifer, J.A. Tallon, W.T. Shih, B.L. Tarasen and G.B. Engle, *6th Int. SAMPE Electronics Conf.*, 1992, pp. 734–747.
[63] W.T. Shih, F.H. Ho and B.B. Burkett, *7th Int. SAMPE Electronics Conf.*, 1994, pp. 296–309.
[64] W. Kowbel, X. Xia and J.C. Withers, *43rd Int. SAMPE Symp.*, 1998, pp. 517–527.
[65] W. Kowbel, X. Xia, C. Bruce and J.C. Withers, *Mater. Res. Soc. Symp. Proc.*, Vol. 515 (Electronic Packaging Materials Science X), Warrendale, PA, 1998, pp. 141–146.

[66] S.K. Datta, S.M. Tewari, J.E. Gatica, W. Shih and L. Bentsen, *Met. Mater. Trans. A*, 1999, 30A, 175-181.
[67] W.T. Shih, *Proc. Int. Electronics Packaging Conf.*, 1995, IEPS, Wheaton, IL, pp. 211-219.
[68] J.-M. Ting and M.L. Lake, *Diamond and Related Materials*, 1994, 3(10), 1243-1248.
[69] M.J. Montesano, *Mat. Tech.*, 1996, 11(3), 87-91.
[70] J. Klett, C. Walls and T. Burchell, *Ext. Abstr. Program - 24th Bienn. Conf. Carbon*, 1999, pp. 132-133.
[71] L. Hozer, Y.-M. Chiang, S. Ivanova and I. Bar-On, *J. Mater. Res.*, 1997, 12(7), 1785-1789.
[72] P.N. Kumta, *J. Mater. Sci.*, 1996, 31(23), 6229-6240.
[73] P.N. Kumta, T. Mah, P.D. Jero and R.J. Kerans, *Mater. Lett.*, 1994, 21(3-4), 329-333.
[74] J.Y. Kim and P.N. Kumta, Proc. *IEEE 1998 National Aerospace and Electronics Conf.*, NAECON 1998, celebrating 50 years, IEEE, New York, 1998, pp. 656-665.
[75] H. Korner, A. Shiota and C.K. Ober, *ANTEC '96: Plastics-Racing into the Future*, Conf. Proc., Society of Plastics Engineers, Technical Papers Series, no. 42, Brookfield, CT, 1996, Vol. 2, pp. 1458-1461.
[76] G.S. Swei and D.J. Arthur, *3rd Int. SAMPE Symp. Exhib.*, SAMPE, Covina, CA, 1989, pp. 1111-1124.
[77] L. Li and D.D.L. Chung, *Composites*, 1991, 22(3), 211-218.
[78] X. Lu and G. Xu, *J. Appl. Polymer Sci.*, 1997, 65(13), 2733-2738.
[79] Y. Xu and D.D.L. Chung, *Composite Interfaces*, 2000, 7(4), 243-256.
[80] Y. Xu, D.D.L. Chung and C. Mroz, *Composites*, Part A, 2001, 32, 1749-1757.
[81] D. Klosterman and L. Li, *J. Electronics Manufacturing*, 1995, 5(4), 277-287.
[82] S.K. Kang, R. Rai and S. Purushothaman, *1996 Proc. 46th Electronic Components & Technology*, IEEE, New York, 1996, pp. 565-570.
[83] G.F.C.M. Lijten, H.M. van Noort and P.J.M. Beris, *J. Electronics Manufacturing*, 1995, 5(4), 253-261.
[84] M.P. Zussman, B. Kirayoglu, S. Sharkey and D.J. Powell, *6th Int. SAMPE Electronics Conf.*, 1992, pp. 437-448.
[85] J.D. Bolt and R.H. French, *Adv. Mater. Processes*, 1988, 134(1), 32-35.
[86] J.D. Bolt, D.P. Button and B.A. Yost, *Mater. Sci. Eng.*, 1989, A109, 207-211.
[87] S.P. Mukerherjee, D. Suryanarayana and D.H. Strope, *J. Non-Crystalline Solids*, 1992, 147-148, 783-791.
[88] M.P. Zussman, B. Kirayoglu, S. Sharkey and D.J. Powell, *6th Int. SAMPE Electronics Conf.*, 1992, pp. 437-448.
[89] J.J. Glatz, R. Morgan and D. Neiswinger, *Int. SAMPE Electronics Conf.*, 1992, Vol. 6, pp. 131-145.
[90] A. Bertram, K. Beasley and W. de la Torre, *Naval Engineers J.*, 1992, 104(3), 276-285.
[91] D. Brookstein and D. Maass, *7th Int. SAMPE Electronics Conf.*, 1994, pp. 310-327.
[92] T.F. Fleming and W.C. Riley, *Proc. SPIE - Int. Soc. for Optical Eng.*, Soc. Photo-Optical Instrumentation Engineers, Bellingham, WA, 1993, Vol. 1997, pp. 136-147.
[93] T.F. Fleming, C.D. Levan and W.C. Riley, *Proc. Technical Conf., Int. Electronics Packaging Conf.*, Wheaton, IL, Int. Electronics Packaging Society, 1995, pp. 493-503.
[94] A.M. Ibrahim, *6th Int. SAMPE Electronics Conf.*, 1992, pp. 556-567.
[95] N. Kiuchi, K. Ozawa, T. Komami, O. Katoh, Y. Arai, T. Watanabe and S. Iwai, *Int. SAMPE Tech. Conf.*, Vol. 30, SAMPE, Covina, CA, 1998, pp. 68-77.
[96] J.W.M. Spicer, D.W. Wilson, R. Osiander, J. Thomas and B.O. Oni, *Proc. SPIE - International Society for Optical Engineering*, Vol. 3700, 1999, pp. 40-47.
[97] D.A. Foster, *SAMPE Q.*, 1989, 21(1), 58-64.
[98] D.A. Foster, *34th Int. SAMPE Symp. Exhib.*, Book 2 (of 2), SAMPE, Covina, CA, 1989, pp. 1401-1410.
[99] W. de la Torre, *6th Int. SAMPE Electronics Conf.*, 1992, pp. 720-733.
[100] X. Luo and D.D.L. Chung, *Int. J. Microcircuits Electronic Packaging*, 2001, 24(2), 141-147.
[101] D. Grivas, D. Frear, L. Quan and J.W. Morris Jr, *J. Electronic Mater.*, 1986, 15(6), 355-359.
[102] S.W. Wilson, A.W. Norris, E.B. Scott and M.R. Costello, *National Electronic Packaging and Production Conference*, Proc. Technical Program, Vol. 2, Reed Exhibition Companies, Norwalk, CT, 1996, pp. 788-796.
[103] A.L. Peterson, *Proc. 40th Electronic Components and Technology Conf.*, IEEE, Piscataway, NJ, Vol. 1, 1990, pp. 613-619.
[104] X. Lu, G. Xu, P.G. Hofstra and R.C. Bajcar, *J. Polymer Sci.*, 1998, 36(13), 2259-2265.
[105] T. Sasaski, K. Hisano, T. Sakamoto, S. Monma, Y. Fijimori, H. Iwasaki and M. Ishizuka, *Japan IEMT Symp. Proc. IEEE/CPMT Int. Electronic Manufacturing Technology (IEMT) Symp.*, IEEE, Piscataway, NJ, 1995, pp. 236-239.

[106] Y. Xu, X. Luo and D.D.L. Chung, *J. Electron. Packaging*, 2000, **122**(2), 128–131.
[107] Y. Xu, X. Luo and D.D.L. Chung, *J. Electron. Packaging*, 2002, **124**(3), 188–191.
[108] Z. Liu and D.D.L. Chung, *J. Electron. Mater.*, 2001, **11**(30), 1458–1465.
[109] X. Luo, R. Chugh, B.C. Biller, Y.M. Hoi and D.D.L. Chung, *J. Electron. Mater.*, 2002, **31**(5), 535–544.
[110] X. Fu and D.D.L. Chung, *Cem. Concr. Res.*, 1997, **27**(12), 1799–1804.

4 Composite materials for electrical applications

4.1 Introduction

Composite materials are traditionally designed for use as structural materials. With the rapid growth of the electronics industry, composite materials are finding more and more electronic applications. Owing to the vast difference in property requirements between structural composites and electronic composites, the design criteria for these two groups of composites are different. While structural composites emphasize high strength and high modulus, electronic composites emphasize high thermal conductivity, low thermal expansion, low dielectric constant, high/low electrical conductivity and/or electromagnetic interference (EMI) shielding effectiveness, depending on the particular electronic application. Low density is desirable for both aerospace structures and aerospace electronics. Structural composites emphasize processability into large parts, such as panels, whereas electronic composites emphasize processability into small parts, such as stand-alone films and coatings. Owing to the small size of the parts, material costs tend to be less of a concern for electronic composites than structural composites. For example, electronic composites can use expensive fillers, such as silver particles, which serve to provide high electrical conductivity.

The electrical applications for composites pertain to microelectronics and resistance heating (such as the deicing of aircraft).

4.2 Applications in microelectronics

The applications of polymer–matrix composites in microelectronics include interconnections, printed circuit boards, substrates, encapsulations, interlayer dielectrics, die attach, electrical contacts, connectors, thermal interface materials, heat sinks, lids and housings. In general, the integrated circuit chips (dies) are attached to a substrate or a printed circuit board on which the interconnection lines have been written (usually by screen printing) on each layer of the multilayer substrate or board. In order to increase the interconnection density, another multilayer involving thinner layers of conductors and interlayer dielectrics may be applied to the substrate before attachment of the chip. By means of soldered joints, wires connect between electrical contact pads on the chip and electrical contact pads on the substrate or board. The chip may be encapsulated with a dielectric for protection. It may also be covered by a thermally conducting (metal) lid. The substrate (or board) is attached to a heat sink. A thermal interface material may be placed between the substrate (or board) and the heat sink to enhance the quality of the thermal contact. The whole assembly may be placed in a thermally conducting housing.

A *printed circuit board* is a sheet for the attachment of chips, whether mounted on substrates, chip carriers or otherwise, and for the drawing of interconnections. It is a polymer–matrix composite that is electrically insulating and has four conductor lines (interconnections) on one or both sides. Multilayer boards have lines on each inside layer so that interconnections on different layers may be connected by short conductor columns called electrical vias. Printed circuit boards (or cards) for the mounting of *pin-inserting-type packages* need to have lead insertion holes punched through the circuit board. Printed circuit boards for the mounting of *surface-mounting-type packages* need no holes. Surface-mounting-type packages, whether with leads (leaded chip carriers) or without leads (*leadless chip carriers*, LLCCs), can be mounted on both sides of a circuit board (i.e., a card), whereas pin-inserting-type packages can only be mounted on one side of a circuit board. In surface mounting technology (SMT), the surfaces of conductor patterns are connected together electrically without employing holes. Solder is typically used to make electrical connections between a surface-mounting-type package (whether leaded or leadless) and a circuit board. A lead insertion hole for pin-inserting-type packages is a plated-through hole, a hole on whose wall a metal is deposited to form a conducting penetrating connection. After pin insertion, the space between the wall and the pin is filled by solder to form a solder joint. Another type of plated-through hole is a via hole, which serves to connect different conductor layers together without the insertion of a lead.

A *substrate*, also called a *chip carrier*, is a sheet on which one or more chips are attached and interconnections are drawn. In the case of a multilayer substrate, interconnections are also drawn on each layer inside the substrate, such that interconnections in different layers are connected, if desired, by electrical vias. A substrate is usually an electrical insulator. Substrate materials include ceramics (e.g., Al_2O_3, AlN, mullite, glass–ceramics), polymers (e.g., polyimide), semiconductors (e.g., silicon) and metals (e.g., aluminum). The most common substrate material is Al_2O_3. As the sintering of Al_2O_3 requires temperatures greater than 1000°C, the metal interconnections need to be refractory, such as tungsten or molybdenum. The disadvantage of tungsten or molybdenum lies in the higher electrical resistivity compared to copper. In order to make use of more conductive metals (e.g., Cu, Au, Ag–Pd) as the interconnections, ceramics that sinter at temperatures below 1000°C ("low temperature") can be used in place of Al_2O_3. The competition between ceramics and polymers for substrates is increasingly keen. Ceramics and polymers are both electrically insulating; ceramics are advantageous in that they tend to have a higher thermal conductivity than polymers; polymers are advantageous in that they tend to have a lower dielectric constant than ceramics. A high thermal conductivity is attractive for heat dissipation; a low dielectric constant is attractive for a smaller capacitive effect, hence a smaller signal delay. Metals are attractive for their very high thermal conductivity compared to ceramics and polymers.

An *interconnection* is a conductor line for signal transmission, power or ground. It is usually in the form of a thick film of thickness > 1 μm. It can be on a chip, a substrate or a printed circuit board. The thick film is made by either screen printing or plating. Thick-film conductor pastes containing silver particles (conductor) and glass frit (binder which functions by the viscous flow of glass upon heating) are widely used to form thick-film conductor lines (interconnections) on substrates by screen printing and subsequent firing. These films suffer from reduction in electrical conductivity by the presence of the glass and porosity in the film after firing. The choice of a metal in a thick film paste depends on the need for withstanding air oxidation in the heating encountered in subsequent processing, which can be the firing of the green thick film together with the green ceramic substrate (a process known as cofiring). It is during cofiring that bonding and sintering take place. Copper is an excellent conductor, but it oxidizes readily when heated in air. The choice of metal also depends on the temperature encountered in subsequent processing. Refractory metals, such as tungsten and molybdenum, are suitable for interconnections heated to high temperatures (>1000°C), for example during Al_2O_3 substrate processing.

A *z-axis anisotropic electrical conductor film* is a film which is electrically conducting only in the z-axis, i.e., in the direction perpendicular to the plane of the film. As one z-axis film can replace a whole array of solder joints, z-axis films are valuable for solder replacement, processing cost reduction and repairability improvement in surface mount technology.

An *interlayer dielectric* is a dielectric film separating the interconnection layers, such that the two kinds of layers alternate and form a thin-film multilayer. The dielectric is a polymer, usually spun on or sprayed; or a ceramic, usually applied by chemical vapor deposition (CVD). The most common multilayer involves polyimide as the dielectric and copper interconnections, plated sputtered, or electron-beam deposited.

A *die attach* is a material for joining a die (a chip) to a substrate. It can be a metal alloy (a solder paste), a polymeric adhesive (a thermoset or a thermoplast) or a glass. Die attach materials are usually applied by screen printing. A solder is attractive in its high thermal conductivity, which enhances heat dissipation. However, its application requires the use of heat and a flux. The flux subsequently needs to be removed chemically. The defluxing process adds costs and is undesirable to the environment (the ozone layer) due to the chlorinated chemicals used. A polymer or glass has poor thermal conductivity, but this problem can be alleviated by the use of a thermally conductive filler, such as silver particles. A thermoplastic provides a reworkable joint, whereas a thermoset does not. Furthermore, a thermoplastic is more ductile than a thermoset. Moreover, a solder suffers from its tendency to experience *thermal fatigue* due to the thermal expansion mismatch between the chip and the substrate and the resulting work hardening and cracking of the solder. In addition, the footprint left by a solder tends to be larger than the footprint left by a polymer, due to the ease of the molten solder to flow.

An *encapsulation* is an electrically insulating conformal coating on a chip for protection against moisture and mobile ions. An encapsulation can be a polymer (e.g., epoxy, polyimide, polyimide siloxane, silicone gel, Parylene and benzocyclobutene), which can be filled with SiO_2, BN, AlN, or other electrically insulating ceramic particles for decreasing thermal expansion and increasing thermal conductivity [1–3]. The decrease in thermal expansion is needed because a neat polymer typically has a much higher coefficient thermal expansion than a semiconductor chip. An encapsulation can also be a ceramic (e.g., SiO_2, Si_3Ni_4, silicon oxynitride). In the process of electronic packaging, encapsulation is a step performed after both die bonding and wire bonding, and before packaging using a molding material. The molding material is typically a polymer, such as epoxy. However, it can also be a ceramic, such as Si_3N_4, cordierite (magnesium silicate), SiO_2 and so on. A ceramic is advantageous (compared to a polymer) in its low coefficient of thermal expansion (CTE) and higher thermal conductivity, but it is much less convenient to apply than a polymer.

A lid is a cover for a chip for physical protection. The chip is typically mounted in a well in a ceramic substrate and the lid covers the well. The lid is preferably a metal because of the need to dissipate heat. It is typically joined to the ceramic substrate by soldering, using a solder preform (e.g., Au–Sn) shaped like a gasket. Because of the low CTE, Kovar, 54Fe–29Ni–17Co is often used for the lid. For the same reason, Kovar is often used for the can (housing or enclosure) in which a substrate is mounted. Although Kovar has a low CTE (5.3×10^{-6} °C^{-1} at 20–200°C), it suffers from a low thermal conductivity of 17 W m^{-1} K^{-1}.

A *heat sink* is a thermal conductor that serves to conduct (mainly) and radiate heat away from the circuitry. It is typically bonded to a printed circuit board. The thermal resistance of the bond and that of the heat sink itself govern the effectiveness of heat dissipation. A heat sink that matches the CTE of the circuit board for resistance to thermal cycling.

Insufficiently fast dissipation of heat is the most critical problem that limits the reliability and performance of microelectronics [4]. The problem becomes more severe as electronics are miniaturized, so the further miniaturization of electronics is hindered by this problem. The problem also accentuates as the power (voltage and current) increases, so power is restricted by the heat dissipation problem. Excessive heating resulting from insufficient heat dissipation causes thermal stress in the electronic package. The stress may cause warpage of the semiconductor chip (die). The problem is compounded by thermal fatigue, which results from cyclic heating and thermal expansion mismatches. Since there are many solder joints in an electronic package, solder joint failure is a big reliability problem. For these various reasons, thermal management has become a key issue within the field of electronic packaging. *Thermal management* refers to the use of materials (heat sinks, thermal interface materials, etc.), devices (fans, heat pipes, etc.) and packaging schemes to attain efficient dissipation of heat. The use of materials is the part that is addressed in this book. (See Chapter 3 in relation to composite materials for thermal applications.)

The use of materials with high thermal conductivity and low thermal expansion for heat sinks, lids, housings, substrates and die attach is an important avenue for alleviating the heat dissipation problem. For this purpose, metal–matrix composites (such as silicon carbide particle aluminum–matrix composites) and polymer–matrix composites (such as silver particle filled epoxy) have been developed. However, another avenue, which has received much less attention, is the improvement of the thermal contact between the various components in an electronic package (e.g., between substrate and heat sink). The higher thermal conductivity of the individual components cannot effectively help heat dissipation unless thermal contact between the components is good. Without good thermal contacts, the use of expensive thermal conducting materials for the components is a waste.

The improvement of a thermal contact involves the use of a *thermal interface material*, such as a thermal fluid, a thermal grease (paste) or a resilient thermal conductor. A thermal fluid or grease is spread on the mating surfaces. A resilient thermal conductor is sandwiched by the mating surfaces and held in place by pressure. Thermal fluids are most commonly mineral oil. Thermal greases (pastes) are most commonly conducting particle (usually metal or metal oxide) filled silicone. Resilient thermal conductors are most commonly conducting particle filled elastomers. Of these three types of thermal interface materials, thermal greases (based on polymers, particularly silicone) are by far the most commonly used. Resilient thermal conductors are not as well developed as thermal fluids or greases.

As the materials to be interfaced are good thermal conductors (such as copper), the effectiveness of a thermal interface material is enhanced by high thermal conductivity and low thickness of the interface material and low thermal contact resistance between the interface material and each mating surface. As the mating surfaces are not perfectly smooth, the interface material must be able to flow or deform, so as to conform to the topology of the mating surfaces. If the interface material is a fluid, grease or paste, it should have a high fluidity (workability) so as to conform and to have a small thickness after mating. On the other hand, the thermal conductivity of the grease or paste increases with increasing filler content and this is accompanied by decrease in workability. Without a filler, as in the case of an oil, thermal conductivity is poor. A thermal interface material in the form of a resilient thermal conductor sheet (e.g., a collection of conducting fibers clinging together without a binder, and a resilient polymer–matrix composite containing a thermally conducting filler) usually cannot be as thin or as conformable as one in the form of a fluid, grease or paste, so its effectiveness requires a very high thermal conductivity within it.

4.3 Applications in resistance heating

Electrical heating includes resistance heating (i.e., Joule heating) and induction heating, in addition to heating by the use of electric heat pumps, plasmas and lasers [5,6]. It is to be distinguished from solar heating [7–11] and the use of fossil fuels such as coal, fuel oil and natural gas [6]. Owing to the environmental problem associated with the use of fossil fuels and because of the high cost of solar heating, electrical heating is becoming increasingly important. Although electric heat pumps are widely used for the electrical heating of buildings, resistance heating is a complementary method which is receiving increasing attention.

Resistance heating involves passing an electric current through a resistor, which is the *heating element*. In relation to the heating of buildings, resistance heating typically involves the embedding of heating elements in structural material, such as concrete [12–14]. The materials of heating elements cannot be too low in electrical resistivity, as this would result in the resistance of the heating element being too low. The materials of heating elements cannot be too high in resistivity either, as this would result in the current in the heating element being too low (unless the voltage is very high). Materials of heating elements include metal alloys (such as nichrome), ceramics (such as silicon carbide [15]), graphite [16–18], carbon fiber mats [19,20], polymer–matrix composites [21–23], carbon–carbon composites [24], asphalt [25] and concrete [26].

Resistance heating is not only useful for the heating of buildings, but also for the deicing of bridge decks [27] and aircraft [21], and for the demolition of concrete structures [28,29]. Flexibility is desirable for the heating element for some applications (e.g., the deicing of aircraft), due to the need of the element to conform to the shape of the part which it heats. For this purpose, flexible graphite (flexible foil which is all graphite) is highly effective [18] and carbon fiber mat (randomly oriented short carbon fibers, optionally metal coated, held together by a small amount of an organic binder) is moderately effective [20]. However, the porosity in the mat allows penetration by a resin, thereby enabling the mat to be used as an interlayer between laminae in a continuous fiber structural composite.

A less common form of resistance heating involves eddy current heating which accompanies induction heating [30]. However, the requirement of induction heating makes this form of resistance heating relatively expensive.

Conventional concrete is electrically conducting, but the resistivity is too high for resistance heating to be effective. The resistivity of concrete can be diminished by the use of an electrically conductive admixture, such as discontinuous carbon fibers [31–35], discontinuous steel fibers [36] and graphite particles [37,38]. It can also be diminished by the use of an alkaline slag binder [26].

4.4 Polymer–matrix composites

Polymer–matrix composites with continuous or discontinuous fillers are used for electronic packaging and thermal management. Composites with continuous fillers (fibers, whether woven or not) are used as substrates, heat sinks and enclosures. Composites with discontinuous fillers (particle or fibers) are used for die attach, electrically/thermally conducting adhesives, encapsulations, thermal interface materials and electrical interconnections (thick film conductors and z-axis conductors). Composites with discontinuous fillers can be in a paste form during processing, thus allowing application by printing (screen printing or jet printing) and injection molding. Composites with continuous fillers cannot undergo paste processing, but the continuous fillers provide lower thermal expansion and higher conductivity than discontinuous fillers.

Composites can have thermoplastic or thermosetting matrices. Thermoplastic matrices have the advantage that a connection can be reworked by heating for the purpose of repair, whereas thermosetting matrices do not allow reworking. On the other hand, controlled-order thermosets are attractive for their thermal stability and dielectric properties [39]. Polymers exhibiting low dielectric constant, low dissipation factor, low CTE and compliance are preferred [40].

Composites can be electrically conducting or electrically insulating; the electrical conductivity is provided by a conductive filler. The composites can be both electrically and thermally conducting, as attained by the use of metal or graphite fillers; they can be electrically insulating but thermally conducting, as attained by the use of diamond, aluminum nitride, boron nitride or alumina fillers [41,42]. An electrically conducting composite can be isotropically conducting [43,44] or anisotropically conducting [45]. A z-axis conductor is an example of an anisotropic conductor.

4.4.1 Polymer–matrix composites with continuous fillers

Epoxy-matrix composites with continuous glass fibers and made by lamination are most commonly used for printed wiring boards, because of the electrically insulating property of glass fibers and the good adhesive behavior and established industrial usage of epoxy. Aramid (Kevlar) fibers can be used instead of glass fibers to provide lower dielectric constant [46]. Alumina (Al_2O_3) fibers can be used for increasing the thermal conductivity [47]. By selecting the fiber orientation and loading in the composite, the dielectric constant can be decreased and thermal conductivity can be increased [48]. By impregnating yarns or fabrics with a silica-based sol and subsequent firing, thermal expansion can be reduced [49]. Matrices other than epoxy can be used. Examples are polyimide and cyanate ester [50].

For heat sinks and enclosures, conducting fibers are used, since the conducting fibers enhance thermal conductivity and the ability to shield *electromagnetic interference* (EMI). EMI shielding is particularly important for enclosures [51]. Carbon fibers are most commonly used for these applications, owing to their conductivity, low thermal expansion and wide availability as a structural reinforcement. For high thermal conductivity, carbon fibers made from mesophase pitch [52-58] or copper-plated carbon fibers are preferred [59-61]. For EMI shielding, both uncoated carbon fibers [62,63] and metal (e.g., nickel, copper) coated carbon fibers [64,65] have been used.

For avionic electronic enclosures, low density (light weight) is essential for saving aircraft fuel. Aluminum is the traditional material for this application. Carbon fiber reinforced epoxy has been judged by consideration of mechanical, electrical, environmental, manufacturing/producibility and design-to-cost criteria to be more attractive than aluminum, glass fiber reinforced epoxy, glass fiber reinforced epoxy with aluminum interlayer, beryllium, aluminum–beryllium and SiC particle reinforced aluminum [66]. A related application is the thermal management of satellites, for which the thermal management materials need to be integrated from the satellite structure down to the electronic device packaging [67]. Continuous carbon fibers are suitable for this application because of their high thermal conductivity, low density, high strength and high modulus.

4.4.2 Polymer–matrix composites with discontinuous fillers

Polymer–matrix composites with discontinuous fillers (particles or short fibers) are widely used in electronics [68], in spite of their poor mechanical properties compared to composites with continuous fibers. This is because materials in electronics do not need to be mechanically strong and discontinuous fillers enable processing through the paste form, which is particularly suitable for making films, whether standalone films or films on a substrate.

Screen printing is a common method for patterning a film on a substrate. In the case of an electrically conducting paste, the pattern is commonly an array of electrical interconnections and electrical contact pads on the substrate. As screen printing involves the paste going through a screen, screen-printable pastes usually contain particles and no fiber and the particles must be sufficiently small, typically less than 10 μm in size. The larger the particles, the poorer is the patternability, i.e., the edge of a printed line is not sufficiently well defined. In applications not requiring patternability, such as thermal interface materials, short fibers are advantageous in that the connectivity of the short fibers is superior to that of particles at the same volume fraction. For a conducting composite, better connectivity of the filler units means higher conductivity for the composite. Instead of using short fibers, one may use elongated particles or flakes for the sake of the connectivity. In general, the higher the aspect ratio, the better is the connectivity for the same volume fraction. The use of elongated particles or flakes can provide an aspect ratio larger than 1, while retaining patternability. Thus, it is an attractive compromise.

In the case of a conducting composite, the greater the volume fraction of the conducting filler, the higher is the conductivity of the composite, since the polymer matrix is usually insulating. However, the greater the filler volume fraction, the higher is the viscosity of the paste and the poorer is the processability of the paste. To attain a high filler volume fraction while maintaining processability, a polymer of low viscosity is preferred and good wettability of the filler by the matrix, as provided by filler surface treatments and/or the use of surfactants, is desirable.

The matrix used in making a polymer–matrix composite can be in the form of a liquid (e.g., a thermosetting resin) or a solid (e.g., a thermoplastic powder) during the mixing of the matrix and the filler. In the case of the matrix in the form of a powder, the distribution of the filler units in the resulting composite depends on the size of the matrix power particles, as the filler units line the interface between adjacent matrix particles and the filler volume fraction needed for *percolation* (i.e., the filler units touching one another to form a continuous path) [69] decreases with increasing matrix particle size. Attaining percolation is accompanied by a large increase in conductivity. However, a large matrix particle size is detrimental to the processability. Therefore, a compromise is needed.

In the case of a matrix in the form of a thermoplastic powder, the percolation attained after mixing the matrix powder and the filler may be degraded or destroyed after subsequent composite fabrication involving flow of the thermoplastic under heat and pressure. Hence, in this case, a thermoplastic that flows less is preferred for attaining high conductivity in the resulting composite [70].

A less common way to attain percolation in a given direction is to apply an electric or magnetic field so as to align the filler units along that direction. For this technique to be possible, the filler units (whether in the bulk or on the surface) must be polarizable electrically or magnetically. Such alignment is one of the techniques used to produce z-axis conductors.

In percolation, the filler units touch one another to form continuous paths, but there is considerable contact resistance at the interface between the touching filler units. To decrease this contact resistance, thereby increasing the conductivity of the composite, one can increase the size of the filler units, so that the amount of interface area is decreased, provided that percolation is maintained. A less common but even more effective way is to bond the filler units together at their junction by using a solid (like solder) that melts and wets the surface of the filler during the composite fabrication. The low melting point solid can be in the form of particles added to the composite mix, or in the form of a coating on the filler units. In this way, a three-dimensionally interconnected conducting network is formed after composite fabrication [71].

An intimate interface between the filler and the matrix is important to the conductivity of a composite, even though the filler is conducting and the matrix may be perfectly insulating. This is because conduction may involve a path from one filler unit to an adjacent one through a thin film of the matrix by means of tunneling. In the case of the matrix being slightly conducting (but not as conducting as the filler), the conduction path involves both the filler and the matrix and the filler–matrix interface is even more important. This interface may be improved by filler surface treatments (by the use of chemicals, heat, plasma, etc.) prior to incorporating the filler in the composite, or by the use of a surfactant [72].

The difference in thermal expansion coefficient between filler and matrix and the fact that composite fabrication occurs at an elevated temperature cause thermal stress during cooling of the fabricated composite. The thermal expansion coefficient of a polymer is usually relatively high, so the filler units are usually under compression after cooling. The compression helps to tighten the filler–matrix interface, although the compressive stress in the filler and the tensile stress in the matrix may degrade the performance and durability of the composite.

In the case of the matrix being conducting, but not as conducting as the filler, as for conducting polymer matrices [73,74], percolation is not essential for the composite to be conducting, although percolation would greatly enhance the conductivity. Below the *percolation threshold* (i.e., the filler volume fraction above which percolation occurs), the conductivity of the composite is enhanced by a uniform distribution of the filler units, since the chance of having a conduction path that involves more filler and less matrix increases as the filler distribution becomes more uniform. Uniformity is never perfect; it is described by the degree of dispersion of the filler. The degree of dispersion can be enhanced by rigorous agitation during mixing of the filler and matrix or by the use of a dispersant (commonly a surfactant). In the case of the matrix in the form of particles that are coarser than the filler units, the addition of fine particles to the mix also helps dispersion of the filler [75].

Because the thermal expansion coefficient of a polymer is relatively high, the polymer matrix expands more than the filler during heating of a polymer–matrix composite. This results in the proximity between adjacent filler units changing with temperature, thus decreasing conductivity of the composite [76]. This phenomenon is detrimental to the thermal stability of the composites.

Corrosion and surface oxidation of the filler are the most common causes of degradation which decrease conductivity of the composite. Thus, oxidation-resistant fillers are essential. Silver and gold are oxidation resistant, but copper is not. Owing to the high cost of silver and gold, the coating of copper, nickel or other lower-cost metal fillers by gold or silver is common for improving oxidation resistance. By far the most common filler is silver particles [77,78].

A z-axis anisotropic electrical conductor film is a film which is electrically conducting in the direction perpendicular to the film, but is insulating in all other directions. This film is technologically valuable for use as an interconnection material in electronic packaging (chip-to-package, package-to-board and board-to-board), as it electrically connects the electrical contact pads touching one side of the film with the corresponding contact pads touching the side of the film directly opposite. Even though the film is in one piece, it contains numerous z-axis conducting paths (not necessarily in a regular array; they can be randomly distributed), so that it can provide numerous interconnections. If each contact pad is large enough to span a few z-axis conducting paths, no alignment is needed between the contact pad array and the z-axis film, whether the conducting paths are ordered or random in their distribution [79–93]. In this situation, in order to attain a high density of interconnections the cross-section of each z-axis conducting path must be small. However, if each contact pad is only large enough to span one z-axis conducting path, alignment is needed between the contact pad array and the z-axis film, and this means that the conducting paths in the z-axis film must be ordered in the same way as the contact pad array [94]. An example of an application of a z-axis conductor film is in the interconnections between the leads from (or contact pads on) a surface mount electronic device and the contact pads on the substrate beneath the device. In this application, one piece of z-axis film can replace a whole array of solder joints, so processing cost can be much reduced. Furthermore, the problem of thermal fatigue of the solder joints can be avoided by this replacement. Another example is in the vertical interconnections in three-dimensional electronic packaging.

A z-axis film is a polymer–matrix composite containing conducting units which form the z-axis conducting paths. The conducting units are usually particles, such as metal particles and metal-coated polymer particles. The particles can be clustered so that each cluster corresponds to one conducting path [79–81,85]. Metal columns [86], metal particle columns (e.g., gold-plated nickel) [84,85] and individual metal-coated polymer particles [83] had been used to provide z-axis conducting paths. Particle columns were formed by magnetic alignment of the particles. Using particle columns, Fulton et al. [80] attained a conducting path width of 400 μm and a pitch (center-to-center distance between adjacent conducting paths) of 290 μm. Also using particle columns, Robinson et al. [84] and Rosen et al. [85] attained a conducting path width of ~10 μm and a pitch of ~100 μm. In general, a large conducting path width is desirable for decreasing the resistance per path, while a small pitch is desirable for high-density interconnection. In contrast to the use of metal wires, metal columns or metal particle columns, Xu and Chung [95] used one metal particle per conducting path (i.e., per connection). The concept of one particle per path had been demonstrated [83] by using metal-coated polymer particles. However, owing to the high resistivity of the metal coating compared to the bulk metal, the z-axis resistivity of the film was high (0.5 Ω.cm for a conducting path). By using metal particles in place of metal-coated polymer particles, Xu and Chung [95] decreased the z-axis resistivity of a conducting path to 10^{-6} Ω.cm. Furthermore, that study did not rely on a polymer (whether the matrix or the particles) for providing resilience, as the resilience is provided by the metal particles, which protrude from both sides of the standalone film. As a result, the problem of stress relaxation of the polymer is eliminated. In addition, the protrusion of the metal particles eliminates the problem of open circuiting the connection upon heating owing to the higher thermal expansion of the polymer compared to the conductor [92].

Most work on z-axis adhesive films [90,91] used an adhesive with randomly dispersed conductive particles (8–12 μm diameter) suspended in it. The particles were phenolic spheres that had been coated with nickel. After bonding under heat (180–190°C) and pressure (1.9 MPa), a particle became oval in shape (4 μm thick). There was one particle per conducting path. The main drawback of this technology is the requirement of heat and pressure for curing the adhesive. Heat and pressure are not desirable in practical use of the z-axis adhesive. Xu and Chung [95] removed the need for heat and pressure through the choice of the polymer.

A different kind of z-axis adhesive film [94] used screening or stenciling to obtain a regular two-dimensional array of silver-filled epoxy conductive dots, but this technology suffers from the large pitch (1500 μm) of the dots and the consequent need for alignment between z-axis film and contact pad array. In the work of Xu and Chung [95], the pitch of the conducting paths in the z-axis adhesive film is as low as 64 μm.

Capacitors require materials with a high dielectric constant. Such materials in the form of thick films allow capacitors to be integrated with the electronic packaging, thereby allowing further miniaturization, in addition to performance and reliability improvements [96]. These thick-film pastes involve ceramic particles with a high dielectric constant, such as barium titanate (BaTiO$_3$), and a polymer (e.g., epoxy) [97,98].

Inductors are needed for transformers, DC/DC converters and other power supply applications. They require magnetic materials. Such materials in the form of thick films allow inductors and transformers to be integrated with the electronic packaging, thereby allowing further miniaturization. These thick-film pastes involve magnetic particles (e.g., ferrite) and a polymer [99,100].

The need for EMI shielding is increasing rapidly due to the interference of radio frequency radiation (such as that from a cellular phone) with digital electronics, and the increasing dependence of society on digital electronics. The associated electronic pollution is an interference problem.

EMI shielding is achieved by using electrical conductors, such as metals and conductive filled polymers [101–109]. *EMI shielding gaskets* [110–121] are resilient conductors. They are needed to electromagnetically seal an enclosure. The resilient conductors are most commonly elastomers (e.g., rubber) that are filled with a conductive filler [122], or elastomers that are coated with a metalized layer. Metalized elastomers suffer from poor durability owing to the tendency of the metal layer to debond from the elastomer. Conductive filled elastomers do not have this problem, but they require the use of a highly conductive filler, such as silver particles, in order to attain a high shielding effectiveness while maintaining resilience. The highly conductive filler tends to be expensive, making the composite expensive. The use of a less conducting filler results in the need for a large volume fraction of the filler in order to attain a high shielding effectiveness; the consequence is diminished resilience or even loss of resilience. Moreover, these composites suffer from degradation of the shielding effectiveness in the presence of moisture or solvents. In addition, the polymer matrix in the composites limits the temperature resistance, and the thermal expansion mismatch between filler and matrix limits the thermal cycling resistance.

Because of the *skin effect* (i.e., electromagnetic radiation at a high frequency interacting with only the near surface region of an electrical conductor), a filler for a polymer–matrix composite for EMI shielding needs to be not only electrically conducting, but also small in unit size. Although connectivity between the filler units is not required for shielding, it helps. Therefore, a filler in the form of a metal fiber of very small diameter is desirable. For this purpose, nickel filaments of diameter 0.4 μm and length > 100 μm, with a carbon core of diameter 0.1 μm, were developed [123]. Their exceptionally small diameter compared to those of existing metal fibers made them outstanding for use as a filler in a polymer for EMI shielding. A shielding effectiveness of 87 dB at 1 GHz was attained in a polyethersulfone–matrix composite with only 7 vol.% nickel filaments [123–125]. The low volume fraction allows resilience in a silicone–matrix composite for EMI gaskets [126].

Electronics are sensitive to electrostatic discharge (ESD), which affects the yield and reliability of integrated circuits [127]. An ESD protective material is an electrical conductor which allows the charge to spread and decay. It is characterized by a low surface resistivity and a low charge decay time.

Polymer-matrix composites containing conductive fibers are used for ESD protection [128,129].

4.5 Ceramic–matrix composites

Controlled resistivity materials are used for controlled electrical conduction, static charge dissipation, lightning protection and EMI shielding in electronic, mechanical, structural, chemical and vacuum applications. In particular, controlled resistivity ceramics, such as alumina–matrix composites containing an electrically conducting particulate filler [130], are used as substrates for handling semiconductor wafers, which require static protection. They are also used in the form of charge-dissipating coatings to improve the breakdown voltage of high-power, high-vacuum devices. In addition, controlled resistivity ceramics in the form of tiles are used for antistatic floors [131–134].

Controlled resistivity materials are mainly in the form of composite materials comprising an electrically insulating matrix and an electrically conductive discontinuous filler, which can be particulate or fibrous. The higher the filler content, the lower is the resistivity of the composite. These composites include those with polymer [135–138], ceramic [130–134] and cement [139] matrices. Polymers and ceramics are usually insulating electrically, but cement is slightly conductive. Polymers that are electrically conductive exist, but they are expensive. Among all these matrices, cement is the least expensive. In addition, the fabrication of cement–matrix composites is inexpensive and takes place at room temperature. The fabrication of ceramic–matrix composites such as alumina– matrix composites is even more expensive than that of polymer–matrix composites, owing to the high processing temperatures. Furthermore, cement–matrix composites, like ceramic–matrix composites, are mechanically more rugged and chemically more resistant than polymer–matrix composites.

In addition to providing a range of resistivity, controlled resistivity materials provide a range of dielectric constant. As the filler content of a ceramic–matrix composite increases, the dielectric constant increases and the resistivity decreases [130]. In the case of an alumina–matrix TiO_2 particulate composite, the relative dielectric constant (1 kHz) is 127,000 (undesirably high) when the resistivity is 5×10^5 Ω.cm and the Al_2O_3 content is >80 wt.%, and the relative dielectric constant (1 kHz) is 26.6 when the resistivity is 1×10^9 Ω.cm and the Al_2O_3 content is >94 wt.% [130]. It would be desirable to have the combination of low resistivity (for charge dissipation and related functions) and low dielectric constant (for avoiding a capacitive effect). Polymer matrices tend to exhibit lower values of dielectric constant than ceramic or cement matrices, but they tend to be insufficient in mechanical ruggedness and in chemical and temperature resistance.

Low-cost and mechanically rugged cement-based controlled resistivity materials exhibiting low values of relative dielectric constant are cement paste containing short electrically conducting fibers [139]. With steel fibers (0.1 vol.%), the resistivity and relative dielectric constant (10 kHz) are 8×10^4 Ω.cm and 20 respectively. With carbon fibers (1.0 vol.%) and silica fume, these quantities are 8×10^2 Ω.cm and 49 respectively.

4.6 Conclusion

Composite materials for microelectronics are typically designed for high thermal conductivity, low CTE, low dielectric constant, high, low or intermediate electrical conductivity and processability (e.g., printability). They include composites with polymer, metal and cement matrices. Applications include heat sinks, housings, printed wiring boards, substrates, lids, die attach, encapsulation, interconnections, thermal interface materials, EMI shielding and ESD protection. Combinations of properties are usually required. For example, for heat sinks and substrates, the combination of high thermal conductivity and low CTE is required for the purpose of heat dissipation and thermal stress reduction. In the case of aerospace electronics, low density is also desired. Polymer–matrix composites for microelectronics include those with continuous and discontinuous fillers. They can be in the form of an adhesive film, a standalone film or a bulk material. They can be isotropic or anisotropic electrically.

Review questions

1. Which parts of an electronic package can benefit from using a metal–matrix composite of low thermal expansion and high thermal conductivity?
2. What is meant by a surface mount component in relation to electronic packaging?
3. What is the function of glass frit (i.e., glass particles) in a thick-film electrical conductor paste?
4. Describe a process for making a multilayer ceramic chip carrier.
5. What is the main application of a z-axis conductor film?
6. What are the functions of electronic packaging?
7. Thermally conducting but electrically insulating polymer–matrix composites are useful for what aspects of electronic packaging?
8. Give an example of each of the following: (a) a z-axis conductor; (b) an electrically conductive thick-film paste.
9. The following materials are used in electronic packaging. For each material, describe the properties which make it useful for electronic packaging: (a) Kovar; (b) silver-epoxy; (c) BN-epoxy.
10. What is the main advantage of a surface-mounting-type electronic package compared to a pin-inserting-type package?
11. What are the three main ingredients of a printed circuit board?
12. What are the two attractive properties of metal–matrix composites which make them useful for electronic packaging applications?
13. Why are ceramics that can be sintered at temperatures below 1000°C attractive for electronic packaging?
14. What are the two main problems with soldered joints in an electronic package?
15. What are the two main criteria that govern the effectiveness of a z-axis conductor?

References

[1] V. Sarihan and T. Fang, *Structural Analysis in Microelectronic and Fiber Optic Systems*, American Society of Mechanical Engineers, EEP, Vol. 12, ASME, New York, 1995, pp. 1–4.

[2] W.E. Marsh, K. Kanakarajan and G.D. Osborn, *Polymer/Inorganic Interfaces*, Materials Research Soc. Symp. Proc., Materials Research Soc., Pittsburgh, PA, Vol. 304, 1993, pp. 91–96.
[3] L.T. Nguyen and I.C. Noyan, *Polymer Eng. Sci.*, 1988, **28**(16), 1013–1025.
[4] W.-J. Yang and K. Kudo, *Proc. Int. Symp. Heat Transfer Science and Technology*, Beijing, China, Hemisphere Publ. Corp., Washington, DC, 1988, pp. 14–30.
[5] A. Perkins and A. Guthrie, *Metallurgia*, 1982, **19**(12), 605–621.
[6] K. Ebeling, *Betonwerk und Fertigteil – Technik*, 1994, **60**(12), 70–76.
[7] F. Lazzari and G. Raffellini, *Int. J. Ambient Energy*, 1981, **2**(3), 141–149.
[8] A. Kumar, U. Singh, A. Srivastava and G.N. Tiwari, *Appl. Energy*, 1981, **8**(4), 255–267.
[9] S.P. Seth, M.S. Sodha and A.K. Seth, *Appl. Energy*, 1982, **10**(2), 141–149.
[10] N.D. Kaushik and S.K. Rao, *Appl. Energy*, 1982, **12**(1), 21–36.
[11] A.H. Fanney, B.P. Dougherty and K.P. Dramp, *Proc. 1997 Int. Solar Energy Conf.*, ASME, New York, 1997, pp. 171–182.
[12] B.H. Ramadan, *ASHRAE Transactions*, 1994, **100**(1), 160–167.
[13] B.G. Fomin, *Gidrotekhnicheskoe Stroitel'Stvo*, 1999, (5), 12–14.
[14] M. Sanchez-Romero and P. Alavedra-Ribot, *Building Res. & Information*, 1996, **24**(6), 369–373.
[15] K. Pelissier, T. Chartier and J.M. Laurent, *Ceramics Int.*, 1998, **24**(5), 371–377.
[16] F.S.G. dos Santos and J.W. Swart, *J. Electrochemical Soc.*, 1990, **137**(4), 1252–1255.
[17] M.J. Cattelino, G.V. Miran and B. Smith, *IEEE Transactions on Electron Devices*, 1991, **38**(10), 2239–2243.
[18] R. Chugh and D.D.L. Chung, *Carbon*, 2002, **40**(13), 2285–2289.
[19] O.A. Portnoi, E.S. Shub, G.A. Il'ina, I.M. Stark, V.P. Zosin, S.T. Slavinskii, V.A. Bushtyrkov, V.A. Klyukvin and R.M. Levit, *Fibre Chemistry* (English translation of Khimicheskie Volokna) 1990, **21**(5), 420–422.
[20] T. Kim and D.D.L. Chung, *Carbon*, in press.
[21] C.-C. Hung, M.E. Dillehay and M. Stahl, *J. Aircraft*, 1987, **24**(10), 725–730.
[22] J. Xie, J. Wang, X. Wang and H. Wang, *Hecheng Shuzhi Ji Suliao/Synthetic Resin & Plastics*, 1996, **13**(1), 50–54.
[23] C. Sandberg, W. Whitney, A. Nassar and G. Kuse, *Proc. 1995 IEEE Int. conf. Systems*, Man and Cybernetics, IEEE, Piscataway, NJ, 1995, **4**, 3346–3351.
[24] V.N. Prokushin, A.A. Shubin, V.V. Klejmenov and E.N. Marmer, *Khimicheskie Volkna*, 1992, (2), 50–51.
[25] H.W. Long and G.E. Long, US Patent 6,193,793 (2001).
[26] I.V. Avtonomov and G.A. Pugachev, *Izvestiya Sibirskogo Otdeleniya Akademii Nauk Sssr, Seriya Tekhnicheskikh Nauk*, 1987, **21**, 110–114.
[27] S. Yehia, C.Y. Tuan, D. Ferdon and B. Chen, *ACI Mater. J.*, 2000, **97**(2), 172–181.
[28] Y. Kasai, *Concr. Int.: Design & Construction*, 1989, **11**(3), 33–38.
[29] W. Nakagawa, K. Nishita and Y. Kasai, *Proc. 2nd ASME–JSME Nuclear Eng. Joint Conf.*, ASME, New York, 1993, pp. 871–876.
[30] F.S. Chute, F.E. Vermeulen and M.R. Cervenan, *Canadian Electrical Eng. J.*, 1981, **6**(1), 20–28.
[31] S. Wen and D.D.L. Chung, *Cem. Concr. Res.*, 1999, **29**(6), 961–965.
[32] S. Wen and D.D.L. Chung, *Cem. Concr. Res.*, 1999, **29**(12), 1989–1993.
[33] P. Chen and D.D.L. Chung, *J. Electron Mater.*, 1995, **24**(1), 47–51.
[34] Z. Shui, J. Li, F. Huang and D. Yang, *J. Wuhan Univ. Tech.*, 1995, **10**(4), 37–41.
[35] P.-W. Chen, X. Fu and D.D.L. Chung, *ACI Mater. J.*, 1997, **94**(2), 147–155.
[36] P.-W. Chen and D.D.L. Chung, *ACI Mater. J.*, 1996, **93**(2), 129–133.
[37] P.L. Zaleski, D.J. Derwin and W.H. Flood Jr, US Patent 5,707,171 (1998).
[38] P. Xie, P. Gu, Y. Fu and J.J. Beaudoin, US Patent 5,447,564 (1995).
[39] H. Korner, A. Shiota and C.K. Ober, *ANTEC '96: Plastics – Racing into the Future*, Conf. Proc., Society of Plastics Engineers, Technical Papers Series, no. 42, Brookfield, Conn., Vol. 2, 1996, pp. 1458–1461.
[40] G.S. Swei and D.J. Arthur, *3rd Int. SAMPE Symp. Exhib.*, SAMPE, Covina, CA, 1989, pp. 1111–1124.
[41] L. Li and D.D.L. Chung, *Composites*, 1991, **22**(3), 211–218.
[42] X. Lu and G. Xu, *J. Appl. Polymer Sci.*, 1997, **65**(13), 2733–2738.
[43] D. Klosterman and L. Li, *J. Electronics Manufacturing*, 1995, **5**(4), 277–287.
[44] S.K. Kang, R. Rai and S. Purushothaman, *1996 Proc. 46th Electronic Components & Technology*, IEEE, New York, 1996, pp. 565–570.
[45] G.F.C.M. Lijten, H.M. van Noort and P.J.M. Beris, *J. Electronics Manufacturing*, 1995, **5**(4), 253–261.
[46] M.P. Zussman, B. Kirayoglu, S. Sharkey and D.J. Powell, *6th Int. SAMPE Electronics Conf.*, 1992, pp. 437–448.

[47] J.D. Bolt and R.H. French, *Adv. Mater. Processes*, 1988, **134**(1), 32-35.
[48] J.D. Bolt, D.P. Button and B.A. Yost, *Mater. Sci. Eng.*, 1989, **A109,** 207-211.
[49] S.P. Mukerherjee, D. Suryanarayana and D.H. Strope, *J. Non-Crystalline Solids*, 1992, **147, 148,** 783-791.
[50] M.P. Zussman, B. Kirayoglu, S. Sharkey and D.J. Powell, *6th Int. SAMPE Electronics Conf.*, 1992, pp. 437-448.
[51] J.J. Glatz, R. Morgan and D. Neiswinger, *Int. SAMPE Electronics Conf.*, Vol. 6, 1992, pp. 131-145.
[52] A. Bertram, K. Beasley and W. de la Torre, *Naval Engineers J.*, 1992, **104**(3), 276-285.
[53] D. Brookstein and D. Maass, *7th Int. SAMPE Electronics Conf.*, 1994, pp. 310-327.
[54] T.F. Fleming and W.C. Riley, *Proc. SPIE - Int. Soc. for Optical Eng.*, Soc. Photo-Optical Instrumentation Engineers, Bellingham, WA, 1993, Vol. 1997, pp. 136-147.
[55] T.F. Fleming, C.D. Levan and W.C. Riley, *Proc. Technical Conf.*, Int. Electronics Packaging Conf., Wheaton, IL, Int. Electronics Packaging Society, 1995, pp. 493-503.
[56] A.M. Ibrahim, *6th Int. SAMPE Electronics Conf.*, 1992, pp. 556-567.
[57] N. Kiuchi, K. Ozawa, T. Komami, O. Katoh, Y. Arai, T. Watanabe and S. Iwai, *Int. SAMPE Tech. Conf.*, Vol. 30, 1998, SAMPE, Covina, CA, pp. 68-77.
[58] J.W.M. Spicer, D.W. Wilson, R. Osiander, J. Thomas and B.O. ONI, *Proc. SPIE - International Society for Optical Engineering*, Vol. 3700, 1999, pp. 40-47.
[59] D.A. Foster, *SAMPE Q.* 1989, **21**(1), 58-64.
[60] D.A. Foster, *34th Int. SAMPE Symp. Exhib.*, Book 2 (of 2), SAMPE, Covina, CA, 1989, pp. 1401-1410.
[61] W. de la Torre, *6th Int. SAMPE Electronics Conf.*, 1992, pp. 720-733.
[62] P.D. Wienhold, D.S. Mehoke, J.C. Roberts, G.R. Seylar and D.L. Kirkbride, *Int. SAMPE Tech. Conf.*, Vol. 30, SAMPE, Covina, CA, 1998, pp. 243-255.
[63] X. Luo and D.D.L. Chung, *Composites*: Part B, 1999, **30**(3), 227-231.
[64] L.G. Morin Jr and R.E. Duvall, *Proc. Int. SAMPE Symp. Exhib.*, Vol. 43, No. 1, SAMPE, Covina, CA, 1998, pp. 874-881.
[65] G. Lu, X. Li and H. Jiang, *Composites Sci. Tech.*, 1996, **56,** 193-200.
[66] P.L. Smaldone, *27th Int. SAMPE Technical Conf.*, 1995, pp. 819-829.
[67] J.J. Glatz, D.L. Vrable, T. Schmedake and C. Johnson, *6th Int. SAMPE Electronics Conf.*, 1992, p. 334-346.
[68] R. Crossman, *Northcon/85 - Conf. Rec.*, distributed by Western Periodicals Co., North Hollywood, CA, published by Electronic Conventions Management Inc., Los Angeles, CA, 1985, 21 pp.
[69] D.S. McLachlan, M. Blaszkiewicz and R.E. Newnham, *J. Am. Ceram. Soc.*, 1990, **73**(8), 2187-2203.
[70] L. Li and D.D.L. Chung, *Polymer Composites*, 1993, **14**(6), 467-472.
[71] L. Li, P. Yih and D.D.L. Chung, *J. Electron. Mater.*, 1992, **21**(11), 1065-1071.
[72] B. Guerrero, C. Alemán and R. Garza, *J. Polymer Eng.*, 1997-1998, **17**(2), 95-110.
[73] M. Omastová, J. Pavlinec, J. Pionteck and F. Simon, *Polymer Int.*, 1997, **43**(2), 109-116.
[74] X.B. Chen and J.-P. Issi, M. Cassart, J. Devaux and D. Billaud, *Polymer*, 1994, **35**(24), 5256-5258.
[75] P. Chen and D.D.L. Chung, *J. Electron. Mater.*, 1995, **24**(1), 47-51.
[76] J. Fournier, G. Boiteux, G. Seytre and G. Marichy, *J. Mater. Sci. Lett.*, 1997, **16**(20), 1677-1679.
[77] D. Klosterman and L. Li, *J. Electronics Manufacturing*, 1995, **5**(4), 277-287.
[78] S.K. Kang, R. Rai and S. Purushothaman, *1996 Proc. 46th Electronic Components & Technology*, IEEE, New York, 1996, pp. 565-570.
[79] W.R. Lambert and W.H. Knausenberger, *Proceedings of the Technical Program – National Electronic Packaging and Production Conference*, 1991, **3**, 1512-1526.
[80] J.A. Fulton, D.R. Horton, R.C. Moore, W.R. Lambert and J.J. Mottine, *Proc. 39th Electronic Components Conf.*, IEEE, 1989, p., 71-77.
[81] W.R. Lambert, J.P. Mitchell, J.A. Suchin and J.A. Fulton, *Proc. 39th Electronic Components Conf.*, IEEE, 1989, pp. 99-106.
[82] P.B. Hogerton, J.B. Hall, J.M. Pujol and R.S. Reylek, *Mat. Res. Soc. Symp. Proc.*, 1989, **154,** 415.
[83] L. Li and D.D.L. Chung, *J. Electronic Packaging*, 1997, **119**(4), 255.
[84] P.T. Robinson, V. Florescu, G. Rosen and M.T. Singer, *Annual Connector & Interconnection Technology Symposium*, International Institute of Connector and Interconnection Technology, Deerfield, IL, 1990, pp. 507-515.
[85] G. Rosen, P.T. Robinson, V. Florescu and M.T. Singer, *Proc. 6th IEEE Holm Conf. Electrical Contacts and 15th Int. Conf. Electric Contacts*, IEEE, Piscataway, NJ, 1990, pp. 151-165.
[86] D.D. Johnson, *Proc. 3rd Int. Symp. Advanced Packaging Materials: Processes, Properties and Interfaces*, IEEE, Piscataway, NJ and IMAPS, Reston, VA, 1997, pp. 29-30.
[87] T. Kokogawa, H. Morishita, K. Adachi, H. Otsuki, H. Takasago and T. Yamazaki, *Conference Record of 1991 International Display Conference*, San Diego, IEEE, Piscataway, NJ and Society for Information Display, Playa Del Rey, CA, 1991, pp. 45-48.

[88] N.P. Kreutter, B.K. Grove, P.B. Hogerton and C.R. Jensen, *7th Electronic Materials and Processing Congress*, Cambridge, MA, Aug. 1992.
[89] H. Yoshigahara, Y. Sagami, T. Yamazaki, A. Burkhart and M. Edwards, *Proc. Technical Program: National Electronic Packaging and Production Conf.*, NEPCON West '91, Cahners Exposition Group, Des Plaines, IL, 1991, **1**, 213–219.
[90] D.M. Bruner, *Int. J. Microcircuits and Electronic Packaging*, 1995, **18**(3) 311.
[91] J.J. Crea and P.B. Hogerton, *Proc. Technical Program: National Electronic Packaging and Production Conf.*, NEPCON West '91, Cahners Exposition Group, Des Plaines, IL, Vol. 1, 1991, pp. 251–259.
[92] K. Gilleo, *Proc. Electricon '94, Electronics Manufacturing Productivity Facility*, Indianapolis, IN, 1994, pp. 11/1–11/12.
[93] K. Chung, G. Dreier, P. Fitzgerald, A. Boyle, M. Lin and J. Sager, *Proc. 41st Electronic Components Conf.*, May 1991, pp. 345.
[94] J.C. Bolger and J.M. Czarnowski, *1995 Japan IEMT Symp. Proc. 1995 Int. Electronic Manufacturing Technology Symp.*, IEEE, New York, 1996, pp. 476–481.
[95] Y. Xu and D.D.L. Chung, *J. Electron. Mater.*, 1999, **28**(11), 1307.
[96] G.Y. Chin, *Adv. Mater. Proc.*, 1990, **137**(1), 47, 50, 86.
[97] S. Liang, S.R. Chong and E.P. Giannelis, *Proc. Electronic Components & Technology Conf.*, IEEE, New York, Vol. 48, 1998, pp. 171–175.
[98] V. Agarwal, P. Chahal, R.R. Tummala and M.G. Allen, *Proc. Electronic Components and Technology Conf.*, IEEE, New York, Vol. 48, 1998, pp. 165–170.
[99] J.Y. Park and M.G. Allen, *Conf. Proc. - IEEE Applied Power Electronics Conf. and Exposition - APEC*, Vol. 1, 1997, pp. 361–367.
[100] J.Y. Park, L.K. Lagorce and M.G. Allen, *IEEE Trans. Magnetics*, 1997, **33**(5), pt. 1, 3322–3324.
[101] R. Charbonneau, *Proc. Int. SAMPE Symp. Exhib.*, Vol. 43, No. 1, SAMPE, Covina, CA, 1998, pp. 833–844.
[102] J. Wang, V.V. Varadan and V.K. Varadan, *SAMPE J.*, 1996, **32**(6), 18–22.
[103] S. Maugdal and S. Sankaran, *Recent Trends in Carbon*, Proc. National Conf., ed. O.P. Bahl, Shipra Publ., Delhi, India, 1997, pp. 12–19.
[104] J.T. Hoback and J.T. Reilly, *J. Elastomers Plastics*, 1988, **20**(1), 54–69.
[105] G. Lu, X. Li and H. Jiang, *Composites Sci. Tech.*, 1996, **56**, 193–200.
[106] D.W. Radford, *J. Adv. Mater.*, Oct. 1994, pp. 45–53.
[107] C. Huang and J. Pai, *J. Appl. Polymer Sci.*, 1997, **63**(1), 115–123.
[108] D.W. Radford and B.C. Cheng, *J. Testing & Evaluation*, 1993, **21**(5), 396–401.
[109] S.R. Gerteisen, *Northcon/85 - Conf. Rec.*, distributed by Western Periodicals Co., North Hollywood, CA, Paper 6.1, published by Electronic Conventions Management Inc., Los Angeles, CA, 1985, 9 pp.
[110] P. O'Shea, *Evaluation Engineering*, 1995, **34**(8), 84–93.
[111] P. O'Shea, *Evaluation Engineering*, 1996, **35**(8), 56–61.
[112] R.A. Rothenberg and D.C. Inman, *1994 Int. Symp. Electromagnetic Compatibility*, Technical Group on EMC of the Institute of Electronics, Information and Communication Engineers, and Technical Group on EMC of the Institute of Electrical Engineers of Japan, 1994, 818 pp.
[113] J.W.M. Child, *Electronic Production*, Oct. 1986, pp. 41–47.
[114] A.K. Subramanian, D.C. Pande and K. Boaz, *Proc. 1995 Int. Conf. Electromagnetic Interference and Compatibility*, Soc. EMC Engineers, Madras, India, 1995, pp. 139–147.
[115] W. Hoge, *Evaluation Engineering*, 1995, **34**(1), 84–86.
[116] J.F. Walther, *IEEE 1989 Int. Symp. Electromagnetic Compatibility: Symp. Record*, IEEE, New York, 1989, pp. 40–45.
[117] H.W. Denny and K.R. Shouse, *IEEE 1990 Int. Symp. Electromagnetic Compatibility: Symp. Record*, IEEE, New York, 1990, pp. 20–24.
[118] R. Bates, S. Spence, J. Rowan and J. Hanrahan, *8th Int. Conf. Electromagnetic Compatibility, Electronics Division*, Institution of Electrical Engineers, London, 1992, pp. 246–250.
[119] J.A. Catrysse, *8th Int. Conf. Electromagnetic Compatibility*, Electronics Division, Institution of Electrical Engineers, London, 1992, pp. 251–255.
[120] A.N. Faught, *IEEE Int. Symp. Electromagnetic Compatibility*, IEEE, New York, 1982, pp. 38–44.
[121] G. Kunkel, *IEEE Int. Symp. Electromagnetic Compatibility*, IEEE, New York, 1980, pp. 211–216.
[122] K.P. Sau, T.K. Chaki, A. Chakraborty and D. Khastgir, *Plastics Rubber & Composites Processing & Applications*, 1997, **26**(7), 291–297.
[123] X. Shui and D.D.L. Chung, *J. Electron Mater.*, 1995, **24**(2), 107–113.
[124] X. Shui and D.D.L. Chung, *J. Electron. Mater.*, 1996, **25**(6), 930–934.
[125] X. Shui and D.D.L. Chung, *J. Electron. Mater.*, 1997, **26**(8), 928–934.

[126] X. Shui and D.D.L. Chung, *J. Electron. Packaging,* 1997, **119**(4), 236–238.
[127] H. Gieser and E. Worley, *International Integrated Reliability Workshop Final Report 1998*, IEEE, Piscataway, NJ, 1998, p. 94–96.
[128] R.B. Rosner, *Electrical Overstress/Electrostatic Discharge Symposium Proceedings 2000*, ESD Assoc., Rome, NY, 2000, p. 121–131.
[129] K.B. Cheng, S. Ramakrishna and K.C. Lee, *Polym. Compos.,* 2001, **22**(2), 185–196.
[130] Data sheet, Controlled Resistivity Alumina for Static Charge Dissipation, WESGO Technical Ceramics, Belmont, CA.
[131] A.V. Zemlyanukhin, V.D. Alekseev, T.V. Timofeeva and Y.V. Nikifirov, *Steklo i Keramika*, 1992, (6), 21.
[132] T.M. Zhdanova, A.V. Zemlyanukhin and S.N. Neumeecheva, *Glass & Ceramics*, 1991, **48**(3–4), 168.
[133] T.M. Zhdanova, A.V. Zemlyanukhin and S.N. Neumeecheva, *Steklo i Keramika*, 1991, (4), 24.
[134] P. O'Shea, *Evaluation Eng.*, 1997, **36**(2), 6.
[135] M.M. Mateev and D.L. Totev, *Dautschuk & Gummi Kunststoffe*, 1996, **49**(6), 427.
[136] S.E. Artemenko, L.P. Nikulina, T.P. Ustinova, D.N. Akbarov, E.P. Krajnov and V.I. Dubkova, *Khimicheskie Volokna*, 1992, (4), 39.
[137] J.E. Travis, *Soc. Plastics Eng. 8th Annual Pacific Technical Conference and Technical Displays*, Society of Plastics Engineers, Brookfield Center, CT, 1985, p. 98.
[138] A.I. Medalia, *Rubber Chem. & Tech.*, 1986, **59**(3), 432.
[139] S. Wen and D.D.L. Chung, *J. Electron. Mater.*, 2001, **30**(11), 1448.

5 Composite materials for electromagnetic applications

5.1 Introduction

This chapter covers composite materials for electromagnetic applications, including structural and non-structural composite materials. Among the structural composite materials, both polymer–matrix and cement–matrix composites are addressed.

Electromagnetic functions include the shielding, reflection and absorption of electromagnetic radiation. In particular, *electromagnetic interference (EMI) shielding* refers to the reflection and/or absorption of electromagnetic radiation by a material, which thereby acts as a shield against the penetration of the radiation through the shield. As electromagnetic radiation, particularly that at high frequencies (e.g., radio waves, such as those emanating from cellular phones) tend to interfere with electronics (e.g., computers), EMI shielding of both electronics and radiation source is needed and is increasingly required by governments around the world. The importance of EMI shielding relates to the high demand of today's society on the reliability of electronics and the rapid growth of radio frequency radiation sources [1–9]. EMI shielding is to be distinguished from magnetic shielding, which refers to the shielding of magnetic fields at low frequencies (e.g., 60 Hz). Materials for EMI shielding are different from those for magnetic fielding.

In contrast to shielding is the transmission of electromagnetic radiation, as needed for low-observable aircraft and radomes. *Electromagnetic observability* of an object refers to the ability of the object to be observed or detected by electromagnetic waves, particularly microwaves associated with a radar. Low observability is desirable for military aircraft and ships [10–18].

5.2 Mechanisms behind electromagnetic functions

The primary mechanism of EMI shielding is usually *reflection*. For reflection of radiation by the shield, the latter must have mobile charge carriers (electrons or holes) which interact with the electromagnetic fields in the radiation. As a result, the shield tends to be electrically conducting, although a high conductivity is not required. For example, a volume resistivity of the order of 1 Ω.cm is typically sufficient. However, electrical conductivity is not the scientific criterion for shielding, as conduction requires connectivity in the conduction path (percolation in the case of a composite material containing a conductive filler), whereas shielding does not. Although shielding does not require connectivity, it is enhanced by connectivity. Metals are by far the most common materials for EMI shielding. They function mainly by reflection due to the free electrons in them. Metal sheets are bulky, so metal coatings made by electroplating, electroless plating or vacuum deposition are commonly used for shielding [19–34]. The coating may be on bulk materials, fibers or particles. Coatings tend to suffer from poor wear or scratch resistance.

A secondary mechanism of EMI shielding is usually *absorption*. For significant absorption of radiation by the shield, the latter should have electric and/or magnetic dipoles which interact with the electromagnetic fields in the radiation. The electric dipoles may be provided by $BaTiO_3$ or other materials having a high value of dielectric constant. The magnetic dipoles may be provided by Fe_3O_4 or other materials having a high value of the magnetic permeability [19], which may be enhanced by reducing the number of magnetic domain walls through the use of a multilayer of magnetic films [35,36].

Absorption loss is a function of the product $\sigma_r \mu_r$, whereas reflection loss is a function of the ratio σ_r/μ_r, where σ_r is the electrical conductivity relative to copper and μ_r is the relative magnetic permeability. Table 5.1 shows these factors for various materials. Silver, copper, gold and aluminum are excellent for reflection owing to their high conductivity. Superpermalloy and mumetal are excellent for absorption owing to their high magnetic permeability. Reflection loss decreases with increasing frequency, whereas absorption loss increases with increasing frequency.

Other than reflection and absorption, a mechanism of shielding is *multiple reflections*, which refers to the reflections at various surfaces or interfaces in the shield. This mechanism requires the presence of a large surface area or interface area in the shield. An example of a shield with a large surface area is a porous or foam material. An example of a shield with a large interface area is a composite material containing a filler which has a large surface area. The loss due to multiple reflections can be neglected when the distance between the reflecting surfaces or interfaces is large compared to the skin depth.

The losses, whether due to reflection, absorption or multiple reflections, are commonly expressed in dB. The sum of all the losses is the shielding effectiveness (in dB). The absorption loss is proportional to the thickness of the shield.

Table 5.1 Electrical conductivity relative to copper (σ_r) and relative magnetic permeability (μ_r) of selected materials.

Material	σ_r	μ_r	$\sigma_r \mu_r$	σ_r/μ_r
Silver	1.05	1	1.05	1.05
Copper	1	1	1	1
Gold	0.7	1	0.7	0.7
Aluminum	0.61	1	0.61	0.61
Brass	0.26	1	0.26	0.26
Bronze	0.18	1	0.18	0.18
Tin	0.15	1	0.15	0.15
Lead	0.08	1	0.08	0.08
Nickel	0.2	100	20	2×10^{-3}
Stainless steel (430)	0.02	500	10	4×10^{-5}
Mumetal (at 1 kHz)	0.03	20,000	600	1.5×10^{-6}
Superpermalloy (at 1 kHz)	0.03	100,000	3000	3×10^{-7}

From Ref. 131.

Electromagnetic radiation at high frequencies penetrates only the near surface region of an electrical conductor. This is known as the skin effect. The electric field of a plane wave penetrating a conductor drops exponentially with increasing depth into the conductor. The depth at which the field drops to 1/e of the incident value is called the *skin depth* (δ), which is given by

$$\delta = \frac{1}{\sqrt{\pi f \mu \sigma}}, \quad (5.1)$$

where f = frequency, μ = magnetic permeability = $\mu_0 \mu_r$, μ_r = relative magnetic permeability, $\mu_0 = 4\pi \times 10^{-7}$ H/m, and σ = electrical conductivity in Ω^{-1} m^{-1}.

Hence, the skin depth decreases with increasing frequency and with increasing conductivity or permeability. For copper, $\mu_r = 1$, $\sigma = 5.8 \times 10^7$ Ω^{-1} m^{-1}, so δ is 2.09 μm at a frequency of 1 GHz. For nickel of $\mu_r = 100$, $\sigma = 1.15 \times 10^7$ Ω^{-1} m^{-1}, so δ is 0.47 μm at 1 GHz. The small value of δ for nickel compared to copper is mainly due to the ferromagnetic nature of nickel.

5.3 Composite materials for electromagnetic functions

Because of the skin effect, a composite material having a conductive filler with a small unit size of filler is more effective than one having a conductive filler with a large unit size of filler. For effective use of the entire cross-section of a filler unit for shielding, the unit size of the filler should be comparable to or less than the skin depth. Therefore, a filler of unit size 1 μm or less is typically preferred, although such a small unit size is not commonly available for most fillers and the dispersion of the filler is more difficult when the filler unit size decreases. Metal-coated polymer fibers or particles are used as fillers for shielding, but they suffer from the fact that the polymer interior of each fiber or particle does not contribute to shielding.

5.3.1 Composite materials with discontinuous fillers

Polymer–matrix composites containing conductive fillers are attractive for shielding [37–68] owing to their processability (e.g., moldability), which helps to reduce or eliminate the seams in the housing that is the shield. The seams are commonly encountered in the case of metal sheets as the shield and they tend to cause leakage of radiation and diminish the effectiveness of the shield. In addition, polymer–matrix composites are attractive because of their low density. The polymer matrix is commonly electrically insulating and does not contribute to shielding, although it can affect the connectivity of the conductive filler and connectivity enhances the shielding effectiveness. In addition, the polymer matrix affects the processability.

Electrically conducting polymers [69–88] are becoming increasingly available, but they are not common and tend to be poor in their processability and mechanical properties. Nevertheless, electrically conducting polymers do not require a conductive filler in order to provide shielding, so that they may be used with or without a filler. In the presence of a conductive filler, an electrically conducting polymer matrix has the added advantage of being able to electrically connect the filler units that do not touch one another, thereby enhancing the connectivity.

Cement is slightly conducting, so the use of a cement matrix also allows the conductive filler units in the composite to be electrically connected, even when the filler units do not touch one another. Thus, cement–matrix composites have higher shielding effectiveness than corresponding polymer–matrix composites in which the polymer matrix is insulating [89]. Moreover, cement is less expensive than polymers and cement–matrix composites are useful for the shielding of rooms in a building [90–92]. Similarly, carbon is a superior matrix than polymers for shielding owing to its conductivity, but carbon–matrix composites are expensive [93].

A seam in a housing that serves as an EMI shield needs to be filled with an EMI gasket (i.e., a resilient EMI shielding material), which is commonly a material based on an elastomer, such as rubber or silicone [94–107]. An elastomer is resilient, but is itself not able to shield, unless it is coated with a conductor (e.g., a metal coating called metalization) or is filled with a conductive filler (typically metal particles such as Ag–Cu). The coating suffers from poor wear resistance due to the tendency for the coating to debond from the elastomer. The use of a conductive filler suffers from the resulting decrease in resilience, especially at the high filler volume fraction that is usually required for sufficient shielding effectiveness. As the decrease in resilience becomes more severe as the filler concentration increases, the use of a filler that is effective even at a low volume fraction is desirable. Therefore, the development of EMI gaskets is more challenging than that of EMI shielding materials in general.

For a general EMI shielding material in the form of a composite material, a filler that is effective at a low concentration is also desirable, although it is not as critical as for EMI gaskets. This is because the strength and ductility of a composite tend to decrease with increasing filler content when the filler–matrix bonding is poor. Poor bonding is quite common for thermoplastic polymer matrices. Furthermore, a low filler content is desirable because of the greater processability, which decreases with increasing viscosity. In addition, a low filler content is desirable owing to the cost and weight saving.

In order for a conductive filler to be highly effective, it should preferably have a small unit size (due to the skin effect), a high conductivity (for shielding by reflection and absorption) and a high aspect ratio (for connectivity). Metals are more attractive for shielding than carbons because of their higher conductivity, although carbons are attractive in their oxidation resistance and thermal stability. Fibers are more attractive than particles because of their high aspect ratio. Thus, metal fibers of a small diameter are desirable. Nickel filaments of diameter 0.4 μm (as made by electroplating carbon filaments of diameter 0.1 μm) are particularly effective [107–109]. Nickel is more attractive than copper because of its superior oxidation resistance. The oxide film is poor in conductivity and is thus detrimental to the connectivity among filler units.

5.3.2 Composite materials with continuous fillers

Continuous fiber polymer–matrix structural composites that are capable of EMI shielding are needed for aircraft and electronic enclosures [93,110–118]. The fibers in these composites are typically carbon fibers, which may be coated with a metal (e.g., nickel [119]) or intercalated (i.e., doped) to increase conductivity [120,121]. An alternative design involves the use of glass fibers (not conducting) and conducting interlayers in the composite [122,123]. Yet another design involves the use of polyester fibers (not conducting) and a conducting polymer (e.g., polypyrrole) matrix [124]. Still another design involves the use of activated carbon fibers, the moderately high specific surface area (90 m^2/g) of which results in extensive multiple reflections while the tensile properties are maintained [125].

Carbon fibers are electromagnetically reflective, and are thus undesirably high in observability. To alleviate this problem, an electromagnetically absorbing layer is attached to the carbon fiber polymer–matrix composite substrate. The absorbing layer material can be a ferrite particle epoxy–matrix composite [126], a carbonyl–iron particle polymer–matrix composite [127], a conductive polymer [128] or other related materials which absorb through the interaction of the electric and magnetic dipoles in the absorbing layer with electromagnetic radiation [10–14]. However, the attached layer may fall off due to degradation of the bond between the layer and the substrate. Thus, it is desirable to decrease the observability of the carbon fiber polymer–matrix structural composite itself.

A way to decrease observability is to decrease reflectivity. Reflectivity is related to electrical conductivity. Hence, a way to decrease reflectivity is to decrease conductivity. The conductivity of a carbon fiber polymer–matrix depends not only on that of the fibers themselves, but also on the electrical connectivity among the fibers. This is particularly true for conductivity in the transverse direction (direction perpendicular to the fiber direction). The electrical connectivity can be decreased by using epoxy-coated carbon fibers. In this way, observability is decreased [129].

Another way to decrease reflectivity is to use coiled carbon nanofibers as a filler [130]. The coiled configuration helps interaction with the electromagnetic radiation, thereby decreasing reflectivity.

5.4 Conclusion

Composite materials for electromagnetic functions (particularly EMI shielding) are mainly electrically conducting materials with polymer, cement and carbon matrices. Among multifunctional structural materials, polymer–matrix composites with continuous carbon fibers and cement–matrix composites with discontinuous submicron-diameter carbon filaments are attractive. Among non-structural materials, polymer–matrix composites with submicron-diameter nickel-coated carbon filaments are particularly effective. Reflection is the dominant mechanism behind the interaction between these composites and electromagnetic radiation.

Review questions

1. What is the main criterion that governs the effectiveness of a material for low-observable aircraft?
2. What are the two required properties of an EMI gasket material?

3. What are the two main mechanisms of EMI shielding?
4. Why are nickel filaments of diameter 0.4 μm more effective than nickel fibers of diameter 2 μm for electromagnetic interference shielding?
5. Why are submicron-diameter carbon filaments more effective than short pitch-based carbon fibers as a filler in cement for providing EMI shielding?

References

[1] D. Bjorklof, *Compliance Engineering*, 1998, **15**(5), 10 pp.
[2] R. Brewer and G. Fenical, *Evaluation Engineering*, 1998, **37**(7), S-4–S-10.
[3] P. O'Shea, *Evaluation Engineering*, 1998, **37**(6), 40, 43, 45–46.
[4] R.S.R. Devender, *Proc. of the Int. Conf. on Electromagnetic Interference and Compatibility*, IEEE, Piscataway, NJ, 1997, pp. 459–466.
[5] B. Geddes, *Control* (Chicago, IL), 1996, **9**(10), 4 pp.
[6] S. Hempelmann, *Galvanotechnik*, 1997, **88**(2), 418–424.
[7] W.D. Kimmel and D.D. Gerke, *Medical Device & Diagnostic Industry*, 1995, **17**(7), 112–115.
[8] H.W. Markstein, *Electronic Packaging & Production*, 1995, **35**(2), 4 pp.
[9] K.A. McRae, *Proc. of the Electrical Eng. Congress 1994*, National Conf. Publication – Institution of Engineers, Australia, Vol. 2, No. 94/11, IE Aust, Crows Nest, NSW, 1994, pp. 495–498.
[10] J. Paterson, *J. Aircraft*, 1999, **36**(2), 380–388.
[11] R.A. Stonier, *SAMPE J.*, 1991, **27**(5), 9–18.
[12] A.S. Brown, *Aerospace America*, 1990, **28**(3), 16–20, 36.
[13] A.C. Brown, *Proc. 36th Annual Tech. Conf. – Society of Vacuum Coaters*, Albuquerque, NM, 1993. pp. 3–9.
[14] J. Nicholas and R.D. Strattan, *Naval Engineers J.*, 1996, **108**(5), 49–56.
[15] T.A. Guy, K.B. Sanger and E. Ruskowski, *Proc. 36th AIAA/ASME/ASCE/AHS/ASC Structures, Structural Dynamics, and Materials Conference and AIAA/ASME Adaptive Structures Forum*, Vol. 1, New York, 1995, pp. 1–7.
[16] S.E. Mouring, *Marine Tech. Soc. J.*, 1998, **32**(2), 41–46.
[17] H. Harboe-Hansen, *Naval Architect*, pp. 51–53.
[18] C.H. Goddard, D.G. Kirkpatrick, P.G. Rainey and J.E. Ball, *Naval Engineers J.*, 1996, **108**(3), 105–116.
[19] V.V. Sadchikov and Z.G. Prudnikova, *Stal'.*, 1997, (4), 66–69.
[20] S. Shinagawa, Y. Kumagai and K. Urabe, *J. Porous Mater.*, 1999, **6**(3), 185–190.
[21] B.C. Jackson and G. Shawhan, *Proc. of the 1998 IEEE Int. Symp. on Electromagnetic Compatibility*, Vol. 1, IEEE, Piscataway, NJ, 1998, pp. 567–572.
[22] R. Kumar, A. Kumar and D. Kumar, *Proc. of the Int. Conf. Electromagnetic Interference and Compatibility*, IEEE, Piscataway, NJ, 1997, pp. 447–450.
[23] L.G. Bhatgadde and S. Joseph, *Proc. of the Int. Conf. on Electromagnetic Interference and Compatibility*, IEEE, Piscataway, NJ, 1997, pp. 443–445.
[24] A. Sidhu, J. Reike, U. Michelsen, R. Messinger, E. Habiger and J. Wolf, *Proc. of the 1997 Int. Symp. Electromagnetic Compatibility*, IEEE, Piscataway, NJ, 1997, pp. 102–105.
[25] J. Hajdu, *Trans. Inst. Metal Finishing*, 1997, **75**(pt 1), B7–B10.
[26] G. Klemmer, *Proc. of the 1996 54th Annual Tech. Conf. – ANTEC*, Conf. Proc., Vol. 3, Soc. Plastics Engineers, Brookfield, CT, 1996, pp. 3430–3432.
[27] D. Gwinner, P. Scheyrer and W. Fernandez, *Proc. Annual Tech. Conf. – Soc. Vacuum Coaters*, Soc. Vacuum Coaters, Albuquerque, NM, 1996, 336 pp.
[28] M.S. Bhatia, *Proc. of the 1995 4th Int. Conf. Electromagnetic Interference and Compatibility*, IEEE, 1995, pp. 321–324.
[29] L. Zhang, W. Li, J. Liu and B. Ren, *Cailiao Gongcheng/J. Mater. Eng.*, 1995, (7), 38–41.
[30] N.V. Mandich, *Plating & Surface Finishing*, 1994, **81**(10), 60–63.
[31] B.C. Jackson and P. Kuzyk, *Proc. of the 9th Int. Conf. on Electromagnetic Compatibility*, IEE Conf. Publication, No. 396, IEE, Stevenage, Engl., 1994, pp. 119–124.
[32] C. Nagasawa, Y. Kumagai, K. Urabe and S. Shinagawa, *J. Porous Mater.*, 1999, **6**(3), 247–254.
[33] D.S. Dixon and J. Masi, *Proc. of the 1998 IEEE Int. Symp. on Electromagnetic Compatibility*, Vol. 2, IEEE, Piscataway, NJ, 1998, pp. 1035–1040.
[34] P.J.D. Mason, *Proc. of the 37th Annual Tech. Conf. – Soc. Vacuum Coaters*, Soc. Vacuum Coaters, Albuquerque, NM, 1994, pp. 192–197.

[35] C.A. Grimes, *Proc. of the 1994 IEEE Aerospace Applications Conf.*, IEEE, Computer Society Press, Los Alamitos, CA, 1994, pp. 211–221.
[36] W.J. Biter, P.J. Jamnicky and W. Coburn, *Proc. of the 1994 7th Int. SAMPE Electronics Conf.*, Vol. 7, SAMPE, Covina, CA, 1994, pp. 234–242.
[37] L. Xing, J. Liu and S. Ren, *Cailiao Gongcheng/J. Mater. Eng.*, 1998, (1), 19–21.
[38] L. Rupprecht and C. Hawkinson, *Medical Device & Diagnostic Industry*, 1999, **21**(1), 8 pp.
[39] S. Tan, M. Zhang and H. Zeng, *Cailiao Gongcheng/J. Mater. Eng.*, 1998, (5), 6–9.
[40] R. Charbonneau, *Proc. of the 1998 43rd Int. SAMPE Symp. and Exhibition*, Vol. 43, No. 1, SAMPE, Covina, CA, 1998, pp. 833–844.
[41] J.M. Kolyer, *Proc. of the 1998 43rd Int. SAMPE Symp. and Exhibition*, Vol. 43, No. 1, Sampe, Covina, CA, pp. 810–822.
[42] S.L. Thompson, *Evaluation Eng.*, 1998, **37**(7), 62–63, 65.
[43] D.A. Olivero and D.W. Radford, *J. Reinforced Plastics & Composites*, 1998, **17**(8), 674–690.
[44] K.P. Sau, T.K. Chaki, A. Chakraborty and D. Khastgir, *Plastics Rubber & Composites Processing & Applications*, 1997, **26**(7), 291–297.
[45] W.B. Genetti, B.P. Grady and E.A. O'Rear, *Proc. of the 1996 MRS Fall Symp.*, Materials Research Society Vol. 445, MRS, Warrendale, PA, 1997, pp. 153–158.
[46] J. Mao, J. Chen, M. Tu, W. Huang and Y. Liu, *Gongneng Cailiao/J. Functional Mater.*, 1997, **28**(2), 137–139.
[47] L. Rupprecht, *Proc. of the 1996 2nd Conf. on Plastics for Portable and Wireless Electronics*, Soc. Plastics Engineers, Brookfield, CT, 1996, pp. 12–20.
[48] B.K. Bachman, *Proc. of the 1996 2nd Conf. on Plastics for Portable and Wireless Electronics*, Soc. Plastics Engineers, Brookfield, CT, 1996, pp. 7–11.
[49] S. Schneider, *Kunststoffe Plast Europe*, 1997, **87**(4), 487–488.
[50] S. Schneider, *Kunstetoffe Plast Europe*, 1997, **87**(4), 26.
[51] C.Y. Huang and J.F. Pai, *J. Appl. Polym. Sci.*, 1997, **63**(1), 115–123.
[52] M.W.K. Rosenow and J.A.E. Bell, *Proc. of the 1997 55th Annual Tech. Conf.*, ANTEC, Vol. 2, Soc. Plastics Engineers, Brookfield, CT, 1997, pp. 1492–1498.
[53] M.W.K. Rosenow and J.A.E. Bell, *Proc. of the 1998 43rd Int. SAMPE Symp. and Exhibition*, Vol. 43, No. 1, SAMPE, Covina, CA, 1998, pp. 854–864.
[54] M.A. Saltzberg, A.L. Neller, C.S. Harvey, T.E. Borninski and R.J. Gordon, *Circuit World*, 1996, **22**(3), 67–68.
[55] J. Wang, V.V. Varadan and V.K. Varadan, *SAMPE J.*, 1996, **32**(6), 18–22.
[56] J.V. Masi and D.S. Dixon, *Proc. of the 1994 7th Int. SAMPE Electronics Conf.*, Vol. 7, SAMPE, Covina, CA, 1994, pp. 243–251.
[57] H. Rahman, J. Dowling and P.K. Saha, *J. Mater. Proc. Tech.*, 1995, **54**(1–4), 21–28.
[58] A.A. Dani and A.A. Ogale, *Proc. of the 26th Int. SAMPE Tech..l Conf. on 50 Years of Progress in Materials and Science Technology*, Vol. 26, SAMPE, Covina, CA, 1994, pp. 689–699.
[59] D.W. Radford, *J. Adv. Mater.*, 1994, **26**(1), 45–53.
[60] C.M. Ma, A.T. Hu and D.K. Chen, *Polymers & Polymer Composites*, 1993, **1**(2), 93–99.
[61] K. Miyashita, Y. Imai, *Int. Progress Urethanes*, 1993, **6**, 195–218.
[62] L. Li, P. Yih and D.D.L. Chung, *J. Electronic Mater.*, 1992, **21**(11), 1065–1071.
[63] L. Li and D.D.L. Chung, *Composites*, 1994, **25**(3), 215–224.
[64] L. Li and D.D.L. Chung, *Polym. Composites*, 1993, **14**(5), 361–366.
[65] L. Li and D.D.L. Chung, *Polym. Composites*, 1993, **14**(6), 467–472.
[66] L. Li and D.D.L. Chung, *Composites*, 1991, **22**(3), 211–218.
[67] M. Zhu and D.D.L. Chung, *J. Electronic Packaging*, 1991, **113**, 417–420.
[68] M. Zhu and D.D.L. Chung, *Composites*, 1992, **23**(5), 355–364.
[69] J.A. Pomposo, J. Rodriguez and H. Grande, *Synth. Metals*, 1999, **104**(2), 107–111.
[70] J.S. Park, S.H. Ryn and O.H. Chung, *Proc. of the 1998 56th Annual Tech. Conf.*, ANTEC, Vol. 2, Soc. Plast. Eng., Brookfield, CT, 1998, pp. 2410–2414.
[71] S. Courric and V.H. Tran, *Polym.*, 1998, **39**(12), 2399–2408.
[72] M. Angelopoulos, *Proc. of the 1997 3rd Annual Conf. on Plastics for Portable and Wireless Electronics*, Soc. Plast. Eng., Brookfield, CT, 1997, pp. 66.
[73] T. Makela, S. Pienimaa, T. Taka, S. Jussila and H. Isotalo, *Synth. Metal*, 1997, **85**(1–3), 1335–1336.
[74] R.S. Kohlman, Y.G. Min, A.G. MacDiarmid and A.J. Epstein, *J. Eng. & Appl. Sci.*, 1996, **2**, 1412–1416.
[75] K. Naishadham, *Proc. of the 1994 7th Int. SAMPE Electronics Conf.*, Vol. 7, SAMPE, Covina, CA, 1994, pp. 252–265.
[76] A. Kaynak, A. Polat and U. Yilmazer, *Mater. Res. Bull.*, 1996, **31**(10), 1195–1206.
[77] A. Kaynak, *Mater. Res. Bull.*, 1996, **31**(7), 845–860.

[78] J. Joo, A.G. MacDiarmid and A.J. Epstein, *Proc. of the 53rd Annual Tech. Conf.*, ANTEC, Vol. 2, Soc. Plast. Eng., Brookfield, CT, 1995, pp. 1672-1677.
[79] C.P.J.H. Borgmans and R.H. Glaser, *Evaluation Eng.*, 1995, **34**(7), S-32-S-37.
[80] P. Yan, *Scientific American*, 1995, **273**(1), 82-87.
[81] P.J. Mooney, *JOM*, 1994, **46**(3), 44-45.
[82] H.H. Kuhn, A.D. Child and W.C. Kimbrell, *Synth. Mat.*, 1995, **71**(1-3), pt. 3, 2139-2142.
[83] M.T. Nguyen and A.F. Diaz, *Adv. Mater.*, 1994, **6**(11), 858-860.
[84] J. Unsworth, C. Conn, Z. Jin, A. Kaynak, R. Ediriweera, P. Innis and N. Booth, *J. Intelligent Mater. Systems Structures*, 1994, **5**(5), 595-604.
[85] A. Kaynak, J. Unsworth, R. Clout, A.S. Mohan and G.E. Beard, *J. Appl. Polym. Sci.*, 1994, **54**(3), 269-278.
[86] A. Kaynak, A.S. Mohan, J. Unsworth and R. Clout R, *J. Mater. Sci. Lett.*, 1994, **13**(15), 1121-1123.
[87] W. Sauerer, *Galvanotechnik*, 1994, **85**(5), 1467-1472.
[88] M.P. Goefe and L.W. Steenbakkers, *Kunststoffe Plast Europe*, 1994, **84**, 16-18.
[89] L. Fu and D.D.L. Chung, *Carbon*, 1998, **36**(4), 459-462.
[90] L. Gnecco, *Evaluation Eng.*, 1999, **38**(3), 3 pp.
[91] S.S. Lin, *SAMPE J.*, 1994, **30**(5), 39-45.
[92] Y. Kurosaki and R. Satake, *Proc. of the IEEE 1994 Int. Symp. on Electromagnetic Compatibility*, IEEE, Piscataway, NJ, 1994, pp. 739-740.
[93] X. Luo and D.D.L. Chung, *Composites: Part B*, 1999, **30**(3), 227-231.
[94] M. Zhu, Y. Qiu and J. Tian, *Gongneng Cailiao/J. Functional Mater.*, 1998, **29**(6), 645-647.
[95] D.A. Case and M.J. Oliver, *Compliance Eng.*, 1999, **16**(2), 40, 42, 44, 46, 48-49.
[96] S. Hudak, *Evaluation Eng.*, 1998, **37**(8), 3 pp.
[97] B.N. Prakash and L.D. Roy, *Proc. of the 1997 5th Int. Conf. on Electromagnetic Interference and Compatibility*, IEEE, Piscataway, NJ, 1997, pp. 1-2.
[98] S.H. Peng and K. Zhang, *Proc. of the 1998 56th Annual Tech. Conf.*, ANTEC, Vol. 2, Soc. Plast. Eng., Brookfield, CT, 1998, pp. 1216-1218.
[99] S.K. Das, J. Nuebel and B. Zand, *Proc. of the 1997 IEEE 14th Int. Symp. on Electromagnetic Compatibility*, IEEE, Piscataway, NJ, 1997, pp. 66-71.
[100] S.H. Peng and W.S.V. Tzeng, *Proc. of the 1997 Int. Symp. on Electromagnetic Compatibility*, IEEE, Piscataway, NJ, 1997, pp. 94-97.
[101] P. O'Shea, *Evaluation Eng.*, 1997, **36**(8), 6 pp.
[102] P. O'Shea, *Evaluation Eng.*, 1996, **35**(8), 4 pp.
[103] Anonymous, *Electronic Eng.*, 1996, **68**(834), 2 pp.
[104] B. Lee, *Engineering*, 1995, **236**(10), 32-33.
[105] R.A. Rothenberg, D.C. Inman and Y. Itani, *Proc. of the 1994 Int. Symp. on Electromagnetic Compatibility*, IEEE, Piscataway, NJ, 1994, pp. 818.
[106] B.D. Mottahed and S. Manoochehri, *Polymer Eng. Sci.*, 1997, **37**(3), 653-666.
[107] X. Shui and D.D.L. Chung, *J. Electron. Packaging*, 1997, **119**(4), 236-238.
[108] X. Shui and D.D.L. Chung, *J. Electron. Mater.*, 1997, **26**(8), 928-934.
[109] X. Shui and D.D.L. Chung, *J. Electron. Mater.*, 1995, **24**(2), 107-113.
[110] Y. Ramadin, S.A. Jawad, S.M. Musameh, M. Ahmad, A.M. Zihlif, A. Paesano, E. Martuscelli and G. Ragosta, *Polym. Int.*, 1994, **34**(2), 145-150.
[111] M.S. Lin, *Proc. of the 1994 Int. Symp. on Electromagnetic Compatibility*, IEEE, Piscataway, NJ, 1994, pp. 112-115.
[112] H.K. Chiu, M.S. Lin and C.H. Chen, *IEEE Transactions on Electromagnetic Compatibility*, 1997, **39**(4), 332-339.
[113] J.C. Roberts and P.D. Weinhold, *J. Composite Mater.*, 1995, **29**(14), 1834-1849.
[114] D.A. Olivero and D.W. Radford, *SAMPE J.*, 1997, **33**(1), 51-57.
[115] M. Choate and G. Broadbent, *Proc. of the 1996 Regional Tech. Conf. of the Soc. of Plastic Engineers on Thermosets: The True Engineering Polymers Technical Papers*, Soc. Plast. Eng., Brookfield, CT, 1996, pp. 69-82.
[116] P.D. Wienhold, D.S. Mehoke, J.C. Roberts, G.R. Seylar and D.L. Kirkbride, *Proc. of the 1998 30th Int. SAMPE Tech. Conf.*, Vol. 30, SAMPE, Covina, CA, 1998, pp. 243-255.
[117] A. Fernyhough and Y. Yokota, *Materials World*, 1997, **5**(4), 202-204.
[118] T. Hiramoto, T. Terauchi and J. Tomibe, *Proc. of the 1998 20th Annual Int. EOS/ESD (Electrical Overstress/Electrostatic Discharge) Symp.*, ESD Assoc., Rome, NY, 1998, pp. 18-21.
[119] L.G. Morin Jr. and R.E. Duvall, *Proc. of the 1998 43rd Int. SAMPE Symp. and Exhibition*, Vol. 43, No. 1, SAMPE, Covina, CA, 1998, pp. 874-881.
[120] J.R. Gaier and J. Terry, *Proc. of the 1994 7th Int. SAMPE Electronics Conf.*, Vol. 7, SAMPE, Covina, CA, 1994, pp. 221-233.

[121] J.R. Gaier, M.L. Davidson and R.K. Shively, *Proc. of the 1996 28th Int. SAMPE Tech. Conf. on Technology Transfer in a Global Community*, Vol. 28, SAMPE, Covina, CA, 1996, pp. 1136-1147.
[122] J.M. Liu, S.N. Vernon, A.D. Hellman and T.A. Campbell, *Proc. of SPIE – Int. Soc. for Optical Eng.*, Vol. 2459, Soc. Photo-Optical Instrumentation Eng., Bellingham, WA, 1995, pp. 60-68.
[123] D.A. Olivero and D.W. Radford, *Proc. of the 1996 28th Int. SAMPE Tech. Conf. on Technology Transfer in a Global Community*, Vol. 28, SAMPE, Covina, CA, 1996, pp. 1110-1121.
[124] M.S. Kim, H.K. Kim, S.W. Byun, S.H. Jeong, Y.K. Hong, J.S. Joo, K.T. Song, J.K. Kim, C.J. Lee and J.Y. Lee, *Synthetic Metals*, 2002, **126**(ER2-3), 233-239.
[125] J. Wu and D.D.L. Chung, *Carbon*, 2002, **40**(ER3), 445-447.
[126] S.S. Kim, G.M. Cheong and B.I. Yoon, *J. de Physique IV*, 1997, 7(1), 425-426.
[127] M. Matsumoto and Y. Miyata, *NTT R&D*, 1999, **48**(3), 343-348.
[128] F. Jousse, *Proc. 1997 42nd Int. SAMPE Symp. and Exhibition*, Vol. 42, no. 2, Covina, CA, 1997, pp. 1552-1558.
[129] J. Wu and D.D.L. Chung, *Composite Interfaces*, 2002, **9**(4), 389-393.
[130] D. Zhao, Z. Shen and J. Yu, *Carbon 2001*, American Carbon Society, Paper 1.24.
[131] C.R. Paul, *Introduction to Electromagnetic Compatibility*, edited by K. Chang, Wiley, 1992, pp. 649.

6 Composite materials for thermoelectric applications

6.1 Introduction

Thermoelectric phenomena involve the transfer of energy between electric power and thermal gradients. They are widely used for cooling and heating, including air conditioning, refrigeration, thermal management and the generation of electrical power from waste heat.

The thermoelectric phenomenon involving the conversion of thermal energy to electrical energy is embodied in the *Seebeck effect*, i.e., the greater concentration of carrier above the Fermi energy at the hot point than the cold point, the consequent movement of mobile carrier from the hot point to the cold point and the resulting voltage difference (called the *Seebeck voltage*) between the hot and cold points. If the mobile carrier is electrons, the hot point is positive in voltage relative to the cold point. If the mobile carrier is holes, the cold point is positive relative to the hot point. Hence, a temperature gradient results in a voltage. The change in Seebeck voltage (hot minus cold) per degree C temperature rise (hot minus cold) is called the *thermoelectric power*, the *thermopower* or the *Seebeck coefficient*. Table 6.1 gives the values of the Seebeck coefficient of various materials. The Seebeck effect is the basis for thermocouples.

For the thermoelectric phenomenon (*Peltier effect*) which involves the conversion of electrical energy to thermal energy (for heating or cooling), the combination of a low thermal conductivity (to reduce heat transfer loss), a high electrical conductivity (to reduce Joule heating) and a high thermoelectric power is required. These three factors are combined in the *thermoelectric figure of merit* Z, which is defined as

$$Z = \frac{\alpha_{AB}^2}{\left[\left(\frac{\kappa}{\sigma}\right)_A^{1/2} + \left(\frac{\kappa}{\sigma}\right)_B^{1/2}\right]^2}, \qquad (6.1)$$

where A and B are the two dissimilar conductors that form a junction, α_{AB} is the Seebeck coefficient difference ($\alpha_{AB} = \alpha_A - \alpha_B$), κ is the thermal conductivity and σ is the electrical conductivity. Values of Z for various junctions are shown in Table 6.2. A junction commonly involves a p-type semiconductor and an n-type semiconductor, as in the last two entries in Table 6.2. In practice a current is

Table 6.1 Seebeck coefficient [64].

Material	Temperature (°C)	Seebeck coefficient (μV/K)
Al	100	−0.20
Cu	100	+3.98
W	100	+3.68
ZnSb	200	+220
Ge	700	−210
$Bi_2Te(Se)_3$	100	−210
TiO_2	725	−200

passed through the junction in order to attain either heating or cooling. A change in current direction causes a change from heating to cooling, or vice versa.

Thermoelectric behavior has been observed in metals, ceramics and semiconductors, as they are electrically conducting. Composite engineering provides a route to develop better thermoelectric materials, as composites with different properties can be combined in a composite in order to achieve a high figure of merit. Metals are usually high in both thermal and electrical conductivities. Since the combination of low thermal conductivity and high electrical conductivity is not common in single-phase materials, the composite route is valuable. Moreover, the composite route can be used to enhance the mechanical properties.

This chapter addresses thermoelectric composite materials, including structural composites and non-structural composites. Not included in this review is work that uses thermoelectric measurement for the purpose of materials characterization.

Table 6.2. Thermoelectric figure of merit Z [64]

Materials that form junction	Z (10^{-3} K^{-1})
Chromel–constantan	0.1
Sb–Bi	0.18
ZnSb–constantan	0.5
PbTe(p)–PbTe(n)	1.3
Bi_2Te_3(p)–Bi_2Te_3(n)	2.0

6.2 Non-structural composites

Thermoelectric non-structural composites are functional materials designed for their thermoelectric properties, with little attention given to structural properties. These composites are useful for thermoelectric devices and equipment, but not for structures.

The development of thermoelectric non-structural composites is centered on increasing the figure of merit Z through decreasing thermal conductivity and increasing electrical conductivity. These composites are mainly ceramic–ceramic composites, such as B_4C–SiB_{14}–Si [1,2], B_4C–YB_6 [3], SiB_6–TiB_2 [4,5], B_4C–TiB_2 [6], B_4C–SiC [7,8], SiB_n–SiB_4 [9], B_4C–W_2B_5 [10], $AgBiTe_2$–Ag_2Te [11], $Bi_{92.5}Sb_{7.5}$–BN–ZrO_2 [12], $CoSb_3$–$FeSb_2$ [13,14], $CoSb_3$–oxide [15], Ni_3B–B_2O_3 [16] and $Ca_2CoO_{3.34}$–CoO_2 [17]. Boron carbide is particularly attractive for high-temperature thermoelectric conversion [18], since it has a high melting temperature. Thermoelectric non-structural composites also include ceramic–metalloid composites, such as Bi_2Te_3–C [19], SiC–Si [20], SiB_{14}–Si [21] and PbTe–SiGe [22].

Composites involving metals [23] are not common for the Peltier effect, as a metal tends to increase the thermal conductivity. However, metals, particularly intermetallic compounds involving rare earth atoms (e.g., $CePd_3$ and $YbAl_3$) [24], are important for the Seebeck effect.

Thermoelectric composites are mainly designed by considering the electrical and thermal properties of the components, as high electrical conductivity tends to be accompanied by high thermoelectric power. For example, the addition of TiB_2 to B_4C to form the B_4C–TiB_2 composite [6] causes Z to increase, due to the increase in electrical conductivity, slight decrease in thermal conductivity and increase in thermoelectric power. Less commonly, composites are designed by considering the interface of the components, as the interface contributes to scattering of the carriers [25].

The thermal stress between a thermoelectric cell and the wall of a heat exchange of a thermoelectric energy conversion system affects thermal coupling as well as durability. To reduce the thermal stress, compliant pads are used at the interface, although the pad acts as a barrier against thermal conduction [26,27]. If the thermoelectric material is itself compliant, a compliant pad will not be necessary.

Metals are in general more compliant than semiconductors, but their Seebeck effect is relatively weak and they tend to suffer from corrosion. Polymers can be more compliant than metals, but they are usually electrically insulating and do not exhibit the Seebeck effect. On the other hand, a polymer containing an electrically conductive discontinuous filler (e.g., carbon black) above the percolation threshold is conductive [28–30]. By using a polymer which is an elastomer (e.g., silicone), the resulting composite is resilient and compliant. A compliant silicone-matrix composite containing carbon black and exhibiting volume electrical resistivity of 2.3 Ω.cm exhibits an absolute thermoelectric power of +2 $\mu V/°C$.

Electrically conductive polymer-matrix composites such as carbon black filled silicone are used for electromagnetic interference (EMI) shielding and for electrostatic discharge protection. The resilience is important for the use as EMI gaskets (Chapter 5). Owing to the heating associated with the operation of microelectronics, which require shielding, the shielding material (particularly that associated with a mixed signal module such as one that involves both data and voice) may encounter a temperature gradient. The thermoelectric effect of the shielding material would result in a voltage, which may affect the performance of the microelectronics. For example, the electrical grounding may be affected. Therefore, the thermoelectric behavior of shielding materials is of concern.

6.3 Structural composites

6.3.1 Introduction to thermoelectric property tailoring by composite engineering

Composite engineering has been widely employed to tailor the strength, modulus, thermal expansion coefficient, electrical resistivity and thermal conductivity of materials. However, it has been little used for tailoring the thermoelectric properties, which are important for electrical energy generation, heating and cooling. Electrical energy generation involves the Seebeck effect, in which a temperature gradient gives rise to a voltage between the hot and cold ends. Heating and cooling involve the Peltier effect, in which heat is evolved (i.e., heating) or absorbed (i.e., cooling) upon passage of an electric current across two dissimilar materials that are electrically connected. The Seebeck effect provides a renewable source of energy, in addition to providing the basis for thermocouples, which are used for temperature measurement. Moreover, it is relevant to the reduction of environmental pollution and global warming. The Peltier effect is relevant to air conditioning, refrigeration and thermal management.

The tailoring of the Seebeck effect involves consideration of mainly a single property, namely the thermoelectric power. On the other hand, the tailoring of the Peltier effect requires consideration of three properties, namely the thermoelectric power, electrical resistivity and thermal conductivity. Composite engineering is valuable for tailoring both Seebeck and Peltier effects, although this paper is focused on the Seebeck effect.

Composite engineering involves artificial combination of different components, such as carbon fibers and a polymer matrix. It is to be distinguished from alloying, which involves diffusion and/or reaction that results in phases governed by thermodynamics. Although the different components of a composite can undergo diffusion or reaction at their interface, the diffusion or reaction is limited in spatial extent and the components largely retain their original compositions during composite fabrication. On the other hand, composite engineering provides numerous parameters that facilitate tailoring of the properties. These parameters include the components (e.g., the matrix and the fillers), the continuity of the components (e.g., the fibrous filler being continuous or discontinuous), the orientation of the components (e.g., the fiber orientation), the relative orientation of the units of a component (e.g., continuous fibers being unidirectional or crossply), the volume fractions of the components, the degree of contact among the units of a component (e.g., the degree of contact among particles of a particulate filler and among the fibers of a fibrous filler), the interface between components (e.g., the fiber–matrix interface), the interface between units of a component (e.g., the interface between laminae in a laminate), etc.

Because fibers are effective as a reinforcement, fibrous composites tend to exhibit good mechanical properties, particularly when the fibers are continuous. Therefore, fibrous composites have been developed for structural applications. The most common structural composites are polymer–matrix composites containing continuous fibers, as used for lightweight structures such as aircraft, sporting goods and machinery. Increasingly common are cement–matrix composites containing short fibers, as used in construction because of their low drying shrinkage and enhanced flexural toughness. Continuous fibers are not common for cement–matrix composites owing to their high cost and the requirement of low cost for a practical concrete.

The approach used in this section involves taking structural composites as a starting point and modifying these composites for the purpose of enhancing the thermoelectric properties. In this way, the resulting composites are multifunctional (i.e., both structural and thermoelectric). Because structural composites are used in large volumes, the rendering of the thermoelectric function to these materials means the availability of large volumes of thermoelectric materials for use in, say, electrical energy generation. Moreover, temperature sensing is useful for structures for the purpose of thermal control, energy saving and hazard mitigation. The rendering of the thermoelectric function to a structural material also means that the structure can sense its own temperature without the need for embedded or attached thermometric devices, which suffer from high cost and poor durability. Embedded devices, in particular, cause degradation of the mechanical properties of the structure.

Section 6.3 addresses the tailoring of the thermoelectric properties of structural composites, specifically cement–matrix composites containing short and randomly oriented fibers, and polymer–matrix composites containing continuous and oriented fibers. In both cases, the fibers serve as the reinforcement.

Two routes are used in the tailoring. One route involves the choice of fibers. The other route, which only applies to polymer–matrix composites, involves the choice of interlaminar filler, which refers to the particulate filler between the laminae. The former route impacts the thermoelectric properties in any direction for a composite containing randomly oriented fibers. In the case of a composite containing oriented continuous fibers, the former route impacts mainly the thermoelectric properties in the fiber direction of the composite, whereas the latter route impacts mainly the thermoelectric properties in the through-thickness direction of the composite.

6.3.2 Tailoring by the choice of fibers

Polymer–matrix composites

Polymer–matrix composites with continuous oriented carbon fibers are widely used for aircraft, sporting goods and other lightweight structures. Glass fibers and polymer fibers are less expensive than carbon fibers, but they are not conductive electrically and are therefore not suitable for rendering the thermoelectric function to the composite.

Carbon fibers can be n-type or p-type even without intercalation. Intercalation greatly increases the carrier concentration, thus making the fibers strongly n-type or strongly p-type, depending on whether the intercalate is an electron donor or an electron acceptor [31]. One of the drawbacks of intercalated graphite is the instability over time, due either to intercalate desorption or reaction with environmental species. For the case of bromine (acceptor) as the intercalate, the instability due to desorption can be overcome by the use of a residue compound, i.e., a compound that has undergone desorption as much as possible so that the remaining intercalate is strongly held, thereby making the compound stable. The stability of bromine-intercalated carbon fibers has been previously demonstrated [32–34]. For the case of an alkali metal such as sodium (donor) as the intercalate, the instability due to reactivity with moisture can be overcome by the use of an alkali metal hydroxide (with the alkali metal ions in excess) as the intercalate [35]. Therefore, this paper uses bromine as the acceptor intercalate and sodium hydroxide (with Na^+ ions in excess) as the donor intercalate.

Although considerable attention has been given to intercalated carbon fibers, little attention has been given to composites that involve these fibers [36–38].

Previous work on these composites has focused on the electrical conductivity, because of the relevance to electromagnetic interference shielding and other applications.

The carbon fibers used were Thornel P-25, P-100 and P-120 2K pitch-based fibers (Amoco Performance Products, Alpharetta, GA) and T-300 PAN-based fibers (in the form of 976 epoxy unidirectional fiber prepregs, Hy-E 1076E, ICI Fiberite, Tempe, AZ).

Intercalation was carried out only for P-100 and P-120 fibers, owing to their relatively high crystallinity. Bromine intercalation involved exposure to bromine vapor in air at room temperature for 10 days, followed by desorption in a fume hood at room temperature for several months. Sodium hydroxide intercalation involved immersion of the fibers in a liquid solution of NaOH and molten sodium contained in a nickel crucible. The atomic ratio of Na to NaOH was 1:100. The procedure is described below. The crucible was placed in a small furnace, which was purged with argon gas. After the furnace had reached 350°C, sodium metal was added to the molten NaOH in the crucible. The fibers (protected by a nickel spring) were then immersed in the liquid solution. The furnace was covered and the temperature of 350°C was maintained for 4 h. After that, the fibers were removed and allowed to cool. Then the fibers (still protected by a nickel spring) were washed by flowing water for 12 h in order to remove the NaOH on the fiber surface. After this, the fibers were dried in a vacuum oven.

Thermocouple junctions were epoxy–matrix composite interlaminar interfaces. In this study, a junction was formed by allowing two laminae to overlap partially and then curing the stack under heat and pressure, as required for the curing of the epoxy matrix. The overlap region served as the junction; the remaining regions served as thermocouple wires. Those junctions involving T-300 fibers used the epoxy in the prepreg as the bonding agent for the junction. Those not involving T-300 fibers used epoxy resin 9405 and curing agent 9470 from Shell Chemical Co. (Houston, TX) as the epoxy matrix as well as bonding agent. Curing of the epoxy in the T-300 prepreg was conducted by heating in a hydraulic hot press at a rate of 2.5°C/min and then maintaining the temperature for 2 h. The curing temperature was 175°C for the epoxy in the T-300 prepregs and 150°C for the other epoxy. The curing pressure was 18 MPa for unidirectional junctions (i.e., the fibers in the two laminae oriented in the same direction) and 16 MPa for crossply junctions (i.e., the fibers in the two laminae oriented at 90°) involving the epoxy in the T-300 prepregs. For junctions involving the other epoxy, the curing pressure was 0.02 MPa.

Thermopower measurement was performed on the fibers (P-25, P-100 and P-120 fiber bundles without matrix, and T-300 prepreg with epoxy matrix) and on the epoxy–matrix composite junctions involving dissimilar fibers. The measurement in the former case involved attaching the two ends of a fiber bundle or prepreg to copper foils using a silver–epoxy conducting adhesive, maintaining one copper foil at a controlled high temperature (up to 200°C) by using a furnace, and maintaining the other copper foil near room temperature. A copper wire was soldered at its end to each of the two copper foils. The copper wires were fed to a Keithley 2001 multimeter for measuring the voltage. T-type thermocouples were used for measuring the temperatures of the hot and cold ends. Voltage and temperature measurements were done simultaneously using the multimeter. The voltage difference (hot minus cold) divided by the temperature difference (hot minus cold) yielded the Seebeck coefficient with copper as the reference, since the copper wires at the two ends of a sample were at different temperatures. This Seebeck coefficient minus the absolute thermoelectric power of copper (+2.34 μV/°C) [39] is the absolute thermoelectric power of the composite. The thermopower measurement in the latter case involved the same configuration, except that the junction was at the hot point and the two ends of the sample away from the junction were attached using silver–epoxy to two copper foils, which were both near room temperature.

Table 6.1 shows the Seebeck coefficient and the absolute thermoelectric power of composites and the thermocouple sensitivity of composite junctions. A positive value of the absolute thermoelectric power indicates p-type behavior; a negative value indicates n-type behavior. Pristine P-25 is slightly n-type; pristine T-300 is strongly n-type. A junction comprising pristine P-25 and pristine T-300 has a positive thermocouple sensitivity that is close to the difference of the Seebeck coefficients (or the absolute thermoelectric powers) of T-300 and P-25, whether the junction is unidirectional or crossply. Pristine P-100 and pristine P-120 are both slightly n-type. Intercalation with sodium causes P-100 and P-120 to become strongly n-type. Intercalation with bromine causes P-100 and P-120 to become strongly p-type. A junction comprising bromine-intercalated P-100 and sodium-intercalated P-100 has a positive thermocouple sensitivity that is close to the sum of the magnitudes of the absolute thermoelectric powers of the bromine-intercalated P-100 and the sodium-intercalated P-100. Similarly, a junction comprising bromine-intercalated P-120 and sodium-intercalated P-120 has a positive thermocouple sensitivity that is close to the sum of the magnitudes of the absolute thermoelectric powers of the bromine-intercalated P-120 and the sodium-intercalated P-120. Figure 6.1 shows the linear relationship of the measured voltage with the temperature difference between hot and cold points for the junction comprising bromine-intercalated P-100 and sodium-intercalated

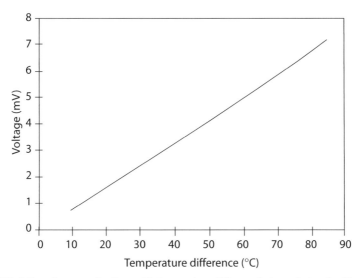

Figure 6.1 Variation of measured voltage with temperature difference between hot and cold points for the epoxy–matrix composite junction comprising bromine-intercalated P-100 and sodium-intercalated P-100.

P-100. By using junctions comprising strongly n-type and strongly p-type partners, a thermocouple sensitivity as high as +82 µV/°C was attained.

It is important to note that the thermocouple junctions do not require any bonding agent other than epoxy, which serves as the matrix of the composite and does not serve as an electrical contact medium (since it is not conducting). In spite of the presence of the epoxy matrix in the junction area, direct contact occurs between a fraction of the fibers of a lamina and a fraction of the fibers of the other lamina, thus resulting in a conduction path in a direction perpendicular to the junction. This conduction path is indicated by direct measurement of the electrical resistance of the junction [40] and enables an electrical contact to be made across the junction. The use of silver paint as an additional bonding agent did not give a better result, as we found experimentally. That the bonding agent did not affect the result is also consistent with the fact that the thermocouple effect is not an interfacial phenomenon. That an additional bonding agent is not necessary facilitates the use of a structural composite as a thermocouple array, as a typical structural composite does not have any extra bonding agent at the interlaminar interface.

Cement–matrix composites

Cement is a low-cost, mechanically rugged and electrically conducting material which can be rendered n-type or p-type by the use of appropriate admixtures, such as short carbon fibers (which contribute holes) for attaining p-type cement and short steel fibers (which contribute electrons) for attaining n-type cement [41–45]. (Cement itself is weakly n-type in relation to electronic/ionic conduction [41].) The fibers also improve the structural properties, such as increasing the flexural strength and toughness and decreasing the drying shrinkage [46–53]. Furthermore, cement-based junctions can be easily made by pouring the dissimilar cement mixes side by side.

The steel fibers used to provide strongly n-type cement paste were made of stainless steel no. 434, as obtained from International Steel Wool Corp. (Springfield, OH). The fibers were cut into pieces of length ~5 mm prior to use in the cement paste in a proportion of 0.5% by mass of cement (i.e., 0.10 vol.%). The properties of the steel fibers are shown in Table 6.3. The mechanical properties of mortars containing these fibers are described in Chen and Chung [47]. However, no aggregate, whether coarse or fine, was used in this work.

The carbon fibers used to provide p-type cement paste were isotropic pitch based, unsized and of length ~5 mm, as obtained from Ashland Petroleum Co. (Ashland, KY). They were used in a proportion of either 0.5% or 0.1% by mass of cement (i.e., either 0.48 or 0.96 vol.% in the case of cement paste with silica fume,

Table 6.3 Properties of steel fibers.

Nominal diameter	60 µm
Tensile strength	970 MPa
Tensile modulus	200 GPa
Elongation at break	3.2%
Volume electrical resistivity	6×10^{-5} Ω.cm
Specific gravity	7.7 g cm^{-3}

and either 0.41 or 0.82 vol.% in the case of cement paste with latex). Silica fume, owing to its fine particulate nature, is particularly effective for enhancing the fiber dispersion [54,55]. The fiber properties are shown in Table 6.4. No aggregate (fine or course) was used. The cement paste with carbon fibers in a proportion of 1.0% by mass of cement was p-type, whereas that with carbon fibers in a proportion of 0.5% by mass of cement was slightly n-type, as shown by thermoelectric power measurement [41].

The cement used in all cases was Portland cement (Type I) from Lafarge Corp. (Southfield, MI). Silica fume (Elkem Materials, Inc., Pittsburgh, PA, EMS 965) was used in a proportion of 15% by mass of cement. The methylcellulose, used in a proportion of 0.4% by mass of cement, was Dow Chemical Corp., Midland, MI, Methocel A15-LV. The defoamer (Colloids, Inc., Marietta, GA, 1010) used whenever methylcellulose was used was in a proportion of 0.13 vol.%. The latex, used in a proportion of 20% by mass of cement, was a styrene butadiene copolymer (Dow Chemical Co., Midland, MI, 460NA) with the polymer making up about 48% for the dispersion and with the styrene and butadiene having a mass ratio of 66:34. The latex was used along with an antifoaming agent (Dow Corning Corp., Midland, MI, no. 2410, 0.5% by mass of latex).

A rotary mixer with a flat beater was used for mixing. Methylcellulose (if applicable) was dissolved in water and then the defoamer was added and stirred by hand for about 2 min. Latex (if applicable) was mixed with the antifoam by hand for about 1 min. The methylcellulose mixture (if applicable), the latex mixture (if applicable), cement, water, silica fume (if applicable), carbon fibers (if applicable) and steel fibers (if applicable) were then mixed in the mixer for 5 min.

A junction between any two types of cement mix was made by pouring the two different mixes into a rectangular mold (160 × 40 × 40 mm) separately, such that the time between the two pours was 10–15 min. The two mixes were poured into two adjacent compartments of the mold and the paper (2 mm thick, without oil on it) separating the compartments was removed immediately after the completion of pouring. Each compartment was roughly half the length of the entire mold.

After pouring into oiled molds, an external electrical vibrator was used to facilitate compaction and decrease the amount of air bubbles. The resulting junction could be seen visually, owing to the color difference between the two halves of a sample. The samples were demolded after 1 day and cured in air at room temperature (relative humidity = 100%) for 28 days.

Five types of cement paste were prepared, namely (i) plain cement paste (weakly n-type, consisting of just cement and water), (ii) steel fiber cement paste (strongly n-type, consisting of cement, water and steel fibers), (iii) carbon fiber silica-fume cement paste (very weakly n-type, consisting of cement, water, silica fume, methylcellulose, defoamer and carbon fibers in a proportion of 0.5% by

Table 6.4 Properties of carbon fibers.

Filament diameter	15 ± 3 μm
Tensile strength	690 MPa
Tensile modulus	48 GPa
Elongation at break	1.4%
Electrical resistivity	3.0×10^{-3} Ω.cm
Specific gravity	1.6 g cm^{-3}
Carbon content	98 wt.%

Table 6.5 Absolute thermoelectric power (μV/°C).

Cement paste	Volume fraction fibers	μV/°C	Type	Ref.
(i) Plain	0	1.99 ± 0.03	weakly n	41
(ii) S_f (0.5*)	0.10%	53.3 ± 4.8	strongly n	45
(iii) C_f (0.5*) + SF	0.48%	0.89 ± 0.09	weakly n	41
(iv) C_f (1.0*) + SF	0.95%	−0.48 ± 0.11	p	41
(v) C_f (0.5*) + L	0.41%	1.14 ± 0.05	weakly n	41

Note: SF = silica fume; L = latex.
*% by mass of cement.

mass of cement), (iv) carbon fiber silica-fume cement paste (p-type, consisting of cement, water, silica fume, methylcellulose, defoamer and carbon fibers in a proportion of 1.0% by mass of cement), and (v) carbon fiber latex cement paste (very weakly n-type, consisting of cement, water, latex and carbon fibers). The water/cement ratio was 0.45 for pastes (i), (ii), (iii) and (iv), and 0.25 for paste (v). The absolute thermoelectric power of each paste is shown in Table 6.5 [41,45].

Three pairs of cement paste were used to make junctions, as described in Table 6.6. For each pair, three specimens were tested in terms of the thermocouple behavior.

Thermocouple testing was conducted by heating the junction by resistance heating, which was provided by nichrome heating wire (wound around the whole perimeter of the sample over a width of 10 mm that was centered at the junction), a transformer and a temperature controller. The voltage difference between the two ends of a sample was measured by using electrical contacts in the form of copper wire wound around the whole perimeter of the sample at each end of the sample. Silver paint was present between the copper wire and the sample surface under the wire. The copper wires from the two ends were fed to a Keithley 2001 multimeter for voltage measurement. A T-type thermocouple was positioned to almost touch the heating wire at the junction. Another T-type thermocouple was attached to one of the two ends of the sample (at essentially room temperature). The difference in temperature between these two locations governs the voltage. Voltage and temperature were measured simultaneously using the multimeter, while the junction temperature was varied through resistance heating. The voltage difference divided by the temperature difference yielded the thermocouple sensitivity.

Figure 6.2 [56] shows plots of the thermocouple voltage versus the temperature difference (relative to essentially room temperature) for junction (a). The thermocouple voltage increases monotonically and reversibly with increasing temperature difference for all junctions. Thermocouple voltage noise decreases and thermocouple sensitivity (Table 6.6 [56]) and reversibility increase in the order (c), (b) and (a). The highest thermocouple sensitivity is 70 ± 7 μV/°C, as attained by junction (a) both during heating and cooling. This value approaches

Table 6.6 Cement junctions.

			Thermocouple sensitivity (μV/°C)	
Junction	Pastes involved	Junction type	Heating	Cooling
(a)	(iv) and (ii)	pn	70 ± 7	70 ± 7
(b)	(iii) and (ii)	nn$^+$	65 ± 5	65 ± 6
(c)	(v) and (ii)	nn$^+$	59 ± 7	58 ± 5

Note: nn+ refers to a junction between a weakly n-type material and a strongly n-type material.

Composite materials for thermoelectric applications 111

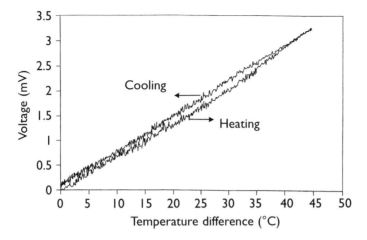

Figure 6.2 Variation of cement-based thermocouple voltage with temperature difference during heating and then cooling for junction (a) of Table 6.6..

that of commercial thermocouples. That junction (a) gives the best thermocouple behavior (in terms of sensitivity, linearity, reversibility and signal-to-noise ratio) is due to the greatest degree of dissimilarity between the materials that make up the junction. The linearity of the plot of thermocouple voltage versus temperature difference is better during cooling than during heating for junction (a).

The values of thermocouple sensitivity (Table 6.6) are higher than (theoretically equal to) the difference in the absolute thermoelectric power of the corresponding two cement pastes that make up the junction (Table 6.5). For example, for junction (a), the difference in the absolute thermoelectric power of pastes (iv) and (ii) is 54 µV/°C, but the thermocouple sensitivity is 70 µV/°C. The reason for this is unclear. Nevertheless, a higher thermocouple sensitivity does correlate with a greater difference in the absolute thermoelectric power.

Table 6.7 Absolute thermoelectric power and volume electrical resistivity of cement pastes (with silica fume except for the paste without fiber) and of steel fiber by itself.

Fiber content		Absolute thermoelectric power (µV/°C)[a]	Resistivity (Ω.cm)
% by mass of cement	Vol. %		
0	0	−1.99 ± 0.04	$(4.7 ± 0.4) × 10^5$
0.5	0.10	−57 ± 4	$(5.6 ± 0.5) × 10^4$
1.0	0.20	−68 ± 5	$(3.2 ± 0.3) × 10^4$
1.1	0.22	−48 ± 5	$(3.0 ± 0.2) × 10^4$
1.2	0.24	−25 ± 2	$(2.3 ± 0.2) × 10^4$
1.3	0.26	−13 ± 1	$(1.8 ± 0.1) × 10^4$
1.4	0.28	0.0 ± 0.2	$(8.7 ± 0.1) × 10^3$
1.5	0.30	+6.3 ± 1.2	$(5.3 ± 0.4) × 10^3$
2.0	0.40	+20 ± 3	$(1.7 ± 0.1) × 10^3$
2.5	0.50	+26 ± 3	$(1.4 ± 0.2) × 10^3$
/	100[b]	+3.76 ± 0.15	$6 × 10^{-5}$ [c]

[a]Measured during heating.
[b]Steel fiber by itself.
[c]From the manufacturer's data sheet.

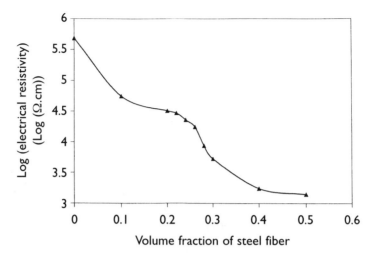

Figure 6.3 Volume electrical resistivity (log scale) of cement pastes containing various volume fractions of steel fiber. All pastes with fibers contained silica fume.

Table 6.7 [57] and Figure 6.3 [57] show that the volume electrical resistivity of cement paste is decreased monotonically by steel fiber addition. The higher the fiber volume fraction, the lower is the resistivity. The absence of an abrupt drop in resistivity as the fiber content increases suggests that all of the fiber volume fractions used are below the percolation threshold, as expected from the previously reported percolation threshold of 0.5–1.0 vol.% for carbon fiber (15 μm diameter) cement paste [54].

Table 6.7 [57] and Figure 6.4 [57] give the absolute thermoelectric power of cement pastes and of the steel fiber by itself. The steel fiber itself has a positive

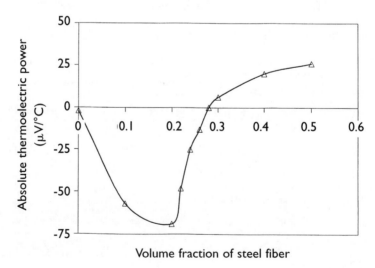

Figure 6.4 Absolute thermoelectric power of cement pastes containing various volume fractions of steel fiber. All pastes with fibers contained silica fume.

value of absolute thermoelectric power, whereas cement paste without fiber has a negative value [58]. The addition of fibers up to 0.20 vol.% makes the value more negative, as reported for the case without silica fume [45]. At the same fiber volume fraction of 0.10%, the use of silica fume changes the absolute thermoelectric power from -53 ± 5 µV/°C [45] to -57 ± 4 µV/°C [57], due to a higher degree of fiber dispersion [54], as shown by the decrease in electrical resistivity from 8×10^4 to 6×10^4 Ω.cm [45]. Increase of the fiber volume fraction from 0.10% to 0.20% causes absolute thermoelectric power to become even more negative, reaching -68 ± 5 µV/°C, which is the highest in magnitude among all cement pastes studied. However, increase of the fiber content beyond 0.20 vol.% makes the value less negative and more positive (as high as $+26 \pm 3$ µV/°C) – even more positive than the value of the fiber by itself ($+3.76 \pm 0.15$ µV/°C). A change in sign occurs at 0.28 vol.%.

It was previously assumed that the steel fiber provides free electrons which would make the absolute thermoelectric power more negative [45]. However, this assumption is incorrect, as the steel fiber itself has a positive value of absolute thermoelectric power (due to scattering of electrons from lattice vibrations within the fiber).

As the steel fiber and the cement paste without fiber have opposite signs of absolute thermoelectric power, the interface between steel fiber and cement paste is a junction of electrically dissimilar materials, like a pn-junction. Carrier scattering at this junction, which is distributed throughout the composite, affects the flow of carriers (electrons and ions) between the hot point and the cold point. Both the negative and positive values of the absolute thermoelectric power of cement pastes containing 0.1–0.5 vol.% steel fibers are attributed to the scattering. A quantitative understanding of the scattering effect requires detailed information on the mean free path and mean free time of the carriers.

The situation is quite different in the case of carbon fiber cement paste. The carbon fiber contributes to hole conduction [42–44], thus making the absolute thermoelectric power of the cement–matrix composite more positive [41]. By using intercalated carbon fiber, which provides even more holes, the absolute thermoelectric power becomes even more positive [59]. Thus, hole conduction rather than scattering dominates the origin of the Seebeck effect in carbon fiber cement paste. In contrast, scattering rather than electron conduction dominates the Seebeck effect in steel fiber cement paste.

6.3.3 Tailoring by the choice of interlaminar filler

In contrast to Section 6.3.2, which involves dissimilar fibers to make thermocouples from structural composites, this section uses dissimilar interlaminar fillers for the two thermocouple legs, which have identical fibers. Furthermore, in contrast to Section 6.3.2.1, which exploits the thermoelectric behavior in the fiber direction of a continuous fiber composite, this section exploits the thermoelectric behavior in the through-thickness direction. The use of dissimilar interlaminar fillers is more suitable for practical application than the use of dissimilar fibers, as the method used to obtain sufficiently dissimilar carbon fibers (i.e., intercalation) is expensive and requires highly crystalline carbon fibers, which are also expensive. Glass fibers and polymer fibers are not suitable, as they are not conducting electrically. Steel fibers are conducting, but their high density makes them unsuitable for lightweight composites. As an interlaminar filler has more effect on the properties (particularly the electrical conduction properties) in the through-thickness direction than those in the fiber direction, this section exploits the thermoelectric behavior in the through-thickness direction.

The thermocouple junction (Section 6.3.2.1) that exploits the thermoelectric behavior in the fiber direction involves two dissimilar laminae that form an interlaminar interface, which is the thermocouple junction. The two laminae can be unidirectional or crossply, but the crossply configuration is attractive in that it provides a two-dimensional array of thermocouple junctions, thereby allowing temperature distribution measurement. Since the thermoelectric behavior in the fiber direction is exploited, only one lamina is needed for each leg. As a consequence, the junction is close to the exposed surface of the laminate and is thus suitable for measuring the temperature of the exposed surface.

On the other hand, for exploiting thermoelectric behavior in the through-thickness direction, each leg needs to be at least a few laminae in thickness. As a result, the junction is not close to the exposed surface and thus cannot be used to measure the temperature of the exposed surface.

An alternative configuration, as illustrated in Figure 6.5, allows the thermocouple junction to be exposed, while each leg comprises multiple laminae. This configuration involves using an electrical interconnection, such as a copper foil that is attached to both laminates. This is the configuration used in this section.

The interlaminar additives used in this section are thermoelectric particles, particularly bismuth, tellurium and bismuth telluride (Bi_2Te_3), which are well known as thermoelectric materials [60]. The thermoplastic polymer is nylon-6 (PA) in the form of unidirectional carbon fiber prepregs of thickness 250 μm.

Sets 1–5 of composite specimens were obtained as described below. Each specimen was made by stacking eight plies of the prepreg in the unidirectional configuration. For Set 1, no interlaminar filler was used. For Sets 2–5, bismuth, tellurium or bismuth telluride particles respectively were spread out manually on each ply as they were laid up. The composition was the same for all seven interlayers in the same composite.

Sets 6–8 of composite specimens were obtained as described below. Each specimen was made by stacking 15 plies of the prepreg in the unidirectional

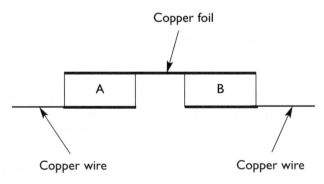

Figure 6.5 Thermocouple involving dissimilar composites A and B and utilizing the thermoelectric behavior of A and B in the through-thickness direction. In this configuration, the thermocouple junction is exposed.

configuration. For Set 6, the 14 interlayers involved alternating layers of tellurium and bismuth telluride particles. For Sets 7 and 8, the composition was the same for all 14 interlayers in the same composite. For Set 7, each interlayer was a mixture of tellurium and bismuth telluride particles (mixed manually before application), such that bismuth telluride particles were in a greater proportion. Set 8 was similar to Set 7, except that tellurium particles were in a greater proportion.

From each of Sets 1–5, two types of specimens were obtained by cutting a laminate. The first type, with a nominal size of 15 × 12 mm, was used to determine the absolute thermoelectric power in the through-thickness direction. The second type, with a nominal size of 35 × 6 mm, was used to determine the value in the longitudinal direction. For each of Sets 6–8, only the type for determining the absolute thermoelectric power in the through-thickness direction was obtained by cutting.

A thermocouple was made using the configuration of Figure 6.5. Each of the two dissimilar laminates (A and B in Figure 6.5) had eight laminae and an interlaminar additive at each of the seven interlaminar interfaces. One laminate (A) had tellurium particles as the additive; the other laminate (B) had bismuth telluride particles as the additive. The copper foil connecting A and B was attached using silver paint, which was applied between the foil and each laminate. For making electrical contact to the ends of A and B away from the copper foil, copper wire, in conjunction with silver paint and a copper foil overlayer, was applied to these ends, as illustrated in Figure 6.5.

Figures 6.6–6.8 [61] show plots of the measured voltage difference (relative to copper) versus the temperature difference obtained during heating for Sets 1, 4 and 5 respectively. In all cases, the data points fall on a nearly straight line through the origin, and the curves for heating and cooling are similar. Table 6.8 [61] shows the absolute thermoelectric power obtained in the through-thickness and longitudinal directions for the composites investigated.

As can be seen, even though the sign of the absolute thermoelectric power is different in the directions considered, the addition of interlaminar thermoelectric particles tends to make it more positive in most cases. The effect is larger in the through-thickness direction than in the longitudinal direction. The thermoelectric behavior of the composite material is influenced by the reinforcing fibers and the interlaminar interfaces. While the fibers govern the behavior in the longitudinal direction, the interlaminar interfaces are encountered in the through-thickness direction. As a result, the effect of the interlaminar particles on thermal or electrical conduction is expected to be larger in the through-thickness direction than in the longitudinal direction.

The use of tellurium particles is more effective than the use of bismuth particles in making absolute thermoelectric power more positive, particularly in the through-thickness direction. This is because in this direction the sign of the absolute thermoelectric power is the same (positive) for the carbon fiber polymer–matrix composite without interlayer and for tellurium (+70 μV/°C [60]), and this is a favorable condition for the enhancement of the thermoelectric effect in composite thermoelectrics [62].

In contrast to bismuth (−72 μV/°C [60]) and tellurium (+70 μV/°C [60]), bismuth telluride (+200 μV/°C [60]) makes the absolute thermoelectric power of the composite more negative, particularly in the through-thickness direction. The origin of this effect is unclear.

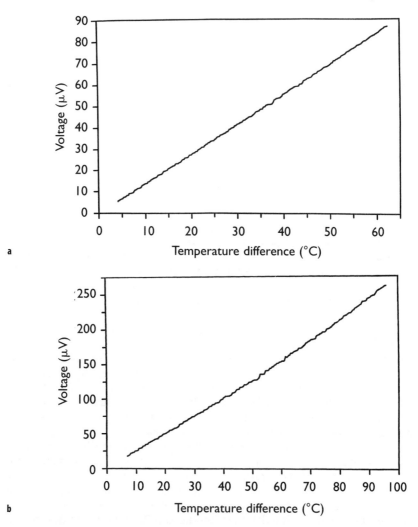

Figure 6.6 Seebeck voltage (relative to copper) vs. temperature difference for a carbon fiber polymer–matrix composite without interlayer during heating (**a**) in the through-thickness direction and (**b**) in the longitudinal direction.

Because tellurium and bismuth telluride as interlaminar fillers give opposite signs of the absolute thermoelectric power, they were used for making a composite thermocouple using the configuration of Figure 6.5. Figure 6.9 shows that the curve of thermocouple voltage versus temperature difference is linear. The thermoelectric sensitivity (e.g., thermocouple voltage per unit temperature difference) is 30 ± 1.5 µV/°C, which, as expected, is roughly equal to the sum of the magnitudes of the absolute thermoelectric power of the two legs in the through-thickness direction (Table 6.8).

Figures 6.10–6.12 show plots of the measured Seebeck voltage (relative to copper) versus the temperature difference obtained during heating for Sets 6–8 respectively. In Figure 6.10 [61], the slope of the curve changed reversibly in sign

Composite materials for thermoelectric applications

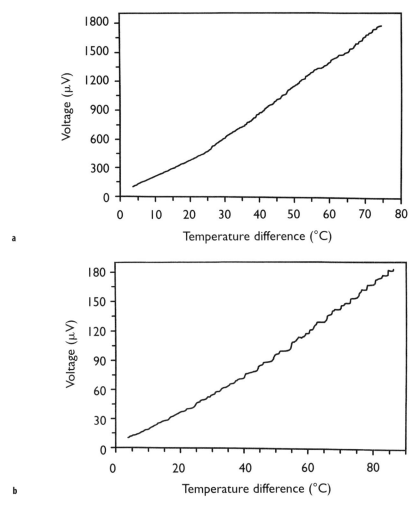

Figure 6.7 Seebeck voltage (relative to copper) vs. temperature difference for a carbon fiber polymer–matrix composite with a tellurium interlayer during heating (**a**) in the through-thickness direction and (**b**) in the longitudinal direction.

as the temperature difference increased. In the temperature difference range of 5–35°C, the absolute thermoelectric power in the through-thickness direction was +8.2 ± 0.3 µV/°C; in the temperature difference range of 55–70°C, the value was −8.5 ± 1.7 µV/°C. In Figure 6.11 [61] and 6.12 [61], there was no change in sign of the slope within a plot, although the slope increased slightly when the temperature difference was high in Figure 6.11 and the slope decreased slightly when the temperature difference was high in Figure 6.12.

Of significance is that the absolute thermoelectric power is strongly negative (−29.4 ± 2.5 µV/°C) for Figure 6.11 (the case in which bismuth telluride dominated) and is strongly positive (+53.6 ± 2.5 µV/°C) for Figure 6.12 (the case in which tellurium dominated). That the case in which bismuth telluride dominated (Set 7) gave a negative value of absolute thermoelectric power is

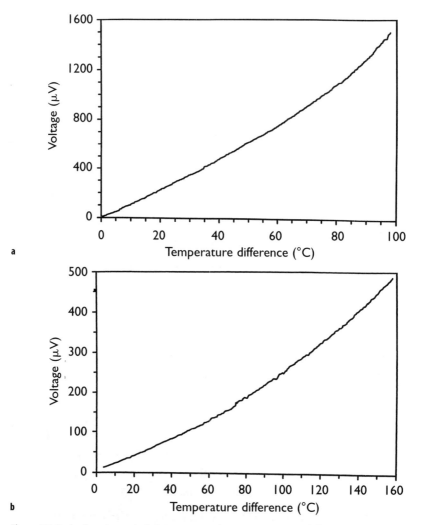

Figure 6.8 Seebeck voltage (relative to copper) vs. temperature difference for a carbon fiber polymer–matrix composite with a bismuth telluride interlayer during heating (**a**) in the through-thickness direction and (**b**) in the longitudinal direction.

consistent with the negative value for the case in which bismuth telluride was the sole interlayer filler (Set 5), but the presence of a small proportion of tellurium along with bismuth telluride caused the absolute thermoelectric power to become much more negative. That the case in which tellurium dominated (Set 8) gave a positive value of absolute thermoelectric power is consistent with the positive value for the case in which tellurium was the sole interlayer filler (Set 4), but the presence of a small proportion of bismuth telluride along with tellurium caused the absolute thermoelectric power to become much more positive.

The origin of the change in sign of the slope in Figure 6.10 is unclear, but it may relate to the dominating effect of bismuth telluride in the high range of temperature difference (so that the absolute thermoelectric power is negative)

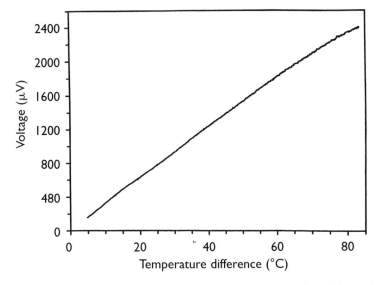

Figure 6.9 Thermocouple voltage vs. temperature difference for a thermocouple with the configuration of Figure 6.3 and involving as the two legs of the thermocouple a composite with tellurium as the interlaminar filler and one with bismuth telluride as the interlaminar filler.

and the dominating effect of tellurium in the low range of temperature difference (so that the absolute thermoelectric power is positive). The increase in slope when the temperature difference is large in Figure 6.11 and the decrease in slope when the temperature difference is large in Figure 6.12 may have origins that are similar to that of the change in sign of the slope in Figure 6.10.

In Set 6, the proportions of tellurium and bismuth telluride were comparable, even though these fillers were not mixed in the same interlayer. The values of absolute thermoelectric power (+8.2 and −8.5 µV/°C, depending on the temperature difference) for Set 6 are intermediate between those for Set 7 (bismuth telluride dominating, −29.4 µV/°C) and Set 8 (tellurium dominating,

Table 6.8 Density, interlayer particle filler volume fraction and absolute thermoelectric power of various thermoplastic-matrix composites.

			Volume fraction	Absolute thermoelectric power (µV/°C)	
Set	Density (g/cm³)	Particle filler	particle filler (%)	Through-thickness	Longitudinal
1	1.4 ± 0.2	None	0	+0.5 ± 0.1	−0.7 ± 0.1
2	1.8 ± 0.2	Bi	4.5 ± 0.5	+0.6 ± 0.1	−0.9 ± 0.1
3	2.3 ± 0.2	Bi	10.5 ± 0.5	+1.2 ± 0.1	−0.3 ± 0.1
4	1.8 ± 0.2	Te	7.3 ± 0.5	+22.3 ± 0.2	−0.1 ± 0.1
5	2.1 ± 0.2	Bi₂Te₃	9.9 ± 0.5	−11.7 ± 1.5	−0.8 ± 0.15
6	2.2 ± 0.2	Te	8.0 ± 0.5	+8.2 ± 0.3[a]	/
		Bi₂Te₃	6.5 ± 0.5	−8.5 ± 1.7[b]	/
7	2.3 ± 0.2	Te	3.6 ± 0.5	−29.4 ± 2.5	/
		Bi₂Te₃	11.5 ± 0.5		
8	2.3 ± 0.2	Te	14.7 ± 0.5	+53.6 ± 2.5	/
		Bi₂Te₃	3.0 ± 0.5		

[a]5–35°C temperature difference.
[b]55–70°C temperature difference.

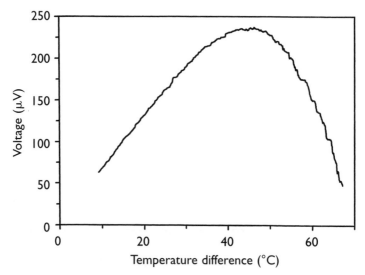

Figure 6.10 Seebeck voltage (relative to copper) vs. temperature difference for Set 6 during heating in the through-thickness direction.

+53.6 μV/°C). Although the case of tellurium and bismuth telluride in comparable proportions and mixed in each interlayer was not investigated, it is likely that this case will give similar results to Set 6.

By using a combination of two thermoelectric materials (combination of Sets 4 and 5 in the through-thickness direction) that exhibit opposite signs of absolute thermoelectric power, such that one material (i.e., one filler in the interlayer) dominated over the other (Sets 7 and 8), we attained magnitudes of absolute

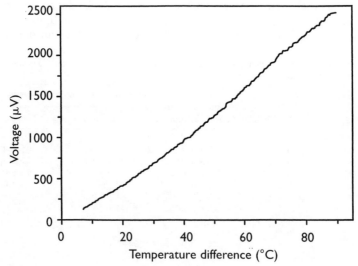

Figure 6.11 Seebeck voltage (relative to copper) vs. temperature difference for Set 7 during heating in the through-thickness direction.

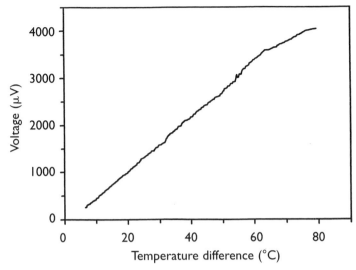

Figure 6.12 Seebeck voltage (relative to copper) vs. temperature difference for Set 8 during heating in the through-thickness direction.

thermoelectric power that are much higher than those of the corresponding materials by themselves (i.e., with only one type of thermoelectric filler in the interlayer). This behavior is consistent with related work on composite thermoelectrics [62,63].

Tellurium (as an element) has an absolute thermoelectric power of +70 μV/°C [60]. By having just 7.3 vol.% Te in a carbon fiber polymer–matrix composite (Set 4), an absolute thermoelectric power of +22 μV/°C (31% of the value of tellurium as an element) was attained in the through-thickness direction. Even though the carbon fiber polymer–matrix composite without an interlayer (Set 1) is negligibly small in its absolute thermoelectric power (+0.5 μV/°C), its presence along with a small proportion of tellurium has a synergistic effect that results in an unexpectedly high value of absolute thermoelectric power in the through-thickness direction. The origin of this synergistic effect is presently not clear, but the effect is consistent with the enhancement of absolute thermoelectric power by a combination of a "high-quality thermoelectric" and a "benign metal" [62].

Set 8 gave a high value of absolute thermoelectric power, i.e., +54 μV/°C. This high value is of technological importance.

6.4 Conclusion

Thermoelectric composite materials include non-structural and structural composites that are designed by consideration of the components' properties, particularly the thermoelectric power, thermal conductivity and electrical conductivity. The non-structural composites are mainly ceramic–ceramic composites. The structural composites are composites with polymer or cement as the matrix, and electrically conducting fibers as the reinforcement.

Tailoring of the thermoelectric properties (the sign and magnitude of the absolute thermoelectric power) was achieved by composite engineering. The techniques involved the choice of reinforcing fibers (continuous or short) in a structural composite and the choice of particulate filler between the laminae in a continuous fiber composite. Tailoring resulted in thermoelectric structural composites, including continuous carbon fiber polymer–matrix composites and short fiber cement–matrix composites. In addition, it resulted in thermocouples in the form of structural composites. The choice of fibers affected thermoelectric behavior in the fiber direction of the composite. The choice of interlaminar filler affected thermoelectric behavior in the through-thickness direction.

The choice of fibers encompassed that between p-type fibers (e.g., pristine and acceptor-intercalated carbon fibers) and n-type fibers (e.g., donor-intercalated carbon fibers). However, in the case where carrier scattering dominates the origin of the Seebeck effect, the use of a p-type fiber (e.g., steel fibers) in an n-type matrix (e.g., cement) may result in a composite with either n-type or p-type behavior, depending on the fiber volume fraction. In the case of continuous fibers, a thermocouple resulted from the use of dissimilar fibers, such that the thermocouple junction was the interface between polymer–matrix laminae of dissimilar fibers. In the case of short fibers, a thermocouple junction was provided by a cementitious bond between cement–matrix composites containing dissimilar fibers.

Interlaminar fillers between continuous carbon fiber polymer–matrix laminae were thermoelectric particulate fillers such as tellurium and bismuth telluride. The use of a mixture of dissimilar fillers, such that one filler dominated over the other, resulted in enhancement of the through-thickness Seebeck effect over that for the case of a single filler (the dominating one) being used. With tellurium dominating over bismuth telluride, an absolute thermoelectric power of +54 µV/°C was attained, compared to the value of +22 µV/°C when tellurium alone was used. With bismuth telluride dominating over tellurium, an absolute thermoelectric power of −29 µV/°C was attained, compared to a value of −12 µV/°C when bismuth telluride alone was used.

The use of dissimilar interlaminar fillers resulted in dissimilar composites, the junction of which served as a thermocouple junction. In this work, the junction was obtained by electrically connecting the top laminae of two dissimilar laminates.

Review questions

1. Describe an application of a thermoelectric material.
2. What is the Seebeck effect?
3. What is the Peltier effect?
4. Describe how a carbon fiber (continuous) polymer–matrix composite can be a two-dimensional array of thermocouples.
5. Describe, in terms of the composition, a cement-based thermocouple.
6. Define intercalation in relation to graphite.

References

[1] J. Li, T. Goto and T. Hirai, *Mater. Trans. Jim*, 1999, **40**(4), 314–319.
[2] J. Li, T. Goto and T. Hirai, *Nippon Seramikkusu Kyokai Gakujutsu Ronbunshi/J. Ceramic Soc. of Japan*, 1998, **106**(1230), 194–197.

[3] J. Li, T. Goto and T. Hirai, *Proc. of the 1998 17th International Conference on Thermoelectrics*, IEEE, Piscataway, NJ, 1998, pp. 587-590.
[4] M. Mukaida, T. Goto and T. Hirai, *Mater. & Manufacturing Processes*, 1992, 7(4), 625-647.
[5] T. Goto, J. Li and T. Hirai, *Proc. of the 1998 17th International Conference on Thermoelectrics*, IEEE, Piscataway, NJ, 1998, pp. 574-577.
[6] T. Goto, J. Li and T. Hirai, *J. Japan Soc. Powder & Powder Metallurgy*, 1997, 44(1), 60-64.
[7] T. Goto, E. Ito, M. Mukaida and T. Hirai, *J. Japan Soc. Powder & Powder Metallurgy*, 1996, 43(3), 311-315.
[8] T. Goto, E. Ito, M. Mukaida and T. Hirai, *J. Japan Soc. Powder & Powder Metallurgy*, 1994, 41(11), 1304-1307.
[9] L. Chen, T. Goto, J. Li and T. Hirai, *J. Japan Soc. Powder & Powder Metallurgy*, 1997, 44(1), 55-59.
[10] K.-F. Cai and C.W. Nan, *Ceramics Int.*, 2000, 26(5), 523-527.
[11] Y. Takigawa, T. Imoto, T. Sakakibara and K. Kurosawa, *Mater. Res. Soc. Symp. Proc.*, 1999, 545, 105-109.
[12] M. Miyajima, K. Takagi, H. Okamura, G.G. Lee, Y. Noda and R. Watanabe, *Proc. of the 1996 15th International Conference on Thermoelectrics*, IEEE, Piscataway, NJ, 1996, pp. 18-21.
[13] S. Katsuyama, Y. Kanayama, M. Ito, K. Majima and H. Nagai, *Proc. of the 1998 17th International Conference on Thermoelectrics*, IEEE, Piscataway, NJ, 1998, pp. 342-345.
[14] S. Katsuyama, *J. Japan Soc. Powder & Powder Metallurgy*, 1999, 46(3), 219-226.
[15] S. Katsuyama, H. Kusaka, M. Ito, K. Majima and H. Nagai, *Proc. of the 1999 18th International Conference on Thermoelectrics*, IEEE, Piscataway, NJ, 1999, pp. 348-351.
[16] Y. Goryachev, V. Dehteruk, M. Siman, L. Fiyalka and E. Shvartsman, *Proc. of the 1998 17th International Conference on Thermoelectrics*, IEEE, Piscataway, NJ, 1998, pp. 109-110.
[17] Y. Miyazaki, K. Kudo, M. Akoshima, Y. Ono, Y. Koike and T. Kajitani, *Japanese J. of Applied Physics Part 2 - Letters*, 2000, 39(6), L531-L533.
[18] F. Thevenot, *J. European Ceramic Soc.*, 1990, 6(4), 205-225.
[19] Y. Goryachev, M. Siman, L. Fiyalka and O. Shvartsman, *Proc. of the 1997 16th International Conference on Thermoelectrics*, IEEE, Piscataway, NJ, 1997, pp. 409-412.
[20] K. Okano and Y. Takagi, *Electrical Eng. in Japan* (English translation of *Denki Gakkai Ronbunshi*), 1996, 117(6), 9-17.
[21] J. Li, T. Goto and T. Hirai, *J. Japan Soc. Powder & Powder Metallurgy*, 1998, 45(6), 581-585.
[22] H. Okamura, M. Miyajima, Y. Noda, A. Kawasaki and R. Watanabe, *J. Japan Soc. Powder & Powder Metallurgy*, 1996, 43(3), 300-305.
[23] L.G. Fel and D.J. Bergman, *Proc. of the 1998 17th International Conference on Thermoelectrics*, IEEE, Piscataway, NJ, 1998, pp. 55-58.
[24] G.D. Mahan, *Proc. of the 1997 16th International Conference on Thermoelectrics*, IEEE, Piscataway, NJ, 1997, pp. 21-27.
[25] R.H. Baughman, A.A. Zakhidov, I.I. Khayrullin, I.A. Udod, C. Cui, G.U. Sumamasedera, L. Grigorian, P.C. Eklund, V. Browning and A. Ehrlich, *Proc. of the 1998 17th International Conference on Thermoelectrics*, IEEE, Piscataway, NJ, 1998, pp. 288-293.
[26] M. Kambe, *Mater. Sci. Forum*, 1999, 308-311, 653-658.
[27] M. Arai, M. Kambe, T. Ogata and Y. Takahashi, *Mippon Kikai Gakkai Ronbunshu, a Hen*, 1996, 62(594), 488-492.
[28] K.G. Princy, R. Joseph and C.S. Kartha, *J. Appl. Polym. Sci.*, 1998, 69(5), 1043-1050.
[29] J. Kost, A. Foux and M. Narkis, *Polym. Eng. & Sci.*, 1994, 34(21), 1628-1634.
[30] J.D. Ajayi and C. Hepburn, *Plastics & Rubber Proc. & Appl.*, 1981, 1(4), 317-326.
[31] J. Tsukamoto, A. Takahashi, J. Tani and T. Ishiguro, *Carbon*, 1989, 27(6), 919.
[32] C.T. Ho and D.D.L. Chung, *Carbon*, 1990, 28(6), 825.
[33] V. Gupta, R.B. Mathur, O.P. Bahl, A. Marchand and S. Flandrois, *Carbon*, 1995, 33(11), 1633.
[34] D.E. Wessbecher, W.C. Forsman and J.R. Gaier, *Synth. Met.*, 1988, 26(2), 185.
[35] C. Hérold, A. Hérold and P. Lagrange, *J. Phys. Chem. Solids*, 1996, 57(6-8), 655.
[36] J.R. Gaier, P.D. Hambourger and M.E. Slabe, *Carbon*, 1991, 29(3), 313.
[37] J.R. Gaier, M.L. Davidson and R.K. Shively, *Tech. Transfer in a Global Community, Int. SAMPE Tech. Conf.*, SAMPE, Covina, CA, 1996, p. 1136.
[38] M. Katsumata, M. Endo, H. Yamanashi and H. Ushijima, *J. Mater. Res.* 1994, 9(7), 1829.
[39] D.D. Pollock, *Thermoelectricity: Theory, Thermometry, Tool*, ASTM Special Technical Publication 852, Philadelphia, PA, 1985, p. 121.
[40] S. Wang and D.D.L. Chung, *Compos. Interfaces*, 1999, 6(6), 497.
[41] S. Wen and D.D.L. Chung, *Cem. Concr. Res.*, 1999, 29(12), 1989.
[42] M. Sun, Z. Li, Q. Mao and D. Shen, *Cem. Concr. Res.*, 1998, 28(4), 549.
[43] M. Sun, Z. Li, Q. Mao and D. Shen, *Cem. Concr. Res.*, 1998, 28(12), 1707.

[44] M. Sun, Z. Li, Q. Mao and D. Shen, *Cem. Concr. Res.*, 1999, **29**(5), 769.
[45] S. Wen and D.D.L. Chung, *Cem. Concr. Res.*, 2000, **30**(4), 661.
[46] P.-W. Chen and D.D.L. Chung, *Composites: Part B*, 1996, **27B**, 269.
[47] P.-W. Chen and D.D.L. Chung, *ACI Mater. J.*, 1996, **93**(2), 129.
[48] A.M. Brandt and L. Kucharska, *Materials for the New Millennium, Proc. Mater. Eng. Conf.*, ASCE, New York, 1996, Vol. 1, p. 271.
[49] N. Banthia and J. Sheng, *Cem. Concr. Comp.*, 1996, **18**(4), 251.
[50] B. Mobasher and C.Y. Li, *ACI Mater. J.*, 1996, **93**(3), 284.
[51] M. Pigeon, M. Azzabi and R. Pleau, *Cem. Concr. Res.*, 1996, **26**(8), 1163.
[52] N. Banthia, C. Yan and K. Sakai, *Cem. Concr. Res.*, 1998, **20**(5), 393.
[53] T. Urano, K. Murakami, Y. Mitsui and H. Sakai, *Compos. - Part A: Applied Science & Manufacturing*, 1996, **27**(3), 183.
[54] P.-W. Chen and D.D.L. Chung, *J. Electron. Mater.*, 1995, **24**(1), 47.
[55] P.-W. Chen, X. Fu and D.D.L. Chung, *ACI Mater. J.*, 1997, **94**(3), 203.
[56] S. Wen and D.D.L. Chung, *J. Mater. Res.*, 2001, **16**(7), 665–667.
[57] S. Wen and D.D.L. Chung, unpublished result.
[58] S. Wen and D.D.L. Chung, *Cem. Concr. Res.*, 2002, **32**(5), 821–823.
[59] S. Wen and D.D.L. Chung, *Cem. Concr. Res.*, 2000, **30**(8), 1295–1298.
[60] R.A. Horne, *J. Appl. Phys.*, 1959, **30**, 393.
[61] V.H. Guerrero, Shoukai Wang, Sihai Wen and D.D.L. Chung, *J. Mater. Sci.*, 2002, **37**(19), 4127.
[62] D.J. Bergman and L.G. Fel, *J. Appl. Phys.*, 1999, **85**(12), 8205.
[63] D.J. Bergman and O. Levy, *J. Appl. Phys.*, 1991, **70**(11), 6821.
[64] M. Ohring, *Engineering Materials Science*, Academic Press, San Diego, 1995, 633 pp.

7 Composite materials for dielectric applications

7.1 Background on dielectric behavior

7.1.1 Dielectric constant

Electrical insulators are also known as dielectrics. Most ionic solids and molecular solids are insulators because of the negligible concentration of conduction electrons or holes.

Consider two metal (conductor) plates connected to the two ends of a battery of voltage V (Figure 7.1). Assuming that the electrical resistance of the connecting wires is negligible, the potential between the two metal plates is V. Assume that the medium between the plates is a vacuum.

There are a lot of conduction electrons in the metal plates and the metal wires. The positive end of the battery attracts conduction electrons, making the left plate positively charged (charge = $+Q$). The negative end of the battery repels conduction electrons, making the right plate negatively charged (charge = $-Q$). Let the area of each plate be A. The magnitude of charge per unit area of each plate is known as the *charge density* (D_o).

$$D_o = \frac{Q}{A}. \tag{7.1}$$

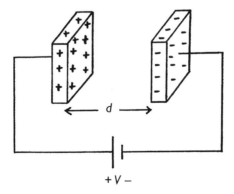

Figure 7.1 A pair of positively charged and negatively charged conductor plates in a vacuum.

The electric field Σ between the plates is given by

$$\Sigma = \frac{V}{d}, \tag{7.2}$$

where d is the separation of the plates. When $\Sigma = 0$, $D_o = 0$. In fact, D is proportional to Σ. Let the proportionality constant be ε_o. Then

$$D_o = \varepsilon_o \Sigma. \tag{7.3}$$

ε_o is called the *permittivity of free space*. It is a universal constant.

$$\varepsilon_o = 8.85 \times 10^{-12} \text{ C/(V.m)}.$$

Equation (7.3) is known as *Gauss's law*.

Just as D_o is proportional to Σ, Q is proportional to V. The plot of Q versus V is a straight line through the origin (Figure 7.2), with

$$\text{Slope} = C_o = \frac{Q}{V} = \frac{\varepsilon_o \Sigma A}{\Sigma d} = \frac{\varepsilon_o A}{d}. \tag{7.4}$$

C_o is known as the *capacitance*. Its unit is coulomb/volt, or *farad* (F). In fact, this is the principle behind a *parallel-plate capacitor*.

Next, consider that the medium between the two plates is not a vacuum, but an insulator whose center of positive charge and center of negative charge coincide when $V = 0$. When $V > 0$, the center of positive charge is shifted toward the negative (right) plate, while the center of negative charge is shifted toward the positive (left) plate. Such displacement of the centers of positive and negative charges is known as *polarization*. In the case of a molecular solid with polarized molecules (e.g., HF), the polarization in the molecular solid is due to the preferred orientation of each molecule, such that the positive end of the molecule is closer to the negative (right) plate. In the case of an ionic solid, polarization is due to the slight movement of cations toward the negative plate and that of anions toward the positive plate. In the case of an atomic solid, the polarization is due to the skewing of the electron clouds towards the positive plate.

When polarization occurs, the center of positive charge sucks more electrons to the negative plate, causing the charge on the negative plate to be $-\kappa Q$, where $\kappa > 1$ (Figure 7.3). Similarly, the center of negative charge repels more electrons away from the positive plate, causing the charge on the positive plate to be κQ. κ is a unitless number called the *relative dielectric constant*. Its value at 1 MHz (10^6 Hz)

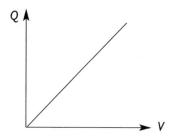

Figure 7.2 Plot of charge Q vs. potential V. The slope = $C_o = \dfrac{\varepsilon_o A}{d}$.

Composite materials for dielectric applications

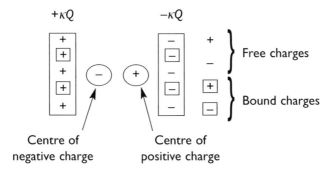

Figure 7.3 A pair of positively charged and negatively charged conductor plates in a medium with relative dielectric constant κ.

is 2.3 for polyethylene, 3.2 for polyvinyl chloride, 6.5 for Al_2O_3, 3000 for $BaTiO_3$, and 78.3 for water.

The charges in the plates when a vacuum is between the plates are called *free charges* (magnitude = Q on each plate). The extra charges in the plates when an insulator is between the plates are called the *bound charges* (magnitude = $\kappa Q - Q$ = $(\kappa - 1)Q$ on each plate).

When an insulator is between the plates, the charge density is given by

$$D_m = \kappa_o = \frac{\kappa Q}{A}. \tag{7.5}$$

Using Equation (7.3), Equation (7.5) becomes

$$D_m = \kappa \varepsilon_o \Sigma = \varepsilon \Sigma, \tag{7.6}$$

where $\varepsilon \equiv \kappa \varepsilon_o$; ε is known as the *dielectric constant*, whereas κ is known as the relative dielectric constant. Hence, ε and ε_o have the same unit.

When an insulator is between the plates, the capacitance is given by

$$C_m = \frac{\kappa Q}{V} = \frac{\kappa \varepsilon_o A \Sigma}{\Sigma d} = \frac{\kappa_o \varepsilon_o A}{d} = \kappa C_o. \tag{7.7}$$

From Equation (7.7), the capacitance is inversely proportional to d, so capacitance measurement provides a way to detect changes in d (i.e., to sense strain).

Mathematically, the *polarization* is defined as the bound charge density, so that it is given by

$$\begin{aligned} P &= D_m - D_o \\ &= \kappa \varepsilon_o \Sigma - \varepsilon_o \Sigma \\ &= (\kappa - 1) \varepsilon_o \Sigma. \end{aligned} \tag{7.8}$$

The plot of P versus Σ is a straight line through the origin, with a slope of $(\kappa - 1)\varepsilon_o$ (Figure 7.4).

The ratio of the bound charge density to the free charge density is given by

$$\frac{\kappa Q - Q}{Q} = \kappa - 1. \tag{7.9}$$

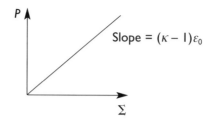

Figure 7.4 Plot of polarization P vs. electric field Σ.

The quantity $\kappa - 1$ is known as the *electric susceptibility* (χ). Hence,

$$\chi = \kappa - 1 = \frac{P}{\varepsilon_0 \Sigma}. \tag{7.10}$$

The *dipole moment* in the polarized insulator is given by

$$(\text{bound charge})d = (\kappa - 1)Qd, \tag{7.11}$$

since the bound charges are induced by the dipole moment in the polarized insulator.

The dipole moment per unit volume of the polarized insulator is thus given by

$$\frac{\text{dipole moment}}{\text{volume}} = \frac{(\kappa-1)Qd}{Ad} = \frac{(\kappa-1)Q}{A} = P. \tag{7.12}$$

Therefore, another meaning of polarization is the dipole moment per unit volume.

Now consider that the applied voltage V is an AC voltage, so that it alternates between positive and negative values. When $V > 0$, $\Sigma > 0$, and the polarization is one way. When $V < 0$, $\Sigma < 0$, and the polarization is in the opposite direction. If the frequency of V is beyond about 10^{10} Hz (Hz = cycles per second), the molecules in the insulator cannot reorient themselves fast enough to respond to V (i.e., *dipole friction* occurs), so κ decreases (Figure 7.5). When the frequency is beyond 10^{15} Hz, even the electron clouds cannot change their skewing direction fast enough to

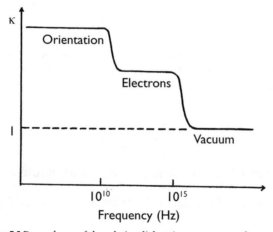

Figure 7.5 Dependence of the relative dielectric constant κ on frequency.

respond to V, so κ decreases further. The minimum value of κ is 1, which is the value for vacuum.

In making capacitors, one prefers to use insulators with very large values of κ, so that C_m is large. However, one should be aware that the value of κ depends on the frequency. It is challenging to make a capacitor that operates at very high frequencies.

The value of κ also changes with temperature and with stress for a given material.

7.1.2 AC loss

In AC condition, it is mathematically more convenient to express Σ and D_m in complex notation, i.e.,

$$\Sigma = \hat{\Sigma}e^{i\omega t} = \hat{\Sigma}\left(\cos\omega t + i\sin\omega t\right) \tag{7.13}$$

and

$$D_m = \hat{D}_m e^{i(\omega t - \delta)} = \hat{D}_m\left[\cos(\omega t - \delta) + i\sin(\omega t - \delta)\right], \tag{7.14}$$

where ω is the angular frequency (which is related to the frequency f by the equation $\omega = 2\pi f$; angular frequency has the unit radians/s, whereas frequency has the unit cycle/s, or Hertz (Hz)), δ is the *dielectric loss angle* (which describes the lag of $\hat{D}_m$ with respect to Σ), $\hat{\Sigma}$ is the amplitude of the Σ wave, and is the amplitude of the D_m wave. If $\delta = 0$, Σ and D_m are in phase and there is no lag. Not only are Σ and D_m complex, ε is complex too. Equation (7.6) then becomes

$$\hat{D}_m e^{i(\omega t - \delta)} = \varepsilon \hat{\Sigma} e^{i\omega t}.$$

Hence,

$$\varepsilon = \frac{\hat{D}_m}{\hat{\Sigma}} e^{-i\delta} = \frac{\hat{D}_m}{\hat{\Sigma}}\left(\cos\delta - i\sin\delta\right). \tag{7.15}$$

The quantity $\hat{D}_m/\hat{\Sigma}$ is known as the *static dielectric constant*. From Equation (7.15), the *loss factor* (or *loss tangent* or *dielectric loss* or *dissipation factor*) tan δ is given by

$$\tan\delta = -\frac{\text{imaginary part of }\varepsilon}{\text{real part of }\varepsilon}. \tag{7.16}$$

In another convention, the dissipation factor is defined as tan δ, while the dielectric loss factor is defined as κ tan δ.

By definition, $\varepsilon = \kappa\varepsilon_o$, so Equation (7.16) can be written as

$$\tan\delta = -\frac{\text{imaginary part of }\kappa}{\text{real part of }\kappa}. \tag{7.17}$$

The loss factor tan δ is more commonly used than δ itself to describe the extent of lag between D_m and Σ. This is because tan δ relates to the energy loss, as explained below. The value of tan δ at 1 MHz (10^6 Hz) is 0.0001 for polyethylene, 0.05 for polyvinyl chloride and 0.001 for Al_2O_3.

Lag occurs when the "insulator" is not a perfect insulator. A non-ideal capacitor can be modeled as a capacitor C in parallel with a resistor (DC) R, as shown in Figure 7.6. If the insulator is perfect, R will be infinite and thus disappears from Figure 7.6. The current i_c through the capacitor is given by

$$i_c = \frac{dQ}{dt} = C\frac{dv}{dt}, \quad (7.18)$$

where v is the AC voltage across both capacitor and resistor. Let

$$v = V \sin \omega t, \quad (7.19)$$

where V is the amplitude of the voltage wave of angular frequency ω, which is related to the frequency f and period T by

$$\omega = 2\pi f = \frac{2\pi}{T}. \quad (7.20)$$

Substituting Equation (7.19) into Equation (7.18),

$$i_c = C\frac{dv}{dt} = \omega C V \cos \omega t \quad (7.21)$$

$$= \frac{V}{1/\omega C} \cos \omega t.$$

Since

$$\sin\left(\omega t + \frac{\pi}{2}\right)$$

$$= \sin \omega t \cos \frac{\pi}{2} + \cos \omega t \sin \frac{\pi}{2}$$

$$= \cos \omega t,$$

Equation (7.21) becomes

$$i_c = \frac{V}{1/\omega C} \sin\left(\omega t + \frac{\pi}{2}\right). \quad (7.22)$$

Comparison of Equation (7.22) with Equation (7.19) shows that i_c leads v by a phase angle of $\pi/2$, or 90°. This pertains to the current through the capacitor only.

The current through the resistor R is

$$i_R = \frac{v}{R} = \frac{V}{R}\sin \omega t. \quad (7.23)$$

i_R and v are in phase.

The phase relationship is conventionally described by a phasor diagram (Figure 7.7), which shows that i_R is in phase with v, i_C is 90° ahead of v and the resultant current i is ahead of v by $\phi = 90° - \delta$. From Figure 7.7,

Composite materials for dielectric applications

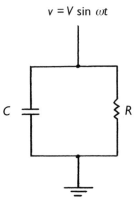

Figure 7.6 A non-ideal capacitor modeled as a capacitor C and a resistor R in parallel.

$$\tan\delta = \frac{V/R}{V\omega C} = \frac{1}{\omega CR}. \tag{7.24}$$

The electrical energy stored in the perfect capacitor is

$$\text{Energy stored} = \int_0^\tau v i_C \, dt$$

$$= \int_0^\tau V^2 \omega C \sin\omega t \cos\omega t \, dt$$

$$= \int_0^\tau \frac{V^2 \omega C}{2} \sin 2\omega t \, dt$$

$$= -\frac{V^2 \omega C}{4\omega} \big[\cos 2\omega t\big]_0^\tau$$

$$= -\frac{1}{4} CV^2 (\cos 2\omega\tau - 1). \tag{7.25}$$

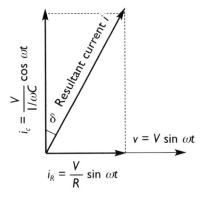

Figure 7.7 Phasor diagram for the non-ideal capacitor of Figure 7.6.

The maximum value of this energy is $\frac{1}{2}CV^2$, and this occurs when $\cos 2\omega t = -1$. The higher is C, the greater is the energy loss.

The loss of energy per cycle due to conduction through the resistor R is

$$\text{energy loss} = \frac{V^2}{R} \int_0^{2\pi/\omega} \sin\omega t \sin\omega t \, dt$$

$$= \frac{V^2}{\omega R} \int_0^{2\pi} \frac{1}{2}(1 - \cos 2\omega t) \, d(\omega t)$$

$$= \frac{V^2}{\omega R} \left[\frac{1}{2}\left(\omega t - \frac{1}{2}\sin 2\omega t\right)\right]_0^{2\pi}$$

$$= \frac{V^2}{\omega R} \left[\frac{1}{2}(2\pi - 0 - 0 + 0)\right]$$

$$= \frac{V^2 \pi}{\omega R}. \tag{7.26}$$

The smaller is R, the greater is the energy loss. When $R = \infty$, R becomes an open circuit and thus disappears from the circuit of Figure 7.6.

$$\frac{\text{energy lost per cycle}}{2\pi \times \text{maximum energy stored}} = \frac{V^2 \pi/\omega R}{2\pi C V^2/2} \tag{7.27}$$

$$= \frac{1}{\omega CR} = \tan\delta.$$

Hence, $\tan \delta$ is related to the energy loss. Energy loss is undesirable for most piezoelectric applications.

Dipole friction causes energy loss, such that the greatest loss occurs at frequencies at which dipole orientation can almost, but not quite, occur, as shown

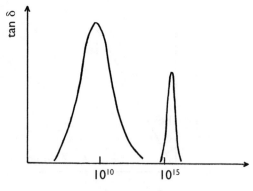

Figure 7.8 Dependence of $\tan \delta$ on frequency.

in Figure 7.8, which corresponds to Figure 7.5. A peak in tan δ occurs at each step decrease in κ as the frequency increases.

7.1.3 Dielectric strength

The maximum electric field that an insulator can withstand before it loses its insulating behavior is known as the *dielectric strength*. It is lower for ceramics (10^4 – 10^7 V/cm) than polymers (10^8 V/cm). In particular, it is 10^4 – 10^5 V/cm for $BaTiO_3$ and $PbZrO_3$. This phenomenon is due to the large electric field providing so much kinetic energy to the few mobile electrons in the insulator that these electrons bombard other atoms, thereby knocking electrons out of these atoms. The additional electrons in turn bombard yet other atoms, resulting in a process called *avalanche breakdown* or *carrier multiplication* (since one electron may knock out two electrons from an atom and each of these two electrons may knock out two electrons from yet another atom, etc). The consequence is a large carrier concentration, which gives rise to an appreciable electrical conductivity. This phenomenon is known as *dielectric breakdown*.

7.2 Piezoelectric behavior

A *piezoelectric* material [1–18] is a material which exhibits one of both of the following effects. In one effect, called the *direct piezoelectric effect*, the relative dielectric constant κ of a material changes in response to stress or strain; the effect allows the sensing of stress or strain through electrical measurement. In the other effect, called the *reverse (or converse) piezoelectric effect*, the strain of a material changes in response to an applied electric field; the effect allows actuation that is controlled electrically. Both effects are reversible. Reversibility is valuable for multiple use (not just once) in sensing or actuation.

Both effects are associated with the electric dipole moment in the material. In the direct piezoelectric effect, the dipole moment per unit volume changes in response to strain, due to the movement of ions or functional groups in the material. The change in dipole moment per unit volume affects the capacitance, which in turn affects the reactance (imaginary part of the complex impedance $Z = R + jX$, where R is the resistance, X is the reactance and $j = \sqrt{-1}$). An RLC meter can be used to measure the capacitance, from which κ can be obtained using Equation (7.7).

In the reverse piezoelectric effect, the dipole moment per unit volume changes in response to an electric field and this change is accompanied by the movement of ions or functional groups, thereby causing strain. The change in dipole moment per unit volume may be due to the change in orientation of electric dipoles in the material. It may also be due to the change in the dipole moment for each dipole.

The most common piezoelectric materials are non-structural materials in the form of ionic crystalline materials whose crystal structure (unit cell) does not have a center of symmetry so that the centers of positive and negative ionic charges do not coincide and an electric dipole (the magnitude of charge multiplied by the distance between the centers of positive and negative charges) results. This situation occurs either with no applied pressure (in the case of barium titanate) or with an applied pressure (in the case of quartz). The dipole moment per unit volume is the polarization P (Equation (7.12)). Every unit cell is associated with a dipole moment. For a column of unit cells in a single crystal, the opposite charges at the adjacent ends of the neighboring unit cells cancel one another, leaving only the charge at the top of the top unit cell and that at the bottom of the bottom unit cell, as illustrated in Figure 7.9. These charges that are not canceled are called surface charges, which give rise to the bound charges in each metal plate sandwiching the piezoelectric material.

Under a mechanical force (i.e., an applied stress), the unit cell is slightly distorted so that the dipole moment per unit volume (i.e., the polarization) is changed and the amount of surface charge is changed (Figure 7.10(a)). This change in surface change amount causes a current pulse to flow from one plate to the other through the external circuit which electrically connects the two plates (Figure 7.10(a) and (b)). The current pulse can be converted to a voltage pulse by using a resistor in the external circuit. If the stress is varying with time at a sufficiently high rate, the current varies with time exactly like the stress varies with time (Figure 7.10(c)), because time does not allow the observation of the tail of a current pulse. Figure 7.10(c) corresponds to a common way in which a piezoelectric sensor is used in practice for sensing dynamic strain. This means that only time-varying stress at a high enough frequency (more than about 0.01 Hz) can be sensed using the method of Figure 7.10(a). On the other hand, if the two plates are not connected electrically, no current can flow and the increase in polarization due to the applied stress causes extra charge on each plate and hence a voltage across the two plates, as illustrated in Figure 7.10(d). A static stress causes a static voltage, whereas a time-varying stress causes a time-varying voltage. Thus, the method of Figure 7.10(d) allows the sensing of both static and dynamic strain. This phenomenon (whether Figure 7.10(a) or (d)) is the direct piezoelectric effect. It allows the conversion (transduction) of mechanical energy to electrical energy for the purpose of sensing. Applications are the detection of mechanical vibrations, force measurement and the detection of audible frequencies.

As a consequence of the polarization (situation in which the positive and negative charge centers do not coincide), an electric field is present in a piezoelectric material. When an electric field is further applied to the material, the dipole moment in the material has to change, thus causing the unit cell dimension to change slightly (i.e., causing a strain) (Figure 7.11). This phenomenon is the reverse piezoelectric effect. It allows the transduction of electrical energy into

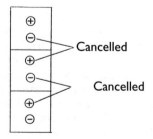

Figure 7.9 A column of three unit cells in a piezoelectric single crystal. Each unit cell has a center of positive charges + and a center of negative charges –. These centers do not coincide.

mechanical energy for the purpose of actuation. Applications are the generation of acoustic or ultrasonic waves (as needed in buzzers and ultrasonic cleaners), micropositioning and vibration-assisted machining.

An example of a piezoelectric material is quartz (SiO_2), which has Si^{4+} and O^{2-} ions. Without an applied stress, the centers of positive and negative charges overlap (Figure 7.12(a)). With an applied stress, the centers of positive and negative charges are displaced from one another, such that the displacement is one way under compression (Figure 7.12(b)) and the other way under tension (Figure 7.12(c)).

The direct piezoelectric effect of Figure 7.10 is described by the equation

$$P = d\sigma, \qquad (7.28)$$

where σ is the applied stress (force per unit area; positive for tensile stress and negative for compressive stress) and d is the *piezoelectric coupling coefficient* (also called the *piezoelectric charge coefficient*). Equation (7.28) means that P is proportional to σ, such that d is the proportionality constant. From Equation (7.28), the unit of d is C/Pa.m², since the unit of P is C/m² and that of σ is Pa. In

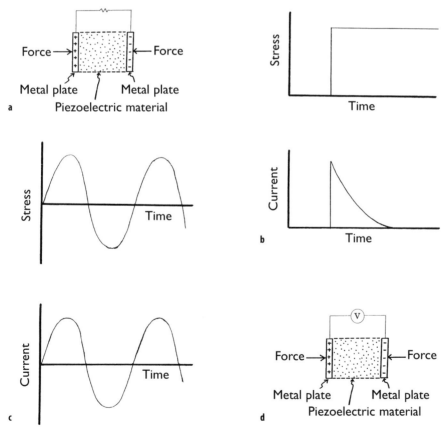

Figure 7.10 The direct piezoelectric effect, in which a current pulse results from an applied stress. (b) Current pulse resulting from static stress. (c) Current wave resulting from stress wave. (d) The direct piezoelectric effect in which a voltage results from an applied stress.

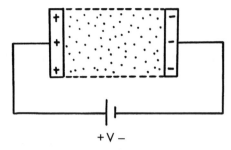

Figure 7.11 The reverse piezoelectric effect, in which a change in dimension results from an applied voltage.

general, P and σ are not necessarily linearly related, so Equation (7.28) should really be written as

$$\partial P = d\partial\sigma \tag{7.29}$$

In the determination of d, the applied electric field Σ is fixed while κ is measured as a function of σ. Hence, from Equations (7.29) and (7.8),

$$d = \varepsilon_o \Sigma \left|\frac{\partial \kappa}{\partial \sigma}\right|. \tag{7.30}$$

The direct piezoelectric effect involves a change in stress causing a change in electric field, thereby causing a change in voltage. Knowing d and κ, one can calculate the change in voltage due to a change in stress $\partial\sigma$, as explained below.

From Equation (7.8),

$$\partial\Sigma = \frac{\partial P}{\varepsilon_o(\kappa-1)}. \tag{7.31}$$

Using Equation (7.29), Equation (7.31) becomes

$$\partial\Sigma = \frac{d\partial\sigma}{\varepsilon_o(\kappa-1)}. \tag{7.32}$$

The change in voltage ∂V is given by

$$\partial V = l\partial\Sigma, \tag{7.33}$$

where l is the length of the specimen in the direction of polarization. From Equations (7.32) and (7.33),

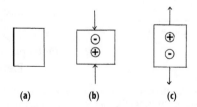

Figure 7.12 Quartz unit cell (a) under no stress, (b) under compression, (c) under tension.

$$\partial V = \frac{l d \partial \sigma}{\varepsilon_o (\kappa - 1)}. \tag{7.34}$$

The *piezoelectric voltage coefficient g* is defined as

$$g = \frac{d}{(\kappa - 1)\varepsilon_o}. \tag{7.35}$$

From Equations (7.34) and (7.35),

$$\partial V = l g \partial \sigma \tag{7.36}$$

For a large value of ∂V, g should be large. From Equation (7.35), for g to be large, d should be large and κ should be small.

Since P and σ do not have to be in the same direction, d is a tensor quantity. Consider three orthogonal axes: x_1, x_2 and x_3. When both P and σ are in the same direction, d is called *longitudinal d*. (When both P and σ are along x_3, the longitudinal d is called d_{33}. Similar definitions apply to d_{22} and d_{11}.) When P and σ are in two orthogonal directions, d is called the *transverse d*. (When P is along x_3 and σ is along x_1, the transverse d is called d_{31}. Similar definitions apply to d_{32}, d_{21}, d_{23}, d_{12} and d_{13}.) When the applied stress is not tensile or compressive, but shear about a certain direction (i.e., forces perpendicular to a certain direction) and P is along a different direction, d is called the *shear d*. (When the shear stress is about x_2 and P is along x_1, the shear d is called d_{15}, in order to distinguish from the transverse d_{12}.) The various d values can be positive or negative. Values of the longitudinal d for selected materials are shown in Table 7.1.

The reverse piezoelectric effect of Figure 7.11 is described by the equation

$$S = d\Sigma, \tag{7.37}$$

where S is the strain (fractional increase in length). From Equation (7.37), the unit of d is m/V, since the unit of Σ is V/m and S has no unit. Note that V/m is the same as C/Pa.m², since $\frac{C}{Pa.m^2} = \frac{C}{N} = \frac{C}{J/m} = \frac{J/V}{J/m} = \frac{m}{V}$. The d in Equation (7.28) is the same as that in Equation (7.37). These two equations are different ways of expressing the same science. In general, S and Σ are not necessarily linearly related, so Equation (7.37) should really be written as

$$\partial S = d \partial \Sigma. \tag{7.38}$$

Dividing Equation (7.37) by Equation (7.28) gives

$$\frac{\Sigma}{\sigma} = \frac{S}{P}. \tag{7.39}$$

Table 7.1 The piezoelectric constant d (longitudinal) for selected materials.

Material	Piezoelectric constant d (C/N = m/V)
Quartz	2.3×10^{-12}
BaTiO$_3$	100×10^{-12}
PbZrTiO$_6$	250×10^{-12}
PbNb$_2$O$_6$	80×10^{-12}

Equivalently, dividing Equation (7.38) by Equation (7.29) gives

$$\frac{\partial \Sigma}{\partial \sigma} = \frac{\partial S}{\partial P}. \tag{7.40}$$

Using Equation (7.8), Equation (7.39) becomes

$$\frac{\Sigma}{\sigma} = \frac{S}{(\kappa-1)\varepsilon_0 \Sigma} \tag{7.41}$$

and Equation (7.40) becomes

$$\frac{\partial \Sigma}{\partial \sigma} = \frac{\partial S}{(\kappa-1)\varepsilon_0 \partial \Sigma}. \tag{7.42}$$

Using Equations (7.37) and (7.41), we get

$$\frac{\Sigma}{\sigma} = \frac{d}{(\kappa-1)\varepsilon_0}. \tag{7.43}$$

Using Equations (7.38) and (7.42), we get

$$\frac{\partial \Sigma}{\partial \sigma} = \frac{d}{(\kappa-1)\varepsilon_0}. \tag{7.44}$$

Using Equation (7.35), Equation (7.43) becomes

$$\Sigma = g\sigma \tag{7.45}$$

and Equation (7.44) becomes

$$\partial \Sigma = g \partial \sigma. \tag{7.46}$$

Like Equation (7.28), Equation (7.45) is used to describe the piezoelectric effect of Figure 7.10. Since g depends on d (Equation (7.35)), g is not an independent parameter.

Stress (σ) and strain (S) are related by Young's modulus (E) through Hooke's law, i.e.,

$$\sigma = ES. \tag{7.47}$$

Hence, Equation (7.45) can be written as

$$\Sigma = gES. \tag{7.48}$$

Rearrangement gives

$$S = \frac{\Sigma}{gE}. \tag{7.49}$$

Combination of Equations (7.49) and (7.37) gives

$$d = \frac{1}{gE}$$

or

Composite materials for dielectric applications

$$E = \frac{1}{gd}. \quad (7.50)$$

Equation (7.50) means that g and d are simply related through E.

A large value of d is desirable for actuators, and a large value of g is desirable for sensors. From Equation (7.35), a low value of κ is favorable for attaining a high value of g.

The *electromechanical coupling factor* or *electromechanical coupling coefficient* (k) is defined as

$$k^2 = \frac{\text{output mechanical energy}}{\text{input electrical energy}}. \quad (7.51)$$

As the effects are reversible,

$$k^2 = \frac{\text{output electrical energy}}{\text{input mechanical energy}}.$$

The factor k describes the efficiency of the energy conversion with a piezoelectric transducer. For a piezoelectric material, k is typically less than 0.1. For a ferroelectric material (Section 7.3), k is typically between 0.4 and 0.7 (0.38 for $BaTiO_3$ and 0.66 for $PbZrO_3$–$PbTiO_3$ solid solution).

The direct piezoelectric effect is useful for strain sensing, which is needed for structural vibration control, traffic monitoring and smart structures in general. The effect is also useful for conversion from mechanical energy to electrical energy, as in converting the stress exerted by a vehicle running on a highway to electricity. This effect is well known for ceramic materials that exhibit the distorted Perovskite structure. These ceramic materials are expensive and are not rugged mechanically, so they are not usually used as structural materials. Devices made from these ceramic materials are commonly embedded in a structure or attached on a structure. Such structures tend to suffer from high cost and poor durability. The use of the structural material itself (e.g., concrete) to provide the piezoelectric effect is desirable, as the structural material is low cost and durable.

The direct piezoelectric effect occurs in cement and is particularly strong when the cement contains short steel fibers and polyvinyl alcohol (PVA) [17]. The longitudinal piezoelectric coupling coefficient d is 2×10^{-11} m/V and the piezoelectric voltage coefficient g is 9×10^{-4} m²/C. These values are comparable to those of commercial ceramic piezoelectric materials such as lead zirconotitanate

Table 7.2 Measured longitudinal piezoelectric coupling coefficient d, measured relative dielectric constant κ, calculated piezoelectric voltage coefficient g and calculated voltage change resulting from a stress change of 1 kPa for a specimen thickness of 1 cm in the direction of polarization.

Material	d (10^{-13} m/V)[a]	κ[b]	g (10^{-4} m²/C)[b]	Voltage change (mV)[b]
Cement paste (plain)	0.659 ± 0.031	35	2.2	2.2
Cement paste with steel fibers and PVA	208 ± 16	2700	8.7	8.7
Cement paste with carbon fibers	3.62 ± 0.40	49	8.5	8.5
PZT	136	1024	15	15

[a] Averaged over the first half of the first stress cycle.
[b] At 10 kHz.

PZT (Table 7.2). The effect in cement is attributed to the movement of the mobile ions in cement.

Another method of attaining a structural composite that is itself piezoelectric is to use piezoelectric fibers as the reinforcement in the composite. However, fine piezoelectric fibers with acceptable mechanical properties are far from being well developed. Their high cost is another problem.

Yet another method to attain a structural composite that is itself piezoelectric is to use conventional reinforcing fibers (e.g., carbon fibers) while exploiting the polymer matrix for the reverse piezoelectric effect. Owing to some ionic character in the covalent bonds, some polymers are expected to polarize in response to an electric field, thereby causing strain. The polarization is enhanced by molecular alignment, which is in turn enhanced by the presence of the fibers. This method involves widely available fibers and matrix and is thereby attractive economically and practically. The reverse piezoelectric effect occurs in the through-thickness direction of a continuous carbon fiber nylon-6 matrix composite [18]. The longitudinal piezoelectric coupling coefficient is 2.2×10^{-6} m/V, as determined up to an electric field of 261 V/m. The coefficient is high compared to values for conventional ceramic piezoelectric materials (e.g., 1×10^{-10} m/V for $BaTiO_3$). This is due to the difference in mechanism behind the effect. The high value is attractive for actuation, as it means that a smaller electric field is needed for the same strain.

Piezoelectric materials in the form of structural composites are in their infancy of development, but they are highly promising.

7.3 Ferroelectric behavior

A subset of piezoelectric materials is ferroelectric. In other words, a *ferroelectric* material is also piezoelectric. However, it has the extra ability to have the electric dipoles in adjacent unit cells interact with one another in such a way that adjacent dipoles tend to align themselves. This phenomenon is called *self-polarization,* which results in *ferroelectric domains* within each of which all the dipoles are in the same direction. As a consequence, even a polycrystalline material can become a single domain.

Quartz is piezoelectric but not ferroelectric. It is a frequently used piezoelectric material due to its availability in large single crystals, its shapeability and its mechanical strength. On the other hand, barium titanate ($BaTiO_3$), lead titanate ($PbTiO_3$) and solid solutions such as lead zirconotitanate ($PbZrO_3$–$PbTiO_3$, abbreviated PZT) and (Pb,La)–(Ti,Zr)O_3 (abbreviated PLZT) are both piezoelectric and ferroelectric.

Barium titanate ($BaTiO_3$) exhibits the *Perovskite structure* (Figure 7.13) above 120°C and a distorted Perovskite structure below 120°C. Above 120°C, the crystal structure of $BaTiO_3$ is cubic; below 120°C, it is slightly tetragonal ($a = b = 3.98$ Å, $c = 4.03$ Å). It is the tetragonal form of $BaTiO_3$ that is ferroelectric, so $BaTiO_3$ is only ferroelectric below 120°C, which is called the *Curie temperature*. (Jacques and Pierre Curie experimentally confirmed the piezoelectric effect over 100 years ago.) In other words, $BaTiO_3$ exhibits a solid–solid phase transformation at 120°C. As shown in Figure 7.14, the relative dielectric constant is low above the Curie temperature, high below the Curie temperature and peaks at the Curie

Composite materials for dielectric applications 141

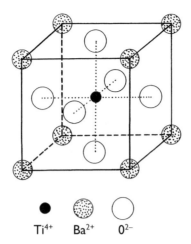

Figure 7.13 Perovskite structure.

temperature (due to the phase transformation). The Curie temperature is 494°C for $PbTiO_3$ and 365°C for PZT (i.e., $PbZrO_3$–$PbTiO_3$ solid solution). It limits the operating temperature range of a piezoelectric material.

In cubic $BaTiO_3$, the centers of positive and negative charges overlap as the ions are symmetrically arranged in the unit cell, so cubic $BaTiO_3$ is not ferroelectric. In tetragonal $BaTiO_3$, the O^{2-} ions are shifted in the negative c-direction, while the Ti^{4+} ions are shifted in the positive c-direction, thus resulting in an electric dipole along the c-axis (Figure 7.15). The dipole moment per unit cell can be calculated from the displacement and charge of each ion and summing the contributions from the ions in the unit cell. There are one Ba^{2+}, one Ti^{4+} ion and three O^{2-} ions per unit cell (Figure 7.13). Consider all displacements with respect to the Ba^{2+} ions. The contribution to the dipole moment of a unit cell by each type of ion is listed in Table 7.3. Hence, the dipole moment per unit cell is 17×10^{-30} C.m. The

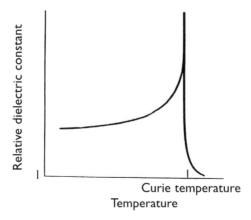

Figure 7.14 The effect of temperature on the relative dielectric constant of a ferroelectric material. The lowest value of this constant is 1.

polarization P is the dipole moment per unit cell, so it is given by the dipole moment per unit cell divided by the volume of a unit cell. Thus,

$$P = \frac{17 \times 10^{-30} \text{C.m}}{4.03 \times 3.98^2 \times 10^{-30} \text{m}^3}$$
$$= 0.27 \text{ C.m}^{-2}.$$

Consider a polycrystalline piece of tetragonal $BaTiO_3$, such that the grains are oriented with the c-axis of each grain along one of six orthogonal or parallel directions (i.e., $+x, -x, +y, -y, +z$ and $-z$). When the applied electric field Σ is zero, the polarization P is non-zero within each grain, but the total polarization of all the grains together is zero, since different grains are oriented differently. Therefore, when $\Sigma = 0, P = 0$.

Because tetragonal $BaTiO_3$ is almost cubic in structure, very slight movement of the ions within a grain can change the c-axis of that grain to a parallel or orthogonal direction. When $\Sigma > 0$, more grains are lined up with the electric dipole moment in the same direction as Σ, so $P > 0$. P increases with Σ much more sharply for a ferroelectric material than one with $P = 0$ within each grain at $\Sigma = 0$. The latter is called a *paraelectric* material (Figure 7.16).

The process of having more and more grains with the dipole moment in the same direction as Σ can be viewed as the movement of the grain boundaries (also called *domain boundaries*) such that the grains with the dipole moment in the same direction as Σ grow while the other grains shrink. These grains are also known as domains (or *ferroelectric domains*).

When all domains have their dipole moment in the same direction as Σ, the ferroelectric material becomes a single domain (a single crystal) and P has reached its maximum, which is called the *saturation polarization*. Its value is 0.26 C/m² for $BaTiO_3$ and ~0.5 C/m² for PZT.

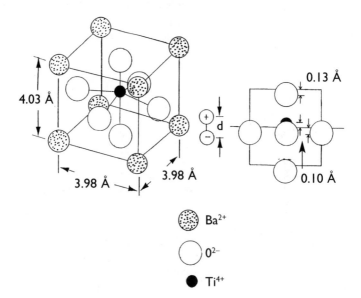

Figure 7.15 Crystal structure of tetragonal $BaTiO_3$.

Composite materials for dielectric applications

Table 7.3 Contribution to dipole moment of a $BaTiO_3$ unit cell by each type of ion.

Ion	Charge (C)	Displacement (m)	Dipole moment (C.m)
Ba^{2+}	$(+2)(1.6 \times 10^{-19})$	0	0
Ti^{4+}	$(+4)(1.6 \times 10^{-19})$	$+0.10(10^{-10})$	6.4×10^{-30}
$2O^{2-}$ (side of cell)	$2(-2)(1.6 \times 10^{-19})$	$-0.10(10^{-10})$	6.4×10^{-30}
O^{2-} (top and bottom of cell)	$(-2)(1.6 \times 10^{-19})$	$-0.13(10^{-10})$	4.2×10^{-30}
			Total = 17×10^{-30}

Upon decreasing Σ after reaching the saturation polarization, domains with dipole moments not in the same direction as Σ appear again and they grow as Σ decreases. At the same time, domains with dipole moments in the same direction as Σ shrink. This process again involves the movement of domain boundaries. In spite of this tendency, P does not return all the way to zero when Σ returns to zero. A *remanent polarization* ($P = P_r > 0$) remains when $\Sigma = 0$ (Figure 7.17).

In order for P to return all the way to zero, an electric field must be applied in the reverse direction. The required electric field is $\Sigma = -\Sigma_c$, where Σ_c is called the *coercive field*.

When Σ is even more negative than $-\Sigma_c$, the domains start to align in the opposite direction until the polarization reaches saturation in the reverse direction. This is known as *polarization reversal*. To bring the negative polarization back to zero, a positive electric field is needed. In this way, the cycling of Σ results in a hysteresis loop in the plot of P versus Σ (Figure 7.18(a)). The corresponding change in strain S associated with the change in polarization is shown in Figure 7.18(b), which is called the *butterfly curve*. The slope of the butterfly curve is d (Equation (7.31)).

A ferroelectric material can be used to store binary information, as a positive remanent polarization can represent "0" while a negative remanent polarization can represent "1". The application of an electric field in the appropriate direction can change the stored information from "0" to "1", or vice versa.

Poling refers to the process in which the domains (i.e., dipoles) are aligned so as to achieve saturation polarization. The process typically involves placing the ferroelectric material in a heated oil bath (e.g., 90°C, below the Curie temperature) and applying an electric field. Heating allows the domains to rotate more easily in the electric field. The electric field is maintained until the oil bath is cooled down

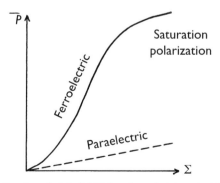

Figure 7.16 Plots of polarization P vs. electric field Σ for a ferroelectric material (continuous line) and a paraelectric material (dashed line).

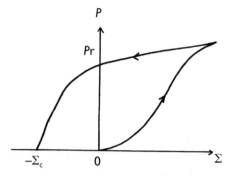

Figure 7.17 Variation of polarization P with electric field Σ during increase of Σ from 0 to a positive value and subsequent decrease of Σ to a negative value ($-\Sigma_c$) at which P returns to 0. Ferroelectric behavior is characterized by a positive value of the remanent polarization and a negative value of $-\Sigma_c$.

to room temperature. An excessive electric field is to be avoided as it can cause dielectric breakdown (Section 7.1.3) in the ferroelectric material.

After poling, *depoling* to a limited extent occurs spontaneously and gradually due to the influence of neighboring domains. This process is known as *aging* or *piezoelectric aging*, which has a logarithmic time dependence. The *aging rate* r is defined as

$$\frac{u_2 - u_1}{u_1} = r \log \frac{t_2}{t_1},$$

where t_1 and t_2 are the numbers of days after polarization, and u_1 and u_2 are the measured parameters, such as capacitance.

Through doping (i.e., substituting for some of the barium and titanium ions in $BaTiO_3$ to form a solid solution), the dielectric constant can be increased, the loss factor (Equation (7.19)) can be decreased and the temperature dependence of the loss factor can be flattened. The Ba^{2+} sites in $BaTiO_3$ are known as A sites; the Ti^{4+}

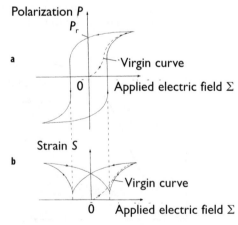

Figure 7.18 (**a**) Plot of polarization P vs. electric field Σ during first application of Σ (dashed curve) and during subsequent cycling of Σ (continuous curve). (**b**) Corresponding plot of strain Σ vs. electric field Σ.

sites are known as B sites. Substitutions for one or both of these sites can occur provided that the overall stoichiometry is maintained. Possible substitutions are shown in Figure 7.19. For example, an A site and a B site can be substituted by an Na^+ ion and an Nb^{5+} ion respectively, since the two sites together should have a charge of 6+. As an another example, three B sites can be substituted by an Mg^{2+} ion and two Nb^{5+} ions, since the three sites together should have a charge of 12+. As a further example, three A sites can be substituted by two La^{3+} ions and a cation vacancy, since the three sites together should have a charge of 6+.

A solute can be an electron acceptor or an electron donor. In the substitution of a B site by Fe^{3+}, Fe^{3+} is an acceptor; by accepting an electron, it changes from Fe^{4+} (unstable, but similar to Ti^{4+} in charge) to Fe^{3+}. However in order to maintain overall charge neutrality, the change of two Fe^{4+} ions to two Fe^{3+} ions must be accompanied by the creation of one O^{2-} vacancy. The O^{2-} vacancies move by diffusion, thereby decreasing the loss factor. Moreover, the Fe^{3+} ion and O^{2-} vacancies form dipoles which align with the polarization of the domain, thus pinning the domain walls. The substitution of two A sites with an La^{3+} ion and an Na^+ ion maintains charge neutrality, such that La^{3+} is the donor and Na^+ is an acceptor. The substitution of an A site by La^{3+} and a B site by Nb^{5+} is also possible; both La^{3+} and Nb^{5+} are donors in this case. The substitution of an A site by La^{3+} without substitution of a B site will require a reduction in oxygen vacancies and/or the creation of cation vacancies in order to maintain charge neutrality. Donors are not as effective as acceptors in pinning the domain walls.

The most common commercial ferroelectric material is PZT (i.e., $PbZrO_3$–$PbTiO_3$ solid solution or lead zirconotitanate). It is also written as $Pb(Ti_{1-z}Zr_z)O_3$ in order to indicate that a fraction of the B sites is substituted by Zr^{4+} ions. Figure 7.20 shows the binary phase diagram of the $PbZrO_3$–$PbTiO_3$ system. The cubic phase at high temperatures is paraelectric. The tetragonal and rhombohedral phases at lower temperatures are ferroelectric. The tetragonal phase is titanium rich; the rhombohedral phase is zirconium rich. The tetragonal phase has larger polarization than the rhombohedral phase. A problem with the tetragonal phase is that the tetragonal long axis is 6% greater than the transverse axis, thus causing stress during polarization (so much stress that the ceramic can be shattered). The substitution of Zr for Ti alleviates this problem. The compositions near the morphotropic phase boundary between the rhombohedral and tetragonal phases are those that pole most efficiently. This is because these

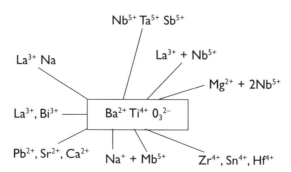

Figure 7.19 Possible substitutions of the A and B sites in $BaTiO_3$.

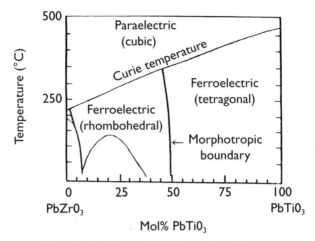

Figure 7.20 The $PbZrO_3$–$PbTiO_3$ binary phase diagram.

compositions are associated with a large number of possible poling directions over a wide temperature range. As a result, the piezoelectric coupling coefficient d is highest near the morphotropic boundary (Figure 7.21).

PZTs are divided into two groups, namely hard PZTs and soft PZTs. This division is according to the difference in piezoelectric properties (not mechanical properties). Hard PZTs generally have low relative dielectric constant, low loss factor and low piezoelectric coefficients, and are relatively difficult to pole and depole. Soft PZTs generally have high relative dielectric constant, high loss factor and high piezoelectric coefficients, and are relatively easy to pole and depole. Table 7.4 lists the properties of two commercial PZTs, namely PZT-5H (soft PZT) and PZT4 (hard PZT). Hard PZTs are doped with acceptors such as Fe^{3+} for Zr^{4+}, thus resulting in oxygen vacancies. Soft PZTs are doped with donors such as La^{3+} for Pb^{2+} and Nb^{5+} for Zr^{4+}, thus resulting in A-site vacancies. Hard PZTs typically have a small grain size (about 2 μm), whereas soft PZTs typically have a larger grain size (about 5 μm).

A relaxor ferroelectric is one having a diffuse ferroelectric phase transition, in addition to having a very strong piezoelectric effect and a very high relative dielectric constant ($\kappa \leq 30,000$, compared to $\kappa \leq 15,000$ for $BaTiO_3$). It has composition $Pb(B_1, B_2)O_3$, where B_1 can be Mg^{2+}, Zn^{2+}, etc., and B_2 can be Nb^{5+},

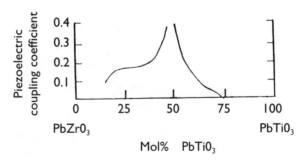

Figure 7.21 A pair of positively charged and negatively charged conductor plates in a vacuum.

Table 7.4 Properties of commercial PZT ceramics.

Property	PZT-5H (soft)	PZT4 (hard)
Permittivity (κ at 1 kHz)	3400	1300
Dielectric loss (tan δ at 1 kHz)	0.02	0.004
Curie temperature (T_c, °C)	193	328
Piezoelectric coefficients (10^{-12} m/V)		
d_{33}	593	289
d_{31}	−274	−123
d_{15}	741	496
Piezoelectric coupling factors		
κ_{33}	0.752	0.70
κ_{31}	−0.388	−0.334
κ_{15}	0.675	0.71

Ta^{5+}, etc. A particularly well-known relaxor ferroelectric is a solid solution of $PbMg_{1/3}Nb_{2/3}O_3$ (abbreviated PMN), and $PbTiO_3$ (abbreviated PT). The solid solution is abbreviated PMNPT. The strong piezoelectric effect stems from the morphotropic boundary, which occurs at 30 mol% $PbTiO_3$ in the PMNPT system. The average Curie temperature is about 0°C.

Another example of a ferroelectric material is poly(vinylidene fluoride) or $(CH_2CF_2)_n$ (abbreviated PVDF), which is a polymer (called a *piezopolymer*, as opposed to a *piezoceramic*), specifically a semi-crystalline (50% crystalline) thermoplastic. Crystallinity is needed for poling, so the ferroelectric behavior depends on the degree of crystallinity. The properties of poled PVDF start to degrade above about 80°C, but the ferroelectric behavior remains up to the crystalline melting temperature of 180°C. The glass transition temperature is −38°C. A mer of this polymer molecule has two fluorine atoms, which are electronegative and non-symmetrically positioned (i.e., at different sides of the molecular chain) (Figure 7.22). Positioning of the fluorine atoms in a molecule into a symmetrical configuration (i.e., on the same side of the molecular chain) results in dipole formation. The dipoles interact to form domains. The different configurations correspond to various isomers, which differ in the positions of the fluorine atoms. As the fluorine atom is more bulky than the carbon and hydrogen atoms, the positions of the fluorine atoms affect the tendency of the macromolecular chain to bend. Owing to the electronegativity of fluorine, the dipole moment of the macromolecule is strongly affected by the positions in the fluorine atoms. The repetitive units for three isomers are shown below.

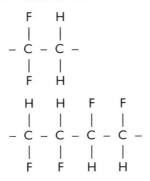

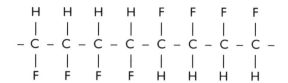

The ferroelectric behavior of PVDF is poorer than that of the oxide single crystals, but PVDF can be processed by polymer processing methods to form large-area lightweight detectors, such as sonar hydrophones and audio transducers. The ferroelectric behavior of PVDF can be enhanced by forming a copolymer, such as polyvinylidene fluoride–trifluoroethylene (abbreviated VF_2–VF_3). In general, piezopolymers suffer from high dielectric loss (particularly at high frequencies), low Curie temperature and low poling efficiency (particularly for specimens with large thickness, >1 mm).

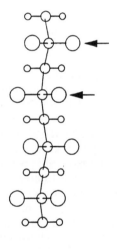

Figure 7.22 A molecular configuration of PVDF.

7.4 Piezoelectric/ferroelectric composite principles

From Equation (7.7), the voltage across a dielectric of relative dielectric constant κ is

Composite materials for dielectric applications

$$V = \frac{\kappa Q}{C_m}. \tag{7.52}$$

From Equations (7.5) and (7.6),

$$\kappa Q = D_m A = \varepsilon \Sigma A. \tag{7.53}$$

From Equation (7.7), the capacitance of a parallel-plate capacitor of thickness x is

$$C_m = \frac{\kappa \varepsilon_0 A}{x} = \frac{\varepsilon A}{x}. \tag{7.54}$$

Using Equations (7.53) and (7.54), Equation (7.52) becomes

$$V = \frac{\varepsilon \Sigma A}{\frac{\varepsilon A}{x}} = \Sigma x. \tag{7.55}$$

Using Equation (7.8), Equation (7.55) becomes

$$V = \frac{Px}{(\kappa - 1)\varepsilon_0}. \tag{7.56}$$

Differentiating,

$$\frac{dV}{d\sigma} = \frac{P}{(\kappa - 1)\varepsilon_0} \frac{dx}{d\sigma} + \frac{x}{(\kappa - 1)\varepsilon_0} \frac{dP}{d\sigma}. \tag{7.57}$$

Equation (7.57) gives the *voltage sensitivity* (dV/dσ), which consists of two terms. The second term involves dP/dσ, which is the piezoelectric coupling coefficient d. The first term involves dx/dσ, which is related to dS/dσ (S = strain), which is the compliance. Hence, the greater the compliance (i.e., the smaller the modulus), the higher the voltage sensitivity.

The polymer PVDF is relatively high in compliance. The rhombohedral form of PZT (Figure 7.20) has greater compliance than the tetragonal form; both forms are ferroelectric. The most common way to increase the compliance is to incorporate a piezoelectric/ferroelectric material in a polymer to form a composite. The mechanical properties in various directions of the composite can be adjusted so as to benefit the piezoelectric/ferroelectric behavior. For example, a composite can be in the form of aligned PZT rods in a polymer matrix (Figure 7.23). Such a composite has the PZT phase connected in only one direction and the polymer phase connected in all three directions. As a result, this composite is referred to as a 1–3 composite (i.e., the active phase having connectivity in one dimension, and the passive phase having connectivity in three dimensions). When a hydrostatic stress is applied to a 1–3 composite, the transverse stress (perpendicular to the rods) is absorbed by the polymer while the longitudinal stress (parallel to the rods) is applied mainly to the PZT rods. As a consequence, the piezoelectric response to the hydrostatic stress is enhanced. Furthermore, the polymer decreases the relative dielectric constant κ, resulting in an increase of g, the voltage coefficient (Equation (7.35)).

There are 32 different methods of connection (32 connectivities) in a two-phase material. Another example of connectivity is 0–3, which refers to particles

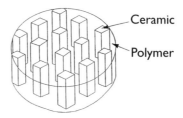

Figure 7.23 A 1–3 piezoelectric/ferroelectric ceramic polymer–matrix composite.

in a matrix. The particles are not continuous in any direction, while the matrix is continuous in all three directions. To make poling of a 0–3 composite more efficient (i.e., requiring less electric field), a conductive filler such as carbon black can be added to the composite so as to increase the electrical conductivity and have more of the poling field applied to the piezoelectric/ferroelectric particles.

The piezoelectric/ferroelectric materials are not cost-effective for reinforcement in a composite. Therefore, a structural composite capable of sensing or actuation is most commonly in the form of a continuous fiber polymer–matrix composite laminate with one or more piezoelectric/ferroelectric elements embedded between the plies in the composite during composite fabrication, such that the leads (wires) from the elements come out of the composite. Although a piezoelectric/ferroelectric element is typically in a sheet form (of thickness similar to that of a prepreg tape), its presence tends to degrade the mechanical properties of the laminate.

A piezoelectric/ferroelectric element for actuation is commonly in the form of two strips of piezoelectric/ferroelectric material, such that one strip is bonded on top of the other and the bi-strip (called a *bimorph*) is fixed at one end to form a cantilever beam configuration. Furthermore, the polarization of the two strips are in opposite directions, such that the application of an electric field across the thickness of the bi-strip causes one strip to extend axially (in the plane of the strip) while the other strip contracts axially, thereby causing the bi-strip to deflect at the tip far from the fixed end (Figure 7.24(a)). In contrast, if the strips are polarized in the same direction, both strips extend axially upon application of the electric field and the resulting movement (Figure 7.24(b)) is much smaller than that for the case of the strips polarized in opposite directions.

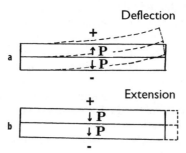

Figure 7.24 Cantilever beam configuration for actuation using a bi-strip of piezoelectric/ferroelectric material. (**a**) Polarization P of the two strips in the bi-strip are in opposite directions, thereby causing deflection at the tip of the beam upon electric field application. (**b**) Polarization P of the two strips are in the same direction, thereby causing extension of the beam upon electric field application.

Composite materials for dielectric applications

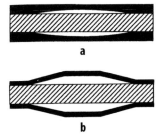

Figure 7.25 Composites in the form of a piezoelectric/ferroelectric material sandwiched by metal faceplates. (a) "Moonie" structure involving air cavities in the shape of a half moon. (b) "Cymbal" structure involving air cavities in the shape of a truncated cone.

A continuous fiber polymer–matrix composite laminate for actuation in the form of bending can be achieved by embedding two piezoelectric/ferroelectric strips in the two sides relative to the plane of symmetry of the laminate, such that one strip extends while the other strip contracts upon application of an electric field. A laminate for actuation in the form of axial extension can be achieved by embedding two strips, such that both strips extend upon application of an electric field.

Macroscopic composites in the form of a piezoelectric/ferroelectric material sandwiched by metal faceplates (Figure 7.25) are attractive for enhancing the piezoelectric coupling coefficient. The air gap between the metal faceplate and the piezoelectric/ferroelectric material allows the metal faceplate to serve as a mechanical transformer for transforming and amplifying a part of the applied axial stress into tangential and radial stresses. The metal faceplate transfers the applied stress to the piezoelectric/ferroelectric material, in addition to transferring the displacement to the medium.

7.5 Pyroelectric behavior

The *pyroelectric* effect refers to the change in polarization in a material due to a change in temperature. The change in polarization gives rise to a change in voltage across the material in the direction of the polarization. The situation is similar to that in Figure 7.10(d), except that heat rather than stress is the input. In this way, thermal energy is converted to electrical energy. The pyroelectric effect is used for temperature sensing, particularly for infrared detection, which is used, for example, in intruder alarms, as the infrared radiation from the warm human or animal body raises the temperature of the pyroelectric detector, which generates a voltage that actuates an alarm.

The pyroelectric coefficient p is given by

$$p = \frac{dP}{dT} = \varepsilon_o \Sigma \frac{d\kappa}{dT}, \qquad (7.58)$$

where P is the polarization, T is the temperature, ε_o is the permittivity of free space (8.85×10^{-12} C/V.m) and Σ is the electric field. Table 7.5 gives the values of p for various materials.

Table 7.5 Pyroelectric coefficient (10^{-6} C/m^2.K).

BaTiO$_3$	20
PZT	380
PVDF	27
Cement paste	0.002

7.6 Introduction to composite materials for dielectric applications

Composite materials are well known for structural applications, owing to the effectiveness of various fibrous and particulate fillers in providing reinforcement, in addition to the low density of many composites. However, composite materials are increasingly used for dielectric applications, i.e., applications that make use of the dielectric (electrically insulating or nearly insulating) behavior. This is because of the need of the electronics industry for dielectric materials in electrical insulation, encapsulation, substrates, interlayer dielectrics in a multilayer ceramic chip carrier, printed circuit boards and capacitors, and because of the rising importance of smart structures which use dielectric materials for piezoelectric, ferroelectric and pyroelectric devices that provide sensing and actuation.

The dielectric behavior of a material can be tailored through the composite route. A dielectric property that is amenable to such tailoring is the relative dielectric constant (κ, equal to 1 in the case of a vacuum). A low value of κ is desirable for electrical insulation, substrates, interlayer dielectrics and printed circuit boards, owing to the capacitor effect associated with a high κ value and the consequent delay in signal propagation in the conductor line adjacent to the dielectric material. A low value of κ is also desirable for piezoelectric, ferroelectric and pyroelectric applications, as the voltage generated is larger when κ is lower. On the other hand, a high value of κ is needed for capacitors. Table 7.6 lists the values of κ for various materials at 1 kHz. Polymers tend to exhibit lower κ than ceramics, although there are exceptions. Owing to electric dipole friction, κ decreases with increasing AC frequency. As a result, the attainment of a high value of κ at a high frequency (as needed for high-speed computers) is a challenge.

Various fundamental studies have been conducted in modeling and understanding the dielectric constant of composite materials [19–28]. However, this section addresses the applications rather than the theory of composite dielectric materials.

As metals are electrically conducting, whereas polymers and ceramics tend to be not conducting, composite dielectric materials are mainly polymer–matrix and ceramic–matrix composites. The chemical bonding in these materials

Table 7.6 Values of the relative dielectric constant κ of various dielectric materials at 1 kHz.

Material	κ
Al$_2$O$_3$ (99.5%)	9.8
BeO (99.5%)	6.7
Cordierite	4.1–5.3
Nylon-66 reinforced with glass fibers	3.7
Polyester	3.7

Data from Ceramic Source '86, American Ceramic Society, Columbus, Ohio, 1985, and Design Handbook for DuPont Engineering Plastics.

involves covalent and ionic bonding. The polymer–matrix composites are mainly polymer–ceramic composites. The ceramic–matrix composites are mainly ceramic–ceramic composites.

The composites can be in bulk, thick-film (typically 1–50 μm thick) and thin-film (typically less than 2000 Å thick) forms. The bulk form is needed for cable jackets and printed circuit boards. Both bulk and thick-film forms are needed for substrates and piezoelectric/ferroelectric/pyroelectric devices. Both thick-film and thin-film forms are needed for wire insulation, interlayer dielectrics and capacitors. Thick films are typically made by the casting of pastes. Thin films are typically made by vapor deposition.

7.7 Composites for electrical insulation

Composites for electrical insulation include those for cable jackets, wire insulation, substrates, interlayer dielectrics, encapsulations and printed circuit boards. Dielectric composites are tailored to attain low κ, high dielectric strength (high value of the electric field at breakdown), low AC loss (low loss tangent, tan δ, which relates to the energy loss), a low value of the coefficient of thermal expansion (CTE; a low value is needed to match those of semiconductors and other components in an electronic package) and preferably high thermal conductivity (for heat dissipation from microelectronics) as well.

7.7.1 Polymer–matrix composites

Compared to ceramics, polymers tend to have lower κ, higher CTE, lower thermal conductivity and lower stiffness. Therefore ceramics are used as fillers in polymer–matrix composites to decrease CTE, increase thermal conductivity and increase stiffness.

The most widely used composites for electrical insulation are those containing continuous fibers (glass, quartz, aramid, etc.), which serve as a reinforcement in the composite laminate [29–31]. These composites are mainly used for printed circuit boards. The fibers are usually woven, although non-woven ones are used also [32]. Short fibers are even less commonly used than continuous fibers for printed circuit boards, but they are used for other forms of insulation [33]. In order to lower κ, fibers of low dielectric constant, such as hollow glass fibers [34], are used. Epoxy and unsaturated polyester resin are commonly used for the matrix of printed wiring boards [30,35], but polyimide is used for high-temperature applications [33].

Particulate fillers are often used in dielectric composites other than laminates. In particular, hollow microspheres made of glass, ceramics or polymer are used as fillers to reduce κ [36,37]. Silica particles, as generated from tetraethoxy silane (TEOS) via a sol–gel process, are used as a filler to lower CTE [38,39].

Composites that are thermally conducting but electrically insulating are critically needed for heat dissipation from microelectronic packages. Aluminum nitride (AlN) and boron nitride (BN) particles are used as fillers to increase thermal conductivity and lower CTE [40–43]. A high filler volume fraction is needed to attain a high thermal conductivity, but it reduces the workability of the paste, which is the form most commonly used in electronic packaging. Therefore, the development of such composites is challenging.

Polymer concrete, which is a low-cost polymer–matrix composite containing sand and gravel as fillers, is a dielectric structural material. It is an alternative to porcelain for outdoor and indoor electrical insulation applications [44,45].

Dielectric composites also include network composites, such as those involving gel silica [46], foams [47] and interpenetrating polymer networks [48].

7.7.2 Ceramic–matrix composites

Different ceramic materials differ in their processability, κ, CTE and thermal conductivity. Thus, ceramic–ceramic composites are designed to improve the processability, decrease κ, decrease CTE or increase the thermal conductivity.

Dielectric ceramic–matrix composites are ceramic–ceramic composites. They include glass–matrix composites [49], composites involving a low-softening-point borosilicate glass and a high-softening-point high-silica glass [50–53], composites with boron nitride (BN) [54] and aluminum nitride (AlN) [55] matrices which are attractive for their high thermal conductivity, AlN–cordierite composites [56] and alumina–matrix composites [57]. Cordierite (Table 7.6) is attractive because of its low κ and relatively low processing temperature.

7.8 Composites for capacitors

The capacitors of this section are dielectric capacitors rather than double-layer capacitors. The former can operate at high frequencies, whereas the latter cannot.

Conventional capacitors involve thick-film dielectrics, whereas integrated capacitors involve thin-film dielectrics. Polymer–matrix composites are more common than ceramic–matrix composites, owing to the ease of processing.

7.8.1 Polymer–matrix composites

Polymers tend to have better processability, lower κ, higher CTE and lower thermal conductivity than ceramics, so polymer–ceramic composites are designed to increase κ, lower CTE or increase thermal conductivity. These composites are polymer–matrix composites containing ceramic particles that exhibit high κ. The most common ceramic particles are the perovskite ceramics, such as $BaTiO_3$, lead magnesium niobate (PMN) and lead magnesium niobate–lead titanate (PMN–PT) [58–68]. Kraft paper, which is not a ceramic and not high in κ, is also commonly used, because of its low cost and amenability to composite fabrication by simply winding the paper [69]. For particulate composites, a small particle size is preferred [70].

7.8.2 Ceramic–matrix composites

Ceramic–ceramic composites for capacitors involve a ceramic with high κ (e.g., $BaTiO_3$ and Ta_2O_5) and another ceramic (e.g., SiC and Al_2O_3) chosen for processability, surface passivation or other attributes [71,72]. Ceramic–matrix metal particle (e.g., nickel) composites are also available for capacitor use, due to the increase of κ by the metal particle addition [73].

7.9 Composites for piezoelectric, ferroelectric and pyroelectric functions

Piezoelectricity and ferroelectricity pertain to the conversion between mechanical energy and electrical energy. Pyroelectricity refers to the conversion from thermal energy to electrical energy. The applications are in sensors and actuators. The sensing relates to strain/stress sensing in the case of piezoelectricity and ferroelectricity, and relates to temperature sensing in the case of pyroelectricity.

Composites for piezoelectric, ferroelectric and pyroelectric functions are designed to improve these particular functions (e.g., through tailoring of the modulus in various directions, κ and the Curie temperature), and to improve processability (e.g., sinterability) and mechanical behavior (e.g., fracture toughness). These composites can involve two or more components that are all capable of providing the piezoelectric, ferroelectric or pyroelectric function, such that the combination of these components results in enhancement of the function. The composites can also involve a component that is capable of providing the function and another component that is not capable of providing the function, such that the combination results in enhancement of the function.

Although perovskite ceramics are the main materials for these functions, the polymer polyvinylidene fluoride (PVDF) is also capable of these functions. Therefore, both ceramic–ceramic and polymer–ceramic composites are relevant.

Among composites in which a component (e.g., epoxy) is not able to provide the function, polymer-matrix composites dominate, owing to their processability. These composites are classified according to the connectivity of each component, as the connectivity affects the functional behavior in various directions and different applications require a different set of directional characteristics.

Among the 32 different connectivities in a two-component material, the 1–3 [74–77], 2–2 [77] and 0–3 [78,79] connectivities are particularly common. In this notation, the first number describes the connectivity of the functional (active) component and the second number describes the connectivity of the non-functional (passive) component. For example, in a 1–3 composite, the functional component has connectivity in one dimension and the non-functional component has connectivity in three dimensions. A composite in the form of aligned lead zirconate titanate (PZT) rods in a polymer matrix has 1–3 connectivity.

7.9.1 Polymer–matrix composites

Polymer–matrix composites for piezoelectric, ferroelectric and pyroelectric functions mainly involve a polymer matrix, which itself cannot provide the function, as the majority component. The polymer serves to increase the compliance, which relates to voltage sensitivity. Such polymers include epoxy

[80], polytetrafluoroethylene (PTFE) [81], polyethylene [78], polypropylene [78], polyvinyl chloride [82] and nylon [83]. However, the composites can involve PVDF or other functional polymers (being able to provide the function) as the matrix [78,84–86] or as a component within the matrix [81].

The functional behavior of PVDF is poorer than that of the ceramic single crystals, but PVDF can be processed by polymer processing methods to form large-area lightweight detectors, such as sonar hydrophones and audio transducers. The ferroelectric behavior of PVDF can be enhanced by forming a copolymer, such as polyvinylidene fluoride–trifluoroethylene (abbreviated VF_2–VF_3). In general, piezopolymers suffer from high AC loss (particularly at high frequencies), low Curie temperature and low poling efficiency (particularly for specimens with large thickness, >1 mm).

The filler in the composites is typically a perovskite ceramic (such as PZT), which can be in the form of wires, fibers or particles, as needed for connectivity. The unit size (e.g., particle size) of the filler affects the functional behavior, partly due to the surface layer on a filler unit [87]. A less common ferroelectric/pyroelectric ceramic is triglycine sulfate (TGS) [82–89].

Although bulk composites have received most attention, composite coatings [90] are increasingly important, owing to the material cost saving and the need for coating particles, fibers, wires and other non-functional surfaces.

7.9.2 Ceramic–matrix composites

Ceramic–matrix composites in this category are ceramic–ceramic composites in which either one or more components is (are) piezoelectric, ferroelectric or pyroelectric. Composites in which both components are functional include $PbZrO_3$/$PbTiO_3$ [91] and related oxide–oxide systems [92], although these materials may be considered alloys (as governed by the phase diagram) rather than composites.

Composites in which only one component is functional include barium strontium titanate matrix composites containing metal oxide fillers such as Al_2O_3 [93], MgO [93] and others [94,95]. They also include PZT matrix composites containing glass, which serves as a sintering aid [96], as a binder in the case of thick-film composites [97], as a barrier layer to prevent reaction between certain constituents [98], and as a component to decrease κ and to increase the Curie temperature [99]. In addition, they include PZT matrix composites containing silver particles, which serve as a sintering aid [100,101].

7.10 Composites for microwave switching and electric field grading

Composites that exhibit dielectric constant that can be tuned by voltage variation [102,103] are of use to switching devices which reversibly change their microwave properties under DC bias. These composites are mainly polymer blends.

The electric field encountered by an electrically insulating part may be non-uniform, so that dielectric breakdown occurs at the locations of high electric field. The field distribution can be homogenized by using a dielectric material which exhibits dielectric constant that varies non-linearly with the electric field [104]. This is known as electric field grading.

7.11 Composites for electromagnetic windows

Electromagnetic windows require transparency in certain frequency ranges, particularly the microwave range, owing to the relevance to radomes. These materials are dielectrics, as conductors tend to be strong reflectors and are thus not transparent. To enhance the mechanical properties, which are needed for impact/ballistic protection, these materials are commonly polymer–matrix composites containing dielectric reinforcing fibers, such as ultra-high-strength polyethylene fibers [105].

7.12 Composites for solid electrolytes

Solid electrolytes that allow ionic conduction rather than electronic conduction are needed for batteries and fuel cells. Ceramic–ceramic [106] and polymer–ceramic [107] composites have been designed to enhance the ionic conductivity. An example is Li_2CO_3–ABO_3, where A = Li, K or Ba, and B = Nb or Ti [106].

7.13 Conclusion

Composite materials for dielectric applications include polymer–matrix and ceramic–matrix composites for electrical insulation, capacitors, piezoelectric/ferroelectric/pyroelectric devices, microwave switching, electric field grading, electromagnetic windows and solid electrolytes. The polymer–ceramic composites are mainly polymer–ceramic composites. The ceramic–matrix composites are mainly ceramic–ceramic composites. The composites are designed to attain a low or high value of κ, in addition to attaining attributes such as high piezoelectric/ferroelectric/pyroelectric response, low CTE, high thermal conductivity, fracture toughness, ionic conductivity and processability, as required for the applications.

Example problems

1. The relative dielectric constant κ is 3000 $BaTiO_3$. What is the dielectric constant ε?

   Solution

   $$\varepsilon = \kappa\varepsilon_o = (3000)\,(8.85 \times 10^{-12}\ C/V.m)$$
   $$= 2.7 \times 10^{-8}\ C/V.m.$$

2. The relative dielectric constant κ is 6.5 for Al_2O_3. What is the electric susceptibility χ?

Solution

From Equation (7.10),

$$\chi = \kappa - 1 = 6.5 - 1 = 5.5.$$

3. The polarization is 0.17 C/m² in a material of thickness 40 µm and diameter 600 µm. What is the dipole moment?

Solution

From Equation (7.12),

$$\text{dipole moment} = \text{polarization} \times \text{volume}$$
$$= (0.17 \text{ C/m}^2)(4 \times 10^{-5} \text{ m})\pi(3 \times 10^{-4} \text{ m})^2$$
$$= 1.9 \times 10^{-12} \text{ C.m.}$$

4. An electric field of 5.4×10^6 V/m is applied to $BaTiO_3$ of a relative dielectric constant κ 3000. How much polarization results?

Solution

From Equation (7.8),

$$P = (\kappa - 1)\varepsilon_o \Sigma$$
$$= (3000 - 1)(8.85 \times 10^{-12} \text{ C/V.m})(5.4 \times 10^6 \text{ V/m})$$
$$= 0.14 \text{ C/m}^2.$$

5. The capacitance is 0.0214 µF for a capacitor with a relative dielectric constant $\kappa = 4700$. If the dielectric is replaced with one with $\kappa = 7600$, what is the capacitance?

Solution

From Equation (7.7), the capacitance is proportional to κ. Hence,

$$\text{capacitance} = \frac{7600}{4700}(0.0214 \text{ µF})$$
$$= 0.0346 \text{ µF.}$$

6. A parallel-plate capacitor of capacitance 0.0375 µF has mica of thickness 50 µm as the dielectric. What area is required for the capacitor if only a single layer of dielectric is used? The relative dielectric constant of mica is 7.0 at 1 MHz.

Solution

From Equation (7.7),

$$C = \frac{\kappa \varepsilon_0 A}{d},$$

so

$$A = \frac{Cd}{\kappa \varepsilon_0}$$

$$= \frac{(3.75 \times 10^{-8} \, C/V)(5 \times 10^{-5} \, m)}{(7.0)(8.85 \times 10^{-12} \, C/(V.m))}$$

$$= 3.0 \times 10^{-2} \, m^2.$$

7. A PbZrTiO$_6$ wafer of thickness 50 μm and diameter 500 μm is subjected to a force of 10 kg in a direction perpendicular to the wafer. How much polarization is produced across the wafer?

Solution

From Table 7.1, the longitudinal d is 250×10^{-12} C/Pa.m^2 (m/V) for PbZrTiO$_6$.

The stress is

$$\sigma = \frac{\text{force}}{\text{area}} = \frac{(10 \text{ kg})(9.807 \text{ N/kg})}{\pi (2.5 \times 10^{-4} \, m)^2} = 5.0 \times 10^8 \, \text{Pa}.$$

From Equation (7.28), the polarization produced is

$$P = d\sigma$$
$$= (2.5 \times 10^{-10} \, C/Pa.m^2)(5.0 \times 10^8 \, Pa)$$
$$= 1.2 \times 10^{-1} \, C/m^2.$$

8. A piezoelectric material with piezoelectric coupling coefficient d of 120×10^{-12} m/V is subjected to an electric field of 5×10^6 V/m. How much strain results?

Solution

From Equation (7.37), the strain is

$$S = d\Sigma$$
$$= (120 \times 10^{-12} \, m/V)(5 \times 10^6 \, V/m)$$
$$= 6 \times 10^{-4}.$$

9. A piezoelectric material has relative dielectric constant $\kappa = 5500$ and piezoelectric coupling coefficient $d = 90 \times 10^{-12}$ m/V. What is its voltage coefficient g?

Solution

From Equation (7.35),

$$g = \frac{d}{(\kappa-1)\varepsilon_o} = \frac{90 \times 10^{-12}\,\text{m/V}}{(5500-1)(8.85 \times 10^{-12}\,\text{C/V.m})}$$
$$= 1.8 \times 10^{-3}\,\text{m}^2/\text{C}.$$

10. A piezoelectric material with voltage coefficient g of 0.22 m²/C is subjected to a stress of 48 MPa. How much electric field (in V/m) is generated?

Solution

From Equation (7.45), the electric field is

$$\Sigma = g\sigma$$
$$= (0.22\,\text{m}^2/\text{C})\,(4.8 \times 10^7\,\text{Pa})$$
$$= 1.1 \times 10^7\,\text{N/C}.$$

Since C/Pa.m² = m/V (Table 4.2) and Pa = N/m²,

$$\text{C/N} = \text{m/V}.$$

Hence,

$$\Sigma = 1.1 \times 10^7\,\text{V/m}.$$

11. The piezoelectric coupling coefficient d is 100×10^{-12} C/Pa.m² (m/V) for BaTiO$_3$. The elastic modulus is 69 GPa. What is the value of the voltage coefficient g for BaTiO$_3$?

Solution

From Equation (7.50),

$$g = \frac{1}{Ed} = \frac{1}{(6.9 \times 10^{10}\,\text{Pa})(1.0 \times 10^{-10}\,\text{C/Pa.m}^2)}$$
$$= 0.14\,\text{m}^2/\text{C}.$$

12. A ferroelectric material has relative dielectric constant (κ) 6600, thickness 50 μm, diameter 750 μm and longitudinal piezoelectric constant (d) 150×10^{-12} m/V (150×10^{-12} C/Pa.m²).

(a) Calculate the polarization and the dipole moment in the thickness direction when the material is subjected to an electric field of 4.1×10^6 V/m in the thickness direction.
(b) Calculate the electric susceptibility χ.
(c) Calculate the capacitance.
(d) Calculate the polarization in the thickness direction when the material is subjected to a force of 24 kg in the thickness direction.
(e) Calculate the strain in the thickness direction resulting from an electric field of 4.1×10^6 V/m in the thickness direction.
(f) Calculate the voltage coefficient (g).

(g) Calculate the electric field generated in the thickness direction when the material is subjected to a stress of 68 MPa in the thickness direction.

Solution

(a)
$$P = (\kappa - 1)\varepsilon_o \Sigma$$
$$= (6600 - 1)(8.85 \times 10^{-12} \text{ C/V.m})(4.1 \times 10^6 \text{ V/m})$$
$$= 0.24 \text{ C/m}^2.$$

Dipole moment = polarization × volume
$$= (0.24 \text{ C/m}^2)(5 \times 10^{-5} \text{ m}) \pi (7.5 \times 10^{-4}/2)^2 \text{ m}^2$$
$$= 5.3 \times 10^{-12} \text{ C.m.}$$

(b)
$$\chi = \kappa - 1 = 6600 - 1 = 6599.$$

(c) Capacitance $= \dfrac{\kappa \varepsilon_o A}{d} = \dfrac{6600 (8.85 \times 10^{-12} \text{C/V.m}) \pi (7.5 \times 10^{-4}/2)^2 \text{ m}^2}{5 \times 10^{-5} \text{ m}}$
$$= 5.2 \times 10^{-10} \text{ F.}$$

(d)
$$P = d\sigma.$$
$$= (1.5 \times 10^{-10} \text{ C/Pa.m}^2) \dfrac{(24 \text{ kg})(9.807 \text{ N/kg})}{\pi (7.5 \times 10^{-4}/2)^2 \text{ m}^2}$$
$$= 8.0 \times 10^{-2} \text{ C/m}^2.$$

(e)
$$S = d\Sigma$$
$$= (1.5 \times 10^{-10} \text{ m/V})(4.1 \times 10^6 \text{ V/m})$$
$$= 6.2 \times 10^{-4}.$$

(f)
$$g = \dfrac{d}{(\kappa-1)\varepsilon_o} = \dfrac{1.5 \times 10^{-10} \text{ m/V}}{(6600-1)(8.85 \times 10^{-12} \text{C/V.m})}$$
$$= 2.6 \times 10^{-3} \text{ m}^2/\text{C}.$$

(g)
$$\Sigma = g\sigma$$
$$= (2.6 \times 10^{-3} \text{ m}^2/\text{C})(6.8 \times 10^7 \text{ Pa})$$
$$= 1.8 \times 10^5 \text{ V/m}.$$

13. A pyroelectric material with pyroelectric coefficient (p) 380×10^{-6} C/m².K and relative dielectric constant (κ) 290 is subjected to a temperature change of 10^{-3} K. What is the resulting change in electric field?

Solution

From Equation (7.58),
$$\delta P = p \delta T$$
$$= (380 \times 10^{-6} \text{ C/m}^2.\text{K})(10^{-3} \text{ K})$$
$$= 380 \times 10^{-9} \text{ C/m}^2.$$

From Equation (7.8),
$$\delta P = (\kappa - 1)\varepsilon_o \delta \Sigma.$$

Rearranging,

$$\delta\Sigma = \frac{\delta P}{(\kappa - 1)\varepsilon_0}$$

$$= \frac{380 \times 10^{-9} \, C/m^2}{(290-1)(8.85 \times 10^{-12} \, C/V.m)}$$

$$= 149 \text{ V/m}.$$

If the pyroelectric material has height $h = 0.1$ mm in a direction perpendicular to the plates sandwiching it, the change in voltage is

$$\delta V = (\delta E)h$$
$$= (149 \text{ V/m}) (0.1 \text{ mm})$$
$$= 0.0149 \text{ V}$$
$$= 14.9 \text{ mV}.$$

Thus, δV is substantial even when δT is only 10^{-3} K.

14. A non-ideal dielectric material of relative dielectric constant $\kappa = 5500$ at 100 Hz can be considered a capacitor C (0.01 F) and a resistor R (10^3 Ω) in parallel. The frequency is 100 Hz. The amplitude of the applied voltage is 2.5 V.

   (a) Calculate the amplitude of the current through the resistor R.
   (b) Calculate the amplitude of the current through the capacitor C.
   (c) Calculate the phase angle δ between the resultant current and the current through the capacitor C.
   (d) Calculate the loss of energy per cycle due to the resistor R.
   (e) If the dielectric material is a sheet of thickness 100 μm, what area of this sheet is needed to attain a capacitance of 0.01 F at 100 Hz?
   (f) If the dielectric material is a disk of diameter 50 cm and has electrical resistivity 10^8 Ω.cm, what thickness of the disk is needed to attain a resistance of 10^3 Ω in the direction along the axis of the disk?
   (g) What is the dipole moment per unit volume in the dielectric material when it is subject to an electric field of 5 kV/cm?
   (h) If the dielectric material is a piezoelectric material with piezoelectric coupling coefficient $d = 160 \times 10^{-12}$ m/V and is subject to an electric field of 5 kV/cm, how much strain results? What is the voltage coefficient of this material?
   (i) If the dielectric material has thickness 100 μm and dielectric strength 10^4 V/cm, what is the highest voltage that can be applied across the thickness without dielectric breakdown?

Solution

(a) $$\frac{V}{R} = \frac{2.5 \text{ V}}{10^3 \, \Omega} = 2.5 \text{ mA}$$

(b) $$\frac{V}{1/\omega C} = V\omega C = (2.5 \text{ V})(2\pi \, 10^2 \text{s}^{-1})(10^{-2} \text{ F}) = 1.57 \text{ A}$$

Composite materials for dielectric applications

Note: F = C/V.

(c) $$\tan\delta = \frac{1}{\omega CR} = \frac{1}{(2\pi 10^2)(10^{-2})(10^3)}$$

$$\delta = 0.009°.$$

Note: $F.\Omega = \frac{C}{V}\Omega = \frac{C}{A} = \frac{C}{C/s} = s.$

(d) $$\frac{V^2\pi}{\omega R} = \frac{(2.5)^2 \pi}{(2\pi\, 10^2)(10^3)} = 0.31 \text{ C.V} = \frac{0.31}{1.6\times 10^{-19}} \text{ eV} = 1.95\times 10^{18} \text{ eV}$$

Note: 1 eV = (1.6 × 10⁻¹⁹ C) V.

$$\frac{V^2}{s^{-1}\Omega} = \frac{V^2}{s^{-1}\frac{V}{A}} = V.A.s = V.C$$

(e) $$C = \frac{\varepsilon A}{d} = \frac{\varepsilon_0 \kappa A}{d} = \frac{\left(8.85\times 10^{-12}\,\frac{C}{V.m}\right)(5500)\,A}{10^{-4}\,m} = 0.01\text{ F}$$

$$A = \frac{(0.01)(10^{-4})}{(8.85\times 10^{-12})(5500)}\,m^2 = 20.5\,m^2$$

(f) $$R = \rho\frac{l}{A}$$

$$l = \frac{RA}{\rho} = \frac{(10^3\,\Omega)\,\pi\,(25\text{ cm})^2}{10^8\,\Omega.\text{cm}} = 2.0\times 10^{-2}\text{ cm} = 200\,\mu m$$

(g) $$P = (\kappa\text{-}1)\varepsilon_0 \Sigma = (5500-1)\left(8.85\times 10^{-12}\,\frac{C}{V.m}\right)\left(5\times 10^3\,\frac{V}{cm}\right)\left(10^2\,\frac{cm}{m}\right)$$

$$= 0.024\,\frac{C}{m^2}$$

(h) $$S = d\Sigma = \left(160\times 10^{-12}\,\frac{m}{V}\right)\left(5\times 10^5\,\frac{V}{m}\right) = 8\times 10^{-5}$$

$$g = \frac{d}{(\kappa\text{-}1)\varepsilon_0} = \frac{160\times 10^{-12}\text{ m/V}}{(5500-1)\left(8.85\times 10^{-12}\,\frac{C}{V.m}\right)} = 3.3\times 10^{-3}\,\frac{m^2}{C}$$

(i) $\quad \text{Voltage} = \Sigma x = \left(10^4 \frac{\text{V}}{\text{cm}}\right)\left(10^{-2}\text{cm}\right) = 10^2 \text{V}$

Review questions

1. Why is it difficult to achieve a capacitor that operates at a high frequency and has a high capacitance?
2. Why is a low value of the relative dielectric constant desirable for electronic packaging?
3. Describe an application of a piezoelectric material.
4. What is the direct piezoelectric effect?
5. Why does the relative dielectric constant of a material decrease with increasing frequency?
6. Define polarization in relation to the dielectric behavior of a material.
7. Define "dipole friction".
8. What is the main application of a dielectric material that exhibits a high value of dielectric constant?

References

[1] J.F. Tressler, S. Alkoy and R.E. Newnham, *J. of Electroceramics*, 1998, **2**(4), 257–272.
[2] J.F. Scott, *Integrated Ferroelectrics*, 1998, **20**(1–4), 15–23.
[3] D.J. Jones, S.E. Prasad and J.B. Wallace, *Key Eng. Mater.*, 1996, **122-124**, 71–144.
[4] A. Safari, R.K. Panda and V.F. Janas, *Key Eng. Mater.*, 1996, **122-124**, 35–70.
[5] A.V. Turik and V.Yu. Topolov, *J. Phys. D – Appl. Phys.*, 1997, **30**(11), 1541–1549.
[6] Y. Yamashita and N. Ichinose, *IEEE Int. Symp. on Applications of Ferroelectrics*, IEEE, Piscataway, NJ, 1996, **1**, 71–77.
[7] A.V. Turik and V.Yu. Topolov, *Key Eng. Mater.*, 1997, **132-136**(pt 2), 1088–1091.
[8] V.Ya. Shur and E.L. Rumyantsev, *Ferroelectrics*, 1997, **191**(1–4), 319–333.
[9] R.G.S. Barsoum, *Smart Mater. & Struct.*, 1997, **6**(1), 117–122.
[10] C.J. Dias and D.K. Das-Gupta, *IEEE Transactions on Dielectrics & Electrical Insulation*, 1996, **3**(5), 706–734.
[11] G. Eberle, H. Schmidt and W. Eisenmenger, *IEEE Transactions on Dielectrics & Electrical Insulation*, 1996, **3**(5), 624–646.
[12] J.F. Scott, *J. Phys. & Chem. of Solids*, 1996, **57**(10), 1439–1443.
[13] H.B. Harrison, Z-Q. Yao and S. Dimitrijev, *Integrated Ferroelectrics*, 1995, **9**(1–3), 105–113.
[14] W. Zhong, D. Vanderbilt, R.D. King-Smith and K. Rabe, *Ferroelectrics*, 1995, **164**(1–3), 291–301.
[15] Q.X. Chen and P.A. Payne, *Measurement Sci. & Tech.*, 1995, **6**(3), 249–267.
[16] D.J. Jones, S.E. Prasad and J.B. Wallace, *Key Eng. Mater.*, 1996, **122-124**, 71–144.
[17] S. Wen and D.D.L. Chung, *Cem. Concr. Res.* 2002, **32**(3), 335–339.
[18] Z. Mei, V.H. Guerrero, D.P. Kowalik and D.D.L. Chung, *Polym. Compos.*, 2002, **23**(5), 697–701.
[19] Mikrajuddin, K. Okuyama, F.G. Shi and H.K. Kim, *Conf. Record of IEEE Int. Symp. on Electrical Insulation*, Piscataway, NJ, 2000, pp. 180–183.
[20] C. Pecharroman and J.S. Moya, *Adv. Mater.*, 2000, **12**(4), 294–297.
[21] E. Tuncer and S.M. Gubanski, *Proc. of SPIE – Int. Soc. for Optical Eng.*, 1999, **4017**, 136–142.
[22] C. Park and R.E. Robertson, *J. Mater. Sci.*, 1998, **33**(14), 3541–3553.
[23] B. Sareni, L. Krahenbuhl, A. Beroual and C. Brosseau, *J. Appl. Phys.*, 1997, **81**(5), 2375–2383.
[24] C. Liu and Q. Zou, *Modelling & Simulation in Mater. Sci. & Eng.*, 1996, **4**(1), 55–71.
[25] Y. Ishibashi and M. Iwata, *Japanese J. Appl. Phys. Part I – Regular Papers Short Notes & Review Papers*, 1996, **35**(9B), 5157–5159.
[26] R. Strumpler, J. Glatz-Reichenbach and F. Greuter, *Proc. of the 1995 MRS Fall Meeting on Electrically Based Microstructural Characterization*, Pittsburgh, PA, 1996, pp. 393–398.
[27] T.W. Smith, M.A. Abkowitz, G.C. Conway, D.J. Luca, J.M. Serpico and G.E. Wnek, *Macromolecules*, 1996, **29**(14), 5042–5045.
[28] S.-L. Wu and I.-C. Tung, *Polym. Compos.*, 1995, **16**(3), 233–239.

[29] Anonymous, *Microwave J.*, 1998, **41**(12), 3 pp.
[30] S. Jain and R. Kumar, *Mater. & Manufacturing Processes*, 1997, **12**(5), 837–847.
[31] K. Nawa and M. Ohkita, *Sumitomo Metals*, 1997, **49**(3), 29–34.
[32] M.P. Zussman, B. Kirayoglu, S. Sharkey and D.J. Powell, *Int. SAMPE Electronics Conf.*, 1992, **6**, 437–448.
[33] S.Q. Gao, X.C. Wang, A.J. Hu, Y.L. Zhang and S.Y. Yang, *High Perf. Polym.*, 2000, **12**(3), 405–417.
[34] S.M. Bleay and L. Humberstone, *Compos. Sci. & Tech.*, 1999, **59**(9), 1321–1329.
[35] R.J. Konsowitz, *Proc. of the Tech. Program – National Electronic Packaging and Production Conf.*, Des Plaines, IL, 1992, **2**, 699–707.
[36] R.M. Japp and K.I. Papathomas, *Proc. of the 1994 MRS Fall Meeting on Hollow and Solid Spheres and Microspheres: Science and Technology Associated with their Fabrication and Application*, Pittsburgh, PA, 1995, **372**, 221–229.
[37] J.P. Ansermet and E. Baeriswyl, *J. Mater. Sci.*, 1994, **29**(11), 2841–2846.
[38] Y. Kim, E. Kang, Y.S. Kwon, W.J. Cho, C. Cho, M. Chang, M. Ree, T. Chang, C.S. Ha, *Synthetic Metals*, 1997, **85**(1–3), 1399–1400.
[39] Y. Kim, M. Ree, T. Chang and C.-S. Ha, *Proc. of the IEEE Int. Conf. on Properties and Appl. of Dielectric Mater.*, Piscataway, NJ, 1997, **2**, 882–885.
[40] S. Yu and P. Hing, *J. Appl. Polym. Sci.*, 2000, **78**(7), 1348–1353.
[41] W. Kim and J. Bae, *Proc. of the 1997 55th Ann. Tech. Conf.*, ANTEC. Brookfield, CT, 1997, **2**, 1438–1442.
[42] Y. Xu, D.D.L. Chung and C. Mroz, *Compos., Part A*, 2001, **32**, 1749–1757.
[43] Y. Xu and D.D.L. Chung, *Compos. Interfaces*, 2000, **7**(4), 243–256.
[44] M. Gunasekaran and H.J. Boneti, *Proc. of the 1996 Ann. Conf. on Electrical Insulation and Dielectric Phenomena*, Piscataway, NJ, 1996, **1**, 420–423.
[45] S.-I. Jeon, S. Hwang-bo, D.-Y. Yi, D.-H. Park, H.-J. Jung, W.-D. Kim, C.-S. Huh and M.-K. Han, *Proc. of the 1994 IEEE Int. Conf. on Properties and Appl. of Dielectric Mater.*, Piscataway, NJ, 1994, **2**, 511–514.
[46] M.C. Nobrega, L.C.F. Gomes, G.P. LaTorre and J.K. West, *Mater. Characterization*, 1998, **40**(1), 1–5.
[47] C.J. Hawker, J.L. Hedrick, R.D. Miller and W. Volksen, *MRS Bull.*, 2000, **25**(4), 54–58.
[48] K.-H. Kim, M.-H. Kim, I.-H. Son and J.-H. Kim, *Proc. of the 1998 IEEE Int. Symp. on Electrical Insulation*, Piscataway, NJ, 1998, **1**, 161–164.
[49] J.-H. Jean and S.-C. Lin, *J. Mater. Res.*, 1999, **14**(4), 1359–1363.
[50] J.-H. Jean, *J. Mater. Res.*, 1996, **11**(8), 2098–2103.
[51] J.-H. Jean, C.-R. Chang, R.-L. Chang and T.-H. Kuan, *Mater. Chem. & Phys.*, 1995, **40**(1), 50–55.
[52] J.-H. Jean, T.-H. Kuan and C.-R. Chang, *Mater. Chem. & Phys.*, 1995, **41**(2), 123–127.
[53] J.-H. Jean and T.-H. Kuan, *Japanese J. Appl. Phys. Part 1 – Regular Papers Short Notes & Review Papers*, 1995, **34**(4A), 1901–1905.
[54] D.-P. Kim and C.G. Cofer and J. Economy, *J. Amer. Ceramic Soc.*, 1995, **78**(6), 1546–1552.
[55] P.N. Kumta, *J. Mater. Sci.*, 1996, **31**(23), 6229–6240.
[56] J. Ma, K. Liao and P. Hing, *J. Mater. Sci.*, 2000, **35**(16), 4137–4141.
[57] D.G. Goski and W.F. Caley, *Canadian Metallurgical Quarterly*, 1999, **38**(2), 119–126.
[58] S. Liang, S.R. Chong and E.P. Giannelis, *Proc. of the 1998 48th Electronic Components & Tech. Conf.*, Piscataway, NJ, 1998, pp. 171–175.
[59] Y. Rao, S. Ogitani, P. Kohl and C.P. Wong, *Proc. – 50th Electronic Components and Tech. Conf.*, Piscataway, NJ, 2000, pp. 183–187.
[60] Y. Rao, C.P. Wong, J. Qu and T. Marinis, *Proc. – 50th Electronic Components and Tech. Conf.*, Piscataway, NJ, 2000, pp. 615–618.
[61] Y. Rao, J. Qu, T. Marinis and C.P. Wong, *IEEE Transactions on Components & Packaging Technologies*, 2000, **23**(4), 680–683.
[62] Y. Rao, S. Ogitani, P. Kohl and C.P. Wong, *Proc. of the Int. Symp. and Exhib. on Adv. Packaging Mater. Processes*, Piscataway, NJ, 2000, pp. 32–37.
[63] V. Paunovic, L. Vulicevic, V. Dimic and D. Stefanovic, *Proc. of the Int. Conf. on Microelectronics*, Piscataway, NJ, 1999, **2**, 535–538.
[64] S. Bhattacharya, R.R. Tummala, P. Chahal and G. White, *Proc. of the 3rd Int. Symp. & Exhib. on Adv. Packaging Mater. Process, Properties & Interfaces*, 1997, pp. 68–70.
[65] K. Nagata, S. Kodama, H. Kawasaki, S. Deki and M. Mizuhata, *J. Appl. Polym. Sci.*, 1995, **56**(10), 1313–1321.
[66] H.-I. Hsiang and F.-S. Yen, *Japanese J. Appl. Phys. Part 1 – Regular Papers Short Notes & Review Papers*, 1994, **33**(7A), 3991–3995.
[67] S.L. Namboodri, H. Zhou, A. Aning and R.G. Kander, *Polym.*, 1994, **35**(19), 4088–4091.
[68] A. Belal, M. Amin, H. Hassan, A.A. El-Mongy, B. Kamal and K. Ibrahim, *Physica Status Solidi A – Applied Research*, 1994, **144**(1), K53–K57.

[69] P. Winsor, T. Scholz, M. Hudis and K.M. Slenes, *Digest of Tech. Papers - IEEE Int. Pulsed Power Conf.*, 1999, **1**, 102–105.
[70] V. Agarwal, P. Chahal, R.R. Tummala and M.G. Allen, *Proc. - 48th Electronic Components and Tech. Conf.*, Piscataway, NJ, 1998, pp. 165–170.
[71] H.J. Hwang and K. Niihara, *J. Mater. Res.*, 1998, **13**(10), 2866–2870.
[72] K. Nomura and H. Ogawa, *J. Electrochem. Soc.*, 1991, **138**(12), 3701–3705.
[73] J. Hojo and H. Emoto, *J. Japan Soc. Powder & Powder Metallurgy*, 1993, **40**(7), 682–686.
[74] T.E. McNulty, V.E. Janas, A. Safari, R.L. Loh and R.B. Cass, *J. Amer. Ceramic Soc.*, 1995, **78**(11), 2913–2916.
[75] T. Ritter, K.K. Shung, W.S. Hackenberger, H. Wang and T.R. Shrout, *Proc. of the IEEE Ultrasonics Symp.*, Piscataway, NJ, 1999, **2**, 1295–1298.
[76] J.-Z. Zhao, C.H.F. Alves, D.A. Snook, J.M. Cannata, W.-H. Chen, R.J. Meyer, Jr, S. Ayyappan, T.A. Ritter and K.K. Shung, *Proc. of the IEEE Ultrasonics Symp.*, Piscataway, NJ, 1999, **2**, 1185–1190.
[77] R. Liu, D. Knapik, K.A. Harasiewicz, F.S. Foster, J.G. Flanagan, C.J. Pavlin and G.E. Trope, *Proc. of the IEEE Ultrasonics Symp.*, Piscataway, NJ, 1999, **2**, 973–976.
[78] A.M. Magerramov, *Physica Status Solidi A - Appl. Res.*, 1999, **174**(2), 551–556.
[80] C. Dias and D.K. Das-Gupta, *Proc. of the 1993 MRS Spring Meeting on High Performance Polymers and Polymer Matrix Composites*, Pittsburgh, PA, 1993, **305**, 183–189.
[81] M.H. Slayton and H.S.N. Setty, *1990 IEEE 7th Int. Symp. on Appl. of Ferroelectrics*, Piscataway, NJ, 1990, pp. 90–92.
[82] X. Zou, L. Zhang, X. Yao, L. Wang and F. Zhang, *Yadian Yu Shengguang/Piezoelectrics & Acoustoopics*, **18**(5), 346–348 (1996).
[83] M. Amin, S.A. Khairy, L.S. Balloomal, H.M. Osman and S.S. Ibrahim, *Ferroelectrics*, 1992, **129**(1–4), 1–11.
[84] B. Nayak, P. Talwar and A. Mansingh, *1990 IEEE 7th International Symposium on Applications of Ferroelectrics*, Piscataway, NJ, 1990, pp. 334–336.
[85] C.J. Dias and D.K. Das-Gupta, *Ferroelectrics*, 1994, **157**(1–4), 405–410.
[86] Y. Bai, V. Bharti, Z.-Y. Cheng, H.S. Xu and Q.M. Zhang, *Mater. Res. Soc. Symp. - Proc.*, Warrendale, PA, 2000, **600**, 281–286.
[87] A. Tripathi, A.K. Tripathi and P.K.C. Pillai, *Proc. on the 7th Int. Symp. on Electrets*, Piscataway, NJ, 1991, pp. 501–506.
[88] R.P. Tandon, N.N. Swami and N.C. Soni, *Ferroelectrics*, 1994, **156**(1–4), 61–66.
[89] H. Zhang, P. Xu, C. Feng, X. Liu, J. Wei, Z. Wu and C. Fang, *J. Infrared & Millimeter Waves/Hongwai Yu Haomibo Xuebao*, 1997, **16**(2), 152–156.
[90] P.M. Xu, G.J. Guo, X.L. Liu, J.Z. Wei and Z.S. Wu, *Ferroelectrics*, 1994, **157**(1–4), 411–414.
[91] J.S. Wright and L.F. Francis, *Proc. of the 1996 MRS Spring Symp. on Ferroelectric Thin Films V*, Pittsburgh, PA, 1996, **433**, 357–362.
[92] J. Hamagami, A. Goto, T. Umegaki and K. Yamashita, *Key Eng. Mater.*, 1999, **169**, 149–152.
[93] Z. Li, L. Zhang and X. Yao, *J. Mater. Sci. Lett.*, 1998, **17**(22), 1921–1923.
[94] P.K. Sharma, K.A. Jose, V.V. Varadan and V.K. Varadan, *Mater. Res. Soc. Symp. Proc.*, 2000, **606**, 175–180.
[95] L.C. Sengupta, E. Ngo and J. Synowczynski, *Integrated Ferroelectrics*, 1997, **15**(1–4), 181–190.
[96] X.M. Chen and J.S. Yang, *J. Mater. Sci. - Mater. in Electronics*, 1997, **8**(3), 147–150.
[97] R.P. Tandon, V. Singh, N. Narayanaswami and V.K. Hans, *Ferroelectrics*, 1997, **196**(1–4), 117–120.
[98] S. Sherrit, C.R. Savin, H.D. Wiederick, B.K. Mukherjee and S.E. Prasad, *J. Amer. Ceramic Soc.*, pp. 1973–75.
[99] Z. Yue, X. Wang, L. Zhang and X. Yao, *J. Mater. Sci. Lett.*, 1997, **16**(16), 1354–1356.
[100] A. Govindan, A.K. Tripathi, T.C. Goel and P.K.C. Pillai, *Proc. of the 7th Int. Symp. on Electrets*, Piscataway, NJ, 1991, pp. 524–529.
[101] D.H. Pearce and T.W. Button, *Ferroelectrics*, 1999, **228**(1), 91–98.
[102] H.J. Hwang, K. Watari, M. Sando and M. Toriyama, *J. Amer. Ceramic Soc.*, 1997, **80**(3), 791–793.
[103] S. Negi, K. Gordon, S.M. Khan and I. Khan, *Proc. of the 1997 Boston Meeting on Polym. Preprints*, Washington, DC, 1998, **39**(2), 721–722.
[104] S. Negi, J. Li, S.M. Khan and I.M. Khan, *Proc. of the 1997 Las Vegas ACS Meeting on Polym. Preprints*, Washington, DC, 1997, **38**(2), 562–563.
[105] R. Strumpler, J. Rhyner, F. Greuter and P. Kluge-Weiss, *Smart Mater. & Struct.*, 1995, **4**(3), 215–222.
[106] D.S. Cordova and L.S. Collier, *37th Int. SAMPE Symp. and Exhib. on Mater. Working for You in the 21st Century*, Covina, CA, 1992, **37**, 1406–1420.
[107] S.S. Bhoga and K. Singh, *Solid State Ionics*, 1998, **111**(1–2), 85–92.
[108] M. Siekierski, W. Wieczorek and J. Przyluski, *Electrochimica Acta*, 1998, **43**(10–11), 1339–1342.

8 Composite materials for optical applications

8.1 Background on optical behavior

8.1.1 The electromagnetic spectrum

Light is *electromagnetic radiation* in the visible region, which includes light of colors violet, blue, green, yellow, orange and red, in order of increasing wavelength from 0.4 to 0.7 μm. The visible region only constitutes a small part of the electromagnetic spectrum, which includes γ-rays, X-rays, ultraviolet, visible, infrared, microwave and radio (TV) radiation, in order of increasing wavelength from 10^{-14} to 10^4 m, as shown in Figure 8.1. The *wavelength* λ is related to the

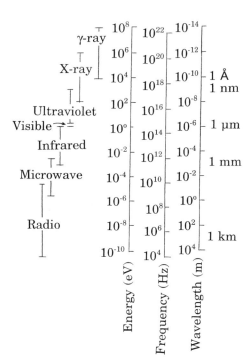

Figure 8.1 The electromagnetic spectrum.

frequency v and *photon energy* E (photon = a quantized packet of electromagnetic radiation) by

$$v = \frac{1}{T} = \frac{v}{\lambda} \qquad (8.1)$$

and

$$E = hv, \qquad (8.2)$$

where T is the *period* (time for the wave to propagate by the distance of a wavelength), v is the *speed* of the wave, and h is a universal constant called Planck's constant ($h = 6.63 \times 10^{-34}$ J.s = 4.14×10^{-15} eV.s). Hence wavelength can alternatively be described in terms of frequency or photon energy. The three alternative scales covering the entire electromagnetic spectrum are shown in Figure 8.1. Note that a large wavelength corresponds to a low frequency as well as a low energy. The speed of electromagnetic radiation depends on the medium. For free space, the speed is c, which equals 3×10^8 m/s.

8.1.2 Interaction of electromagnetic radiation with materials

Electromagnetic radiation is associated with electric and magnetic fields, as illustrated in Figure 8.2, where the electric field is in the y direction, the magnetic field is in the z direction and both electric field wave and magnetic field wave propagate in the x direction. As required by Maxwell's equations (which describe the nature of electromagnetism), the electric and magnetic fields are in a plane perpendicular to the direction of propagation. In other waves, both electric field and magnetic field waves are *transverse waves* (not longitudinal waves). The electric field in an electromagnetic wave may be in any direction in the yz plane, although it is drawn in the y direction in Figure 8.2. If the electric field is not restricted to any particular direction in the yz plane, it is said to be *unpolarized*. If it is restricted to a particular direction the yz plane, it is said to be *linearly polarized*. The waves in Figure 8.2 are said to be *plane waves*, because the wavefront is a plane, which is perpendicular to the direction of propagation. A plane wave is in contrast to a spherical wave, the wavefront of which is a sphere.

The electric and magnetic fields in electromagnetic radiation allow the radiation to interact with nuclei, electrons, molecular vibrations and molecular rotations in a material. Interaction with nuclei causing nuclear reactions requires high-energy radiation (γ-rays or X-rays); interaction with inner electrons causing electronic transitions in atoms requires lower-energy radiation (X-rays or ultraviolet); interaction with outer electrons causing electronic transitions in

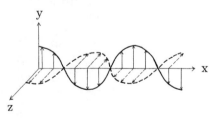

Figure 8.2 Electric field (solid arrows in the y direction) and magnetic field (dashed arrows in the z direction) in an electromagnetic wave moving in the x direction.

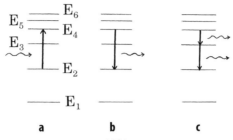

Figure 8.3 Schematic electron energy level diagram for an isolated atom. (**a**) Absorption of incident photon of energy $h\upsilon = E_4 - E_2$ causing electronic transition from E_2 to E_4. (**b**) Emission of photon of energy $h\upsilon = E_4 - E_2$ due to electronic transition from E_4 to E_2. (**c**) Emission of photon of energy $h\upsilon = E_4 - E_3$ and photon of energy $h\upsilon = E_3 - E_2$ due to electronic transition from E_4 to E_3 and that from E_3 to E_2 respectively.

atoms requires even lower-energy radiation (ultraviolet or visible); interaction with molecular vibrations requires still lower-energy radiation (infrared); interaction with molecular rotations requires yet lower energy radiation (infrared or microwave).

Figure 8.3(a) illustrates the excitation of an electron in an isolated atom from energy E_2 to energy E_4 due to absorption of a photon of energy equal to the difference in energy between E_4 and E_2. After absorption, the excited electron spontaneously decays to a lower energy (say, the initial energy E_2) once the incident radiation is removed, thereby emitting a photon of energy equal to the difference in energy between E_4 and E_2 (Figure 8.3(b)). Instead of decaying to the initial energy E_2 in one step, the electron may decay to an intermediate energy (say, E_3), emitting a photon of energy equal to the difference between E_4 and E_3, before continuing to decay to the initial energy E_2 and, in the process, emitting a photon of energy equal to the difference between E_3 and E_2 (Figure 8.3(c)). In other words, electron decay can cause the emission of one photon (Figure 8.3(b)) or multiple photons (Figure 8.3(c)). This emission is called *luminescence*, if the photons are in the visible region.

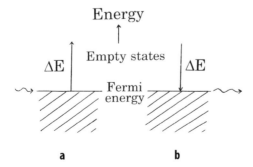

Figure 8.4 Schematic electron energy diagram for a metal, showing filled states below the Fermi energy and empty states above the Fermi energy. (**a**) Absorption of incident photon of energy DE causing electronic transition from Fermi energy to an empty state at energy DE above the Fermi energy. (**b**) Emission of photon of energy DE due to electronic transition from energy DE above the Fermi energy to the Fermi energy.

In the case of a metal instead of an isolated atom, an electron at the Fermi energy (highest energy of the filled electron states) absorbs a photon and is thus excited to an empty state above the Fermi energy (Figure 8.4(a)). The change in energy is equal to the photon energy. After that, the electron spontaneously returns to the initial energy, thereby emitting a photon (Figure 8.4(b)).

In the case of an intrinsic semiconductor, an electron at the top of the valence band absorbs a photon of energy exceeding the energy band gap E_g, and thereby it is excited to an empty state in the conduction band (Figure 8.5(a)). The excited electron is a conduction (free) electron. At the same time, a hole is created at the top of the valence band (where the electron resided). Hence, a pair of conduction electron and hole is created. This results in increases in conduction electron and hole concentrations, so that the electrical conductivity of the semiconductor is increased. This phenomenon in which electrical conductivity is increased by incident electromagnetic radiation is called *photoconduction*. Upon removal of the radiation, the excited electron spontaneously returns to the top of the valence band, thus emitting a photon (Figure 8.5(b)).

In the case of a p-type semiconductor, the absorption of a photon of energy E_A causes the transition of an electron from the top of the valence band to the acceptor level a little above the top of the valence band. In the case of an n-type semiconductor, the absorption of a photon of energy E_D causes the transition of an electron from the donor level a little below the bottom of the conduction band to the bottom of the conduction band (Figure 8.6).

Since a metal has no energy band gap, it absorbs photons of any energy. An intrinsic semiconductor has an energy band gap, so it absorbs photons of energy exceeding E_g. An extrinsic semiconductor has energy levels within the band gap,

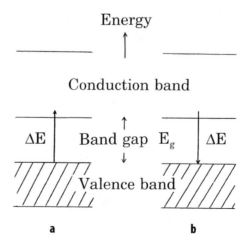

Figure 8.5 Schematic electron energy diagram for an intrinsic semiconductor, showing filled states in the valence band and empty states in the conduction band. The two bands are separated by band gap E_g. (a) Absorption of incident photon of energy $\Delta E > E_g$ causing electronic transition from top of the valence band to the conduction band, thereby generating a hole in the valence band and a conduction electron in the conduction band. (b) Emission of photon of energy $\Delta E > E_g$ due to electronic transition from the conduction band to the top of the valence band. The transition involves recombination of conduction electron and hole.

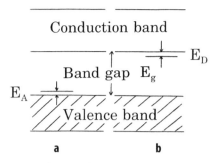

Figure 8.6 Schematic electron energy diagrams for an extrinsic semiconductors: (**a**) p-type; (**b**) n-type.

so it absorbs photons of energy exceeding E_A (for p-type semiconductors) or E_D (for n-type semiconductors).

In metals, there is no energy gap, so luminescence (emission of photons in the visible region) does not occur. In intrinsic semiconductors, there is an energy gap, so luminescence occurs and is called *fluorescence*. In extrinsic semiconductors, the donor or acceptor level within the band gap can trap an electron for a limited time when the electron is making a transition downward across the energy band gap. As a result, two photons are emitted at different times. This emission over a period of time is called *phosphorescence*. In the case of an n-type semiconductor, the first photon emitted is of energy E_D and the second photon is of energy $E_g - E_D$.

Different materials absorb photons of different energies, so the color of different materials is different and measurement of the *absorptivity A* (fraction of intensity of electromagnetic radiation that is absorbed) as a function of photon energy provides useful information on the electronic properties of the material. For a highly absorbing material, reflectivity R (fraction of intensity that is reflected) is usually measured instead of the absorptivity. The law of the conservation of energy dictates that the fraction absorbed (A), the fraction reflected (R) and the fraction transmitted (T) must add up to 1, i.e.,

$$A + R + T = 1. \tag{8.3}$$

The absorptivity A is usually not measured directly, but is obtained from Equation (8.3) after R and T have been measured. The reflectivity of a semiconductor decreases abruptly as the photon energy is increased beyond the energy gap (i.e., as the wavelength is decreased below that corresponding to the energy gap), because of increased absorptivity.

8.1.3 Reflection and refraction

The speed of electromagnetic radiation depends on the medium. If the medium is free space (vacuum), $v = c = 3 \times 10^8$ m/s. If the medium is not a vacuum, $v < c$. The ratio of c to v is defined as the *refractive index n* of the medium. Values of n ranges from 1 (for a vacuum) to 1.00029 for air, 1.46 for fused quartz, 1.55 for polystyrene, 1.77 for sapphire and 2.42 for diamond. The refractive index of a material varies with the wavelength. The above values for n are for a wavelength

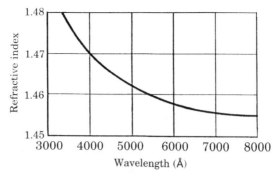

Figure 8.7 Variation of refractive index with wavelength for fused quartz.

of 5890 Å (yellow sodium light). In general, n decreases with increasing wavelength, as shown in Figure 8.7 for fused quartz.

When an incident ray traveling in a medium of refractive index n_1 encounters the interface with a medium of refractive index n_2, as shown in Figure 8.8, it is partly reflected and partly transmitted. The reflected ray is symmetric with the incident ray around the normal to the interface, such that the *angle of incidence* (angle between incident ray and the normal) equals the *angle of reflection* (angle between reflected ray and the normal), as in the case of mirror reflection. The transmitted ray is not in the same direction as the incident ray if n_1 is not equal to n_2. This phenomenon is known as *refraction*. The *angle of refraction* (angle between refracted ray and the normal) is not equal to the angle of incidence. The relationship between the angle of incidence θ_1 and the angle of refraction θ_2 is

$$n_1 \sin \theta_1 = n_2 \sin \theta_2, \tag{8.4}$$

which is known as *Snell's law*. Hence, if $n_1 > n_2$, then $\theta_2 > \theta_1$. Since a larger refractive index means lower speed, $n_1 > n_2$ means $v_2 > v_1$. Thus, the medium with the larger speed is associated with a larger angle between the ray in it and the normal.

When $\theta_2 = 90°$, the refracted ray is along the interface (Figure 8.9). According to Equation (8.4), this occurs when

$$\sin\theta_1 = \frac{n_2}{n_1}. \tag{8.5}$$

The value of θ_1 corresponding to $\theta_2 = 90°$ is called θ_c (the *critical angle*).

When $\theta_1 > \theta_c$, there is no refracted ray and all the incident ray is reflected (Figure 8.10). This is known as *total internal reflection*.

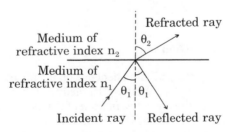

Figure 8.8 Snell's law, which governs the geometry of reflection and refraction at the interface between media of different values of the refractive index.

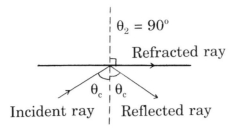

Figure 8.9 When the angle of incidence $\theta1$ equals the critical angle θc, the refracted ray is along the interface between the media of different values of refractive index.

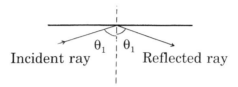

Figure 8.10 When the angle of incidence $\theta1$ exceeds the critical angle θc, there is no refracted ray and total internal reflection occurs.

8.1.4 Optical fiber

An optical fiber guides the light in it so that the light stays inside even when the fiber is bent (Figure 8.11). This is because the fiber has a cladding of refractive index n_2 and a core of refractive index n_1, such that $n_1 > n_2$ and total internal reflection takes place when $\theta_1 > \theta_c$. This means that the incident ray should have an angle of incidence more than θ_c in order to have the light not leak out of the core (Figure 8.12). Hence, incoming rays that are at too large an angle (exceeding

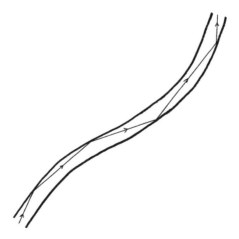

Figure 8.11 The trapping of light within an optical fiber as the light travels through a bent optical fiber due to total internal reflection.

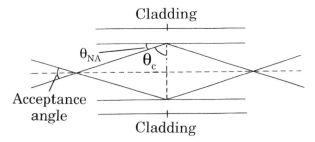

Figure 8.12 Relationship between the critical angle θ_c, the numerical aperture angle θ_{NA} and the acceptance angle.

θ_{NA}) from the axis of the fiber leak. The *acceptance angle* of the fiber is defined as twice θ_{NA}. Rays within the acceptance angle do not leak.

The *numerical aperture* (NA) of the fiber is defined as $n_1 \sin \theta_{NA}$. Since $\theta_{NA} = 90° - \theta_c$,

$$n_1 \sin \theta_{NA} = n_1 \sin(90° - \theta_c) = n_1 \cos \theta_c.$$

Since

$$\sin \theta_c = \frac{n_2}{n_1},$$

$$\cos \theta_c = \sqrt{1 - \left(\frac{n_2}{n_1}\right)^2} = \sqrt{\frac{n_1^2 - n_2^2}{n_1^2}}$$

$$= \frac{\sqrt{n_1^2 - n_2^2}}{n_1}.$$

Hence,

$$\text{numerical aperture} = n_1 \left(\frac{\sqrt{n_1^2 - n_2^2}}{n_1}\right)$$

$$= \sqrt{n_1^2 - n_2^2}. \tag{8.6}$$

An optical fiber (or optical wave guide) has a low-index glass cladding and a normal-index glass core. The refractive index may decrease sharply or gradually from core to cladding (Figure 8.13), depending on how the fiber is made. A sharp decrease in index is obtained in a composite glass fiber; a gradual decrease is obtained in a glass fiber that is doped at the surface to lower the index. A gradual decrease is akin to having a diffuse interface between core and cladding. As a consequence, a ray does not change direction sharply as it is reflected by the interface (Figure 8.13). In contrast, a sharp decrease in index corresponds to a sharp interface and a ray changes direction sharply upon reflection by the interface. A fiber with a sharp change in index is called a *stepped index fiber*. A

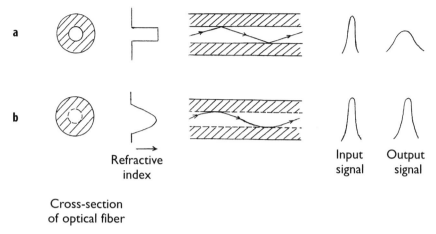

Figure 8.13 (a) A stepped index optical fiber. (b) A graded index optical fiber.

fiber with a gradual change in index is called a *graded index fiber*. A graded index fiber gives a sharper output pulse (i.e., less pulse distortion) in response to an input pulse, compared to a stepped index fiber.

An optical fiber may have different diameters of core. A small core (e.g., 3 μm diameter) means that only rays that are essentially parallel to the fiber axis can go all the way through the fiber, as off-axis rays need to be reflected too many times as they travel through the fiber and, as a result, tend to leak. A large core (e.g., 50–200 μm) means that both on-axis and off-axis rays make their way through the fiber. Thus, a fiber with a large core is called a *multimode fiber*, whereas one with a small core is called *single-mode fiber* (Figure 8.14). A single-mode fiber gives less pulse distortion than a multimode fiber, so it is preferred for long-distance optical communication. However, the intensity of light that can go through a single-mode fiber is smaller than that for a multimode fiber. The NA tends to be around 0.1 for a single-mode glass fiber and around 0.2 for a multimode glass fiber.

A single-mode fiber tends to have the cladding thicker than the core, so that the overall fiber diameter is not too small. For example, the cladding may be 70–150 μm thick, while the core diameter is 3 μm. A multimode fiber tends to have the

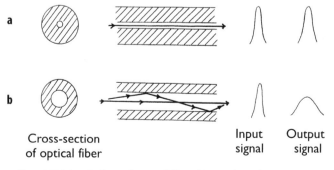

Figure 8.14 (a) A single-mode optical fiber. (b) A multimode optical fiber.

cladding thinner than the core, as the core is already large. For example, the cladding may be 1–50 µm thick, while the core diameter is 50–200 µm.

A single-mode fiber is stepped index, whereas a multimode fiber may be either stepped index or graded index. Thus, there are three basic types of optical fibers: single-mode stepped index, multimode stepped index and multimode graded index. Pulse distortion increases in the order: single-mode stepped index, multimode graded index and multimode stepped index.

Light is absorbed as it travels through any medium (whether solid, liquid or gas), such that the intensity I at distance x is related to the intensity I_o at $x = 0$ by

$$I = I_o e^{-\alpha x}, \qquad (8.7)$$

where α is the *absorption coefficient*, which varies from one medium to another and has the unit m^{-1}. Equation (8.7) is known as the *Beer–Lambert law*. The greater is α, the more severe is the absorption. Equation (8.7) means that the intensity decreases exponentially as light travels through the medium.

Arranging Equation (8.7) and taking the natural logarithm gives

$$\ln \frac{I}{I_o} = -\alpha x. \qquad (8.8)$$

Converting natural logarithm to logarithm to the base 10 gives

$$2.3 \log \frac{I}{I_o} = -\alpha x. \qquad (8.9)$$

As log (I/I_o) is proportional to x, Equation (8.9) is more convenient to use than Equation (8.7). Thus, it is customary to define

$$\text{attenuation loss (in dB)} = -10 \log \frac{I}{I_o}. \qquad (8.10)$$

When $I/I_o = 0.1$, the attenuation loss is 10 dB.
When $I/I_o = 0.01$, the attenuation loss is 20 dB. Note that dB is the abbreviation for *decibel*. From Equations (8.9) and (8.10),

$$\text{attenuation loss (in dB)} = \frac{10\alpha x}{2.3}. \qquad (8.11)$$

Hence, the attenuation loss is proportional to x.

The attenuation loss, also called *optical loss*, per unit length of an optical fiber varies with wavelength, because the absorptivity (Section 8.2) and light scattering (i.e., scattering of light out of the medium) depend on the wavelength. Absorption losses occur when the frequency of the light is resonant with oscillations in the electronic or molecular structure of the fiber material. Various ions and functional groups have characteristic absorption peaks at well-defined frequencies. Scattering that occurs at inhomogeneities in the fiber is linear, i.e., there is no change of frequency. Scattering due to phonon–phonon interaction or Raman scattering is non-linear, i.e., there is a change of frequency. A typical loss for glass fibers is around 1 dB/km. Polymers are not as attractive as glass for use as optical fibers because of their relatively high attenuation loss.

The imperfect coupling between the light source and an optical fiber is another source of loss, called *coupling loss*, which is typically 10–12 dB. This loss is because the light from the source has rays that are at angles greater than the acceptance angle of the optical fiber (Figure 8.12). Even if the light source (a light-emitting diode with rays exiting it within an angle of 100°) is butt directly with the optical fiber, coupling loss still occurs. Less coupling loss occurs if the light source is a laser, since laser light diverges negligibly as it travels.

An optical fiber is most commonly used for the transmission of signals, e.g., for optical communication. Related to this application is the use of an optical fiber as a light guide to connect light to a sensor and to return the light from the sensor to an analyzer. However, an optical fiber can also serve as a sensor [1–28]. For example, it can be used for sensing the strain and damage in a structure in which the optical fiber is embedded. The amplitude, phase and polarization of light that travels through an optical fiber can be affected by the strain and damage in the fiber. These quantities can be monitored, using appropriate instrumentation. Strain in the structure gives rise to strain or bending in the fiber; damage in the structure gives rise to more strain, more bending or even damage in the fiber. For strain sensing, the optical fiber is preferably a ductile material, so polymers are sometimes used in place of glass, although they suffer from high attenuation loss. The intensity of light (related to the amplitude) that goes through an optical fiber is called the *light throughput*, which decreases as the fiber decreases in diameter, as the fiber bends (causing leakage through the cladding) and as the fiber is damaged. (Damage increases the attenuation loss, due to increased absorption or light scattering by the fiber.) An optical fiber may contain partially reflecting (partially transmitting) mirrors at certain points along its length within the fiber. In this way, part of the light is reflected and part is transmitted. By measuring the time it takes for the reflected light to reach the start of the fiber, information can be obtained concerning the location of the strain or damage. This technique is called *time domain reflectometry*.

An optical fiber sensor (also called a *fiber-optic sensor*) can be of one of three types, i.e., *transmission-gap sensor*, *evanescent-wave sensor* and *internal-sensing sensor*. A transmission-gap sensor has a gap between the input fiber and the output fiber (which are end to end except for the gap) and the disturbance at the gap affects the output. The disturbance may be pressure, temperature, etc. When the ends of the fibers delineating the gap are polished to enhance light reflection, a slight change in the gap distance causes a change in phase difference between the light rays reflected from the adjacent ends of the two fibers and traveling in the same direction back toward the light source. This is the principle behind a *Fabry–Perot fiber optic strain sensor*. (Refer to an introductory physics textbook for the technique of phase measurement.) An evanescent-wave sensor has part of the length of an optical fiber stripped of its cladding. The stripped part is the sensor, since the light loss from the stripped part is affected by the refractive index of the medium around the stripped part. Hence, a change in medium is detected by this sensor. For example, an evanescent-wave sensor is used for monitoring the curing of the polymer during polymer–matrix composite fabrication, since the refractive index of the polymer changes as it cures. An internal-sensing sensor is just an unmodified optical fiber; the amplitude and phase of light going through the fiber are affected by the disturbance encountered by the fiber.

An optical fiber's sensing region may be coated with special materials to enhance the response to certain disturbance. A coating in the form of a magnetostrictive material may be used to enhance the response to magnetic fields and one in the form of a piezoelectric material may be used to enhance the response to electric fields. The special material may be in the cladding. For example, the special material is a sensor material which reacts with a chemical

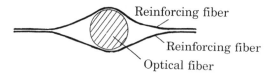

Figure 8.15 Distortion of reinforcing fibers in the vicinity of an optical fiber perpendicular to them.

species and changes its refractive index, thereby changing the light loss from the fiber and enabling the fiber to sense the chemical species.

Optical fibers are often embedded in a structure to form a grid, i.e., a number of fibers in the x direction and a number of fibers in the y direction. The grid allows information to be obtained concerning the location of strain or damage.

In order for an optical fiber embedded in a structure to be truly sensitive to the strain/damage of the structure, the optical fiber must be well adhered to the structure. Thus, the surface of an optical fiber may be treated or coated so as to promote adhesion. The long-term durability of the fiber in the structure should also be considered. In particular, glass slowly dissolves in concrete, which is alkaline.

In the case of embedding an optical fiber between the layers of reinforcing fibers (not optical fibers) in a structural composite material, the large diameter of the optical fiber compared to the reinforcing fibers causes the reinforcing fibers that are perpendicular to the optical fiber to be bent or distorted in the vicinity of an optical fiber (Figure 8.15). This distortion degrades the effectiveness of the reinforcing fibers for reinforcing the composite. Furthermore, the optical fiber is akin to a flaw in the composite; it causes stress concentration and weakens the composite. Thus, the optical fiber is an intrusive sensor.

A fiber-optic sensor system includes a light source, a fiber-optic sensor, a light detector and electronic processing equipment, which are connected in the order given.

8.1.5 Light sources

Light sources include the sun, lamps, light-emitting diodes and lasers. For use with optical fibers, which are small in size, light-emitting diodes and lasers are appropriate.

Light-emitting diodes

A light-emitting diode (LED) is a pn junction under forward bias. The majority of the holes in the p-side move toward the n-side, while the majority of the electrons in the n-side move toward the p-side (Figure 8.16). The holes meet the electrons at the junction, resulting in recombination. As a result, light of photon energy equal to the energy band gap is emitted from the junction. By making one of the sides (say, the p-side) thin and essentially not covered by electrical contacts, light can exit from the junction through the thin side (Figure 8.17). By using semiconductors of different band gaps, light of different wavelengths (different colors) can be obtained. This phenomenon of light emission is known as *electroluminescence*.

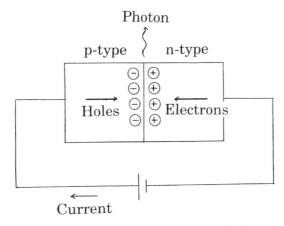

Figure 8.16 A light-emitting diode made from a pn junction under forward bias.

Lasers

Laser is an abbreviation for "light amplification by stimulated emission of radiation". *Stimulated emission*, to be distinguished from *spontaneous emission*, refers to the downward transition of an electron (say, from the conduction band to the valence band or from an excited state to the ground state) occurring due to another electron doing the same thing. In other words, the downward transition of one electron stimulates another electron to do the same thing. In spontaneous emission, no stimulation occurs. Stimulated emission causes a *stimulated photon*, which exists along with the original photon (called the *active photon*), as illustrated in Figure 8.18.

A helium–neon (He–Ne) laser involves a plasma tube containing a mixture of He and Ne gases (90:10 ratio, 3×10^{-3} atm pressure), such that a DC voltage (~2 kV) is applied along the length of the tube (Figure 8.19). The ground state electronic configuration of He is $1s^2$; that of Ne is $1s^2 2s^2 2p^6$. The applied electric field causes ionization of the atoms, thus forming a plasma in the plasma tube. This excitation cause the originally empty 2s energy level of He to be occupied (Figure 8.20). Because the He 2s level is at essentially the same energy as the Ne 3s level (also empty in the ground state of Ne), thermal collision causes the transfer of the electron from He 2s to Ne 3s. The Ne 3s electron stays in the Ne 3s level for ~6×10^{-6} s before relaxing to lower levels. As a result, the Ne 3s level is more

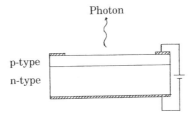

Figure 8.17 A light-emitting diode with light exiting from the exposed thin p-side of the pn junction. The shaded regions are metal thin films serving as electrical contacts.

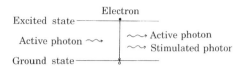

Figure 8.18 Schematic electron energy level diagram illustrating the emission of a stimulated photon by an active photon of the same energy.

populated with electrons than the Ne 2p level – a situation known as *population inversion* (a desirable situation for stimulated emission). The population inversion causes stimulated emission to occur as the electrons relax from the Ne 3s level to the Ne 2p level. The resulting laser light is red, with wavelength 6328 Å. Outside the plasma tube and at its two ends are two parallel mirrors, each with a reflectivity of 0.9999. The emitted light is reflected by each mirror, except for a fraction of 0.0001, which is transmitted through the mirror. The reflected portion re-enters the plasma tube and stimulates other electrons to relax from the Ne 3s level to the Ne 2p level. By reflecting back and forth between the two mirrors, the number of stimulated photons becomes higher and higher. An intense laser beam is then transmitted through either mirror.

In the case of a semiconductor laser, electrons are excited into the conduction band by an applied voltage. An electron recombines with a hole to produce a photon, which stimulates the emission of a second photon by a second recombination. The ends of the semiconductor are mirrored, such that one mirror is totally reflective and the other mirror is partially reflective. Reflection of the photons by the mirrors back into the semiconductor allows these photons to stimulate even more photons. A fraction of the photons is emitted through the partially reflective mirror as a laser beam.

A particularly effective semiconductor laser is a pn junction with very heavily doped n- and p-type regions. Furthermore, the semiconductor is a *direct gap semiconductor*, i.e., a semiconductor in which the momentum of the electrons at the bottom of the conduction band and the momentum of the holes at the top of the valence band are equal. The tendency for recombination is enhanced by having the momenta of the electron and hole equal. Examples of such semiconductors are GaAs and InP, which are compound semiconductors exhibiting the zinc blende crystal structure. In each compound, one element is from Group III of the periodic table whereas the other element is from Group V. Hence, these semiconductors are called *III–V compound semiconductors*. Silicon, an elemental semiconductor, is an *indirect gap semiconductor*, i.e., the momenta of the electrons and holes are not equal, so silicon is not used for lasers. The doping is so heavy that the Fermi energy is above the bottom of the conduction band in the n-side and below the top of the valence band in the p-side (Figure 8.21) in equilibrium (i.e., without bias). Upon sufficient forward bias, population

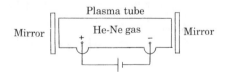

Figure 8.19 He–Ne laser comprising a plasma tube, two mirrors and a DC power supply.

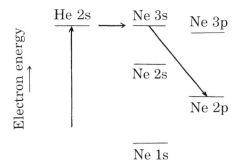

Figure 8.20 Energy-level diagram of the He–Ne system. Arrows indicate electronic transitions.

inversion occurs at the junction, i.e., the electron-rich conduction band of the n-side overlaps the hole-rich valence band of the p-side. Recombination occurs and light is emitted from the junction (the light confinement region).

A laser beam is characterized by its being parallel (not diverging), nearly *monochromatic* (of nearly one wavelength) and *coherent* (with any two points in the laser beam having a predictable phase relationship).

By "nearly monochromatic", one means that the range or band of frequencies (called the *frequency bandwidth* Δv) is narrow. An He–Ne laser (stabilized) has $\Delta v = 10^4$ Hz, whereas spectral lines emitted by gas discharge tubes have $\Delta v = 10^9$ Hz. White light ranges in frequency from 4×10^{14} Hz to 7×10^{14} Hz, so its $\Delta v = 3 \times 10^{14}$ Hz. Light-emitting diodes have much larger Δv than lasers.

To understand coherence, consider predicting the relationship between the phase of a light wave at two different times at the same point in space. Suppose that at time t_o the wave is at a peak. At time $t_o + \Delta t$, the wave would have gone through $v\Delta t$ cycles, where v is the frequency (cycles per second), if the light were truly monochromatic. From the number of cycles, the phase at $t_o + \Delta t$ is obtained. If the light is not monochromatic, but has a frequency range from v to $v + \Delta v$, the number of cycles is between $v\Delta t$ and $(v + \Delta v) \Delta t$. If Δt is small, $\Delta v \Delta t \ll 1$ and the number of cycles is $v\Delta t$, just as in the monochromatic case. The condition for coherence (i.e., having predictable phase relationship) is

$$\Delta v \Delta t \ll 1$$

or

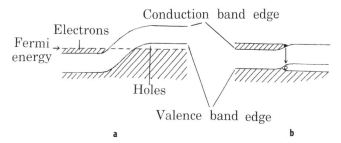

Figure 8.21 Semiconductor laser: (**a**) pn junction without bias; (**b**) pn junction under forward bias. The shaded regions are occupied by electrons.

$$\Delta t \ll \frac{1}{\Delta v}. \qquad (8.12)$$

In a time interval Δt, a wave propagates by a distance Δx, where $\Delta x = c\Delta t$ and c is the speed of light. Thus, comparing the phases at two points a distance of Δx apart at a fixed time is the same as comparing the phases at the same point in space over a time interval Δt. Hence, Equation (8.12) can be written as

$$\Delta x \ll \frac{c}{\Delta v}. \qquad (8.13)$$

The distance $c/\Delta v$ is called the *coherence length* (x_c). In other words,

$$x_c = \frac{c}{\Delta v}$$

and the condition for coherence is

$$\Delta x \ll x_c.$$

The coherence length is the distance within which the light has a predictable phase relationship, i.e., the distance within which the light is coherent.

8.1.6 Light detection

The sensing or detection of light is relevant to smart structures. For example, a laser beam hitting an aircraft in a war is sensed as soon as it hits, and the aircraft is then directed to respond to the threat in an appropriate fashion. As another example, a machine component rotates when light is directed at it and stops when the light is off. In the latter example, light serves as a switch. A *light detector* (also called an *optical detector*) is usually a semiconductor, which can be an elemental semiconductor (e.g., Si) or a compound semiconductor (e.g., GaAs, InP).

Semiconductors (preferably intrinsic) are effective for detecting light of wavelength exceeding the energy band gap, as the light excites electrons (since light is electromagnetic radiation and electrons are charged) from the valence band to the conduction band, thus generating conduction electrons and holes and causing the electrical conductivity to increase. The increase in conductivity results in an increase in current in the circuit. The increased current causes an increased voltage across a resistor in series with the semiconductor in the circuit. The increase in voltage is called the *photovoltage*. This phenomenon is called *photoconduction* or *photoconductivity*. It is a *photovoltaic effect*, which is any effect in which radiation energy is converted to electrical energy. By monitoring the conductivity, light can be detected.

The electrical conductivity σ of a semiconductor is given by

$$\sigma = qn\mu_n + qp\mu_p, \qquad (8.14)$$

where n = conduction electron concentration, p = hole concentration, μ_n = conduction electron mobility and μ_p = hole mobility.

For an intrinsic semiconductor, $n = p$. Thus, Equation (8.14) becomes

$$\sigma = qn(\mu_n + \mu_p).$$

For an intrinsic semiconductor, the conductivity σ_1 in darkness (i.e., no incident light) is given by

$$\sigma_1 = qn_1 (\mu_n + \mu_p), \qquad (8.15)$$

where n_1 is the conduction electron (or hole) concentration in darkness. The conductivity σ_2 in light (i.e., with incident light) is given by

$$\sigma_2 = qn_2 (\mu_n + \mu_p), \qquad (8.16)$$

where n_2 is the conduction electron (or hole) concentration in light. Implicit in Equation (8.16) is the notion that light affects the mobilities negligibly. The ratio of light conductivity σ_2 to dark conductivity σ_1 is called the *photoresponse*, which describes the effectiveness of the light detector. Hence, from Equations (8.15) and (8.16),

$$\text{photoresponse} = \frac{\sigma_2}{\sigma_1} = \frac{n_2}{n_1}. \qquad (8.17)$$

A good light detector has photoresponse as high as 10^3. For a given light detector, the photoresponse increases with light intensity, since n_2 increases with light intensity.

Photoconduction is the basis for detecting electromagnetic radiation of a large range of wavelength. X-ray is electromagnetic radiation of a small wavelength (Figure 8.1). It is emitted from a solid material that has been excited by having a core electron removed from an atom in the material. The emission occurs upon the transition of an electron in the atom from a higher energy level to the core level with an electron removed, although the emission of another electron (called the *Auger electron*) at a high energy level is a process that competes with the X-ray emission. Because the X-ray emitted has a photon energy that is equal to the difference in energy between the energy level from which the electron comes and the core energy level at which the electron lands, analysis of the X-ray wavelength (or energy) gives information on the energy levels, which can be used to identify the element which gives the X-ray emission. This technique of elemental composition analysis is called *X-ray spectroscopy* (or *X-ray spectrometry*).

A widely used detector of X-rays is an intrinsic semiconductor such as silicon. The detector is known as a *semiconductor detector*. As the photon energy of the X-ray is much greater than the energy band gap of a semiconductor, a photon of X-ray hitting the semiconductor causes the excitation of a large number of electrons of the semiconductor from the valence band to the conduction band. Each electronic transition causes the creation of a hole in the valence band, in addition to a conduction electron in the conduction band, i.e., an electron–hole pair. In the presence of an electric field applied to the semiconductor, the conduction electrons move toward the positive end of the voltage gradient, while the holes move toward the negative end of the voltage gradient. This results in a current. The conventional direction of the current is the direction of the flow of holes. The electrons and holes contribute additively to the current. Thus, each photon of X-ray results in a current pulse. The height of each pulse increases with increasing photon energy. The number of pulses over a period of time increases with increasing intensity of the X-ray. Hence, both photon energy and intensity can be determined.

In order to decrease the *dark current* (current in the absence of photons hitting the detector), which is due to thermal excitation of electrons from the valence band

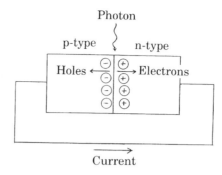

Figure 8.22 A pn junction solar cell.

to the conduction band, the semiconductor detector is cooled to 77 K (boiling point of nitrogen) by using liquid nitrogen. As a result, the detector is housed in a cryostat which has a window (typically made of beryllium) for the X-ray to enter.

A short-circuited pn junction is particularly effective as a light detector (Figure 8.22). When light is directed at the junction, electrons are excited across the band gap and conduction electrons and holes are generated. The potential gradient in the depletion region sweeps the conduction electrons toward the n-side and the holes toward the p-side, thus resulting in a current from the n-side to the p-side within the semiconductor. This is the basis of a pn junction *solar cell* (the most important photovoltaic device).

8.2 Composite materials for optical waveguides

Optical waveguides can be in the form of fibers [29], slabs [30], thick films [31–33] and thin films [34]. A requirement for waveguide materials concerns the refractive index, as explained in Section 8.1.4. Another requirement is transparency (i.e., low optical loss) at the relevant wavelength. Composite material design provides a route for tailoring the refractive index. For attaining transparency in the direction of wave propagation, different components in a composite should exhibit similar values of refractive index. Otherwise, one component should be nano-sized. In the case of an optical fiber, the core and cladding are different in refractive index, by design, while the core must be sufficiently transparent. Examples of transparent composites are polyimide containing nano-sized titanium dioxide particles [35], a TiO_2–SiO_2 composite film [34] and glass particle epoxy–matrix composites [36,37].

8.3 Composite materials for optical filters

Optical filters are for transmission or reflection of selected wavelengths. They are usually in the form of thin films, particularly multilayers. The different layers are designed by consideration of the refractive index and thickness (in relation to the wavelength) of each layer. Materials with both high and low values of refractive index are needed for this purpose. Glasses and polymers tend to have low values of refractive index. Composite engineering provides a route to obtain materials with high values of refractive index. Examples of such composites are poly(ethylene oxide) (PEO) containing iron sulfides [38], and a composite of tantalum oxide and hafnium oxide [39]. Related applications of high refractive index materials pertain to antireflective coatings and lenses [40].

8.4 Composite materials for lasers

Solid-state lasers involve laser-active materials (Section 8.1.5.2). A laser-active material is conventionally used by itself in a monolithic form. However, a composite containing the laser-active material (preferably in the form of nanoparticles to increase its surface area) and a refractive-index-matched polymer matrix can be used instead. The attractions of a composite are processability, such as castability to form thick films, and low cost [41].

8.5 Summary

Composite materials for optical applications are in their early stage of development and usage, but the field is progressing rapidly. This chapter does not address polymer–matrix composites for non-linear optical applications, which include optical data storage, optical image processing, holography, optical computing and pattern recognition [42–44], owing to the complexity of the physics behind non-linear optics and the infancy of the subject.

Example problems

1. What are the wavelength and frequency of electromagnetic radiation of photon energy 0.02 eV in free space?

Solution

From Equation (8.2),

$$\upsilon = \frac{E}{h} = \frac{0.02 \text{ eV}}{4.14 \times 10^{-15} \text{ eV.s}} = 5 \times 10^{12} \text{ s}^{-1}.$$

From Equation (8.1),

$$\lambda = \frac{v}{\upsilon} = \frac{c}{\upsilon} = \frac{3 \times 10^8 \text{ m/s}}{5 \times 10^{12} \text{ s}^{-1}} = 6 \times 10^{-5} \text{ m} = 60 \text{ } \mu\text{m}.$$

2. An optical fiber has a core of refractive index 1.48 and a cladding of refractive index 1.35. (a) What is the critical angle? (b) What is the acceptance angle? (c) What is the numerical aperture?

Solution

(a)
$$\sin\theta_c = \frac{n_2}{n_1} = \frac{1.35}{1.48}$$
$$\theta_c = 66°.$$

(b)
Acceptance angle = $2(90° - \theta_c)$
= $2(90° - 66°)$
= $48°$.

(c) From Equation (8.6),
$$\text{numerical aperture} = \sqrt{n_1^2 - n_2^2}$$
$$= \sqrt{1.48^2 - 1.35^2}$$
$$= 0.61.$$

3. An optical fiber of length 1.8 km has an attenuation loss of 1.5 dB/km. (a) What is the ratio of the intensity of light exiting the fiber to that entering the fiber at the other end? (b) What is the absorption coefficient of the optical fiber material?

Solution

(a)
Attenuation loss = (1.5 dB/km) (1.8 km)
= 2.7 dB.

From Equation (8.10),
$$2.7 = -10 \log \frac{I}{I_o} = 10 \log \frac{I_o}{I}$$

$$\frac{I_o}{I} = 1.86$$

$$\frac{I}{I_o} = 0.54.$$

(b) From Equation (8.11), with $x = 1$ km,
$$1.5 = \frac{10\alpha}{2.3}.$$

Hence,
$$\alpha = 0.35 \text{ km}^{-1}.$$

4. What is the coherence length of a laser with frequency bandwidth (a) $\Delta v = 10^4$ Hz; (b) $\Delta v = 10^8$ Hz?

Solution

(a) From Equation (8.13),

$$x_c = \frac{c}{\Delta v} = \frac{3 \times 10^8 \text{ m/s}}{10^4 \text{ Hz}} = 3 \times 10^4 \text{ m}.$$

5. A semiconductor has a photoresponse of 145. Its carrier concentration in darkness is 5.4×10^{15} cm^{-3}. What is the carrier concentration in light?

Solution

From Equation (8.17),

$$n_2 = n_1 \text{ (photoresponse)}$$
$$= (5.4 \times 10^{15} \text{ cm}^{-3})(145)$$
$$= 7.8 \times 10^{17} \text{ cm}^{-3}.$$

6. An intrinsic semiconductor has energy band gap $E_g = 0.9$ eV and refractive index 2.58.

   (a) What is the maximum wavelength (in μm) of electromagnetic radiation that can be absorbed by this material?
   (b) What is the speed of electromagnetic radiation in this material?
   (c) What is the critical angle θ_c (in any) for electromagnetic radiation traveling in air and incident on this material?
   (d) The intensity of electromagnetic radiation incident on the semiconductor is I_0 and that transmitted through the back side of the semiconductor is I, such that $I/I_0 = 0.38$. What is the attenuation loss in dB?
   (e) This semiconductor is used to make a pn junction to serve as a light-emitting diode. What is the frequency of light emitted?
   (f) The frequency bandwidth of the light-emitting diode is 10^7 Hz. What is the coherence length of light generated by this diode?
   (g) The electrical conductivity of this semiconductor when light is incident on it is 250 times that when it is dark. What is the photoresponse?

Solution

(a)
$$E = hv = \frac{hc}{\lambda} \Rightarrow \lambda = \frac{hc}{E}$$

$$= \frac{(4.14 \times 10^{-15} \text{ eV.s})(3 \times 10^8 \text{ m/s})}{0.9 \text{ eV}} = 1.4 \times 10^{-6} \text{ m}.$$

(b)
$$n = \frac{c}{v} \Rightarrow v = \frac{c}{n} = \frac{3 \times 10^8 \text{ m/s}}{2.58} = 1.2 \times 10^8 \text{ m/s}$$

(c)
$$\sin \theta_c = \frac{n_2}{n_1} = \frac{2.58}{1} \Rightarrow \theta_c \text{ does not exist.}$$

(d)   Attenuation loss $= -10 \log \dfrac{I}{I_o} = -10 \log 0.38 = 4.2$ dB.

(e) $$\upsilon = \dfrac{E}{h} = \dfrac{0.9 \text{ eV}}{4.14 \times 10^{-15} \text{ eV.s}} = 2.2 \times 10^{14} \text{ s}^{-1}.$$

(f) $$x_c = \dfrac{c}{\Delta \upsilon} = \dfrac{3 \times 10^8 \text{ m/s}}{10^7 \text{ Hz}} = 30 \text{ m}.$$

(g) $$\text{Photoresponse} = \dfrac{\sigma_2}{\sigma_1} = 250.$$

7. How many electron–hole pairs are generated in a semiconductor (silicon) detector by a photon of X-ray of energy 1.49 keV, which is the energy associated with the transition of an electron of aluminum from the L energy level to the K energy level? The energy band gap of silicon is 1.1 eV.

Solution

$$\text{No. of electron–hole pairs} = \dfrac{1.49 \text{ keV}}{1.1 \text{ eV}} = \dfrac{1490 \text{ eV}}{1.1 \text{ eV}} = 1355.$$

Review questions

1. What is the mechanism behind photoconduction?
2. How does the absorptivity of a semiconductor vary with the photon energy in the range from one below the energy band gap to one above the energy band gap?
3. How is the refractive index of a medium related to the speed of electromagnetic radiation in the medium?
4. What is the main advantage of a graded index optical fiber compared to a stepped index optical fiber?
5. How can an optical fiber function as a damage sensor?
6. Both light-emitting diode and solar cell involve a pn junction. What is the difference in operation condition?
7. What are the main advantages of a laser compared to a light-emitting diode as a light source?
8. Give an example of a direct gap semiconductor and an example of an indirect gap semiconductor.
9. Give an example of a III–V compound semiconductor.
10. What is the principle behind X-ray spectroscopy?
11. Why should a semiconductor detector be kept cold during use?
12. What are the main requirements for an optical waveguide material?

References

[1] N. Mrad, R.G. Melton and B.T. Kulakowski, *J. of Intelligent Mater. Syst. & Struct.*, 1997, 8(11), 920–928.

[2] L. Yuan and L. Zhou, *Measurement Sci. & Tech.*, 1998, **9**(8), 1174–1179.
[3] V.E. Saouma, D.Z. Anderson, K. Ostrander, B. Lee and V. Slowik, *Mater. & Struct.*, 1998, **31**(208), 259–266.
[4] G. Surace and A. Chiaradia, *Proc. SPIE – Int. Soc. for Optical Eng.*, Society of Photo-Optical Instrumentation Engineers, Bellingham, WA, 1997, **3041**, 644–650.
[5] N. Takeda and T. Kosaka, *Proc. SPIE – Int. Soc. for Optical Eng.*, Society of Photo-Optical Instrumentation Engineers, Bellingham, WA, 1997, **3041**, 635–643.
[6] M. Volanthen, H. Geiger and J.P. Dakin, *IEE Colloquium* (Digest), 1997, (033), 4/1–4/6.
[7] F. Ansari, *Cem. & Concr. Compos.*, 1997, **19**(1), 3–19.
[8] B. Grossman, S. Murshid and G. Cowell, *Instrumentation in the Aerospace Industry: Proc. Int. Symp.*, Instrument Society of America, Research Triangle Park, NC, 1997, **43**, 435–443.
[9] M.A. El-Sherif and J. Radhakrishnan, *J. of Reinforced Plastics & Compos.*, 1997, **16**(2), 144–154.
[10] T.R. Wolinski, A.W. Domanski and P. Galazka, *Proc. SPIE – Int. Soc. for Optical Eng.*, 1996, **2839**, 265–271.
[11] W. Pan, W. Tang and W. Zhou, *Proc. SPIE – Int. Soc. for Optical Eng.*, 1996, **2895**, 166–170.
[12] A. Schena, R. Falciai and R. Fontana, *Conf. on Lasers and Electro-Optics Europe – Tech. Digest*, 1996, 47 pp., CMM6.
[13] S. Bourasseau, M. Dupont, M. Pernice, A. Thiriot, P. Blanquet, T. Demol, C. Delebarre and D. Coutellier, *Proc. SPIE – Int. Soc. for Optical Eng.*, Society of Photo-Optical Instrumentation Engineers, Bellingham, WA, 1996, **2779**, 180–185.
[14] W.B. Spillman Jr, *Proc. 1996 Int. Gas Turbine and Aeroengine Congress & Exhib.*, Birmingham, UK, 1996.
[15] E. Udd, *Proc. IEEE*, 1996, **84**(6), 884–894.
[16] E. Udd, *Optics & Photonics News*, 1996, **7**(5), 16–22.
[17] S.M. Yang and C.W. Chen, *J. of Intelligent Mater. Syst. & Struct.*, 1996, **7**(1), 71–77.
[18] A.S. Voloshin, L. Han and J.P. Coulter, *Proc. SPIE – Int. Soc. for Optical Eng.*, Society of Photo-Optical Instrumentation Engineers, Bellingham, WA, 1995, **2443**, 554–564.
[19] L. Han, A. Voloshin and J. Coulter, *Smart Mater. & Struct.*, 1995, **4**(2), 100–105.
[20] E.J. Friebele, C.G. Askins, M.A. Putnam, G.M. Williams, A.D. Kersey, A.A. Fosha Jr, J. Florio Jr, R.P. Donti and R.G. Blosser, *Proc. SPIE – Int. Soc. for Optical Eng.*, Society of Photo-Optical Instrumentation Engineers, Bellingham, WA, 1995, **2447**, 305–311.
[21] M.M. Thomas, R.A. Glowasky, B.E. McIlroy and T.A. Story, *Proc. SPIE – Int. Soc. for Optical Eng.*, Society of Photo-Optical Instrumentation Engineers, Bellingham, WA, 1995, **2447**, 266–273.
[22] K.D. Bennett and J. Wang, *Proc. SPIE – Int. Soc. for Optical Eng.*, Society of Photo-Optical Instrumentation Engineers, Bellingham, WA, 1995, **2441**, 296–304.
[23] D. Sun, M.K. Burford and R.O. Claus, *Proc. SPIE – Int. Soc. for Optical Eng.*, Society of Photo-Optical Instrumentation Engineers, Bellingham, WA, 1995, **2441**, 291–295.
[24] P.L. Fuhr and D.R. Huston, *Proc. SPIE – Int. Soc. for Optical Eng.*, Society of Photo-Optical Instrumentation Engineers, Bellingham, WA, 1995, **2574**, 6–13.
[25] E. Udd, *Proc. SPIE – Int. Soc. for Optical Eng.*, Society of Photo-Optical Instrumentation Engineers, Bellingham, WA, 1995, **2574**, 2–5.
[26] B. Culshaw, W.C. Michie and P.T. Gardiner, *Proc. SPIE The Int. Soc. for Optical Eng.*, Society of Photo-Optical Instrumentation Engineers, Bellingham, WA, 1994, **2341**, 134–151.
[27] W.B. Spillman Jr, *IEEE LEOS Ann. Meeting – Proc.*, IEEE, Piscataway, NJ, **2**, 230–231.
[28] J.S. Sirkis, *Adaptive Structures and Composite Materials: Analysis and Application*, American Society of Mechanical Engineers, Aerospace Division (Publication) AD. ASME, New York, 1994, **45**, 85–92.
[29] V. Matejec, M. Hayer, M. Pospisilova and I. Kasik, *J. of Sol-Gel Sci. & Tech.*, 1997, **8**(1–3), 889–893.
[30] M. Yoshida and P.N. Prasad, *Proc. of the 1995 MRS Spring Meeting on Thin Films for Integrated Optics Applications*, Materials Research Society, Materials Research Society, Pittsburgh, PA, 1995, **392**, 103–108.
[31] S. Motakef, J.M. Boulton and D.R. Uhlmann, *Optics Letters*, 1994, **19**(15), 1125–1127.
[32] J. Graf, H. Sautter, B. Zeitz, H. Drug and H. Schmidt, *Conf. on Lasers and Electro-Optics Europe – Technical Digest*, 1996, 59 pp.
[33] W. Que, Y. Zhou, Y.L. Lam, Y.C. Chan, H.T. Tan, T.H. Tan and C.H. Kam, *J. of Electronic Mater.*, 2000, **29**(8), 1052–1058.

[34] X. Wang, H. Masumoto, Y. Someno and T. Hirai, *Thin Solid Films*, 1999, **338**(1-2), 105-109.
[35] M. Yoshida, M. Lal, N.D. Kumar and P.N. Prasad, *J. of Mater. Sci.*, 1997, **32**(15), 4047-4051.
[36] T. Naganuma and Y. Kagawa, *Acta Materialia*, 1999, **47**(17), 4321-4327.
[37] Y. Kagawa, H. Iba, M. Tanaka, H. Sato and T. Chang, *Acta Materialia*, 1997, **46**(1), 265-271.
[38] T. Kyprianidou-Leodidou, H.-J. Althaus, Y. Wyser, D. Vetter, M. Buchler, W. Caseri and U.W. Suter, *J. of Mater. Res.*, 1997, **12**(8), 2198-2206.
[39] J. Simpson and A. Pitt, *Proc. of SPIE - Int. Soc. for Optical Eng.*, 1999, **3578**, 555.
[40] L.L. Beecroft and C.K. Ober, *J. of Macromolecular Sci. - Pure & Appl. Chem.*, 1997, (4), 573-586.
[41] L.L. Beecroft, R.T. Leidner, C.K. Ober, D.B. Barber and C.R. Pollock, *Proc. of the 1996 MRS Spring Symp. on Better Ceramics Through Chem. VII: Organic/Inorganic Hybrid Materials*, Materials Research Society, Pittsburgh, PA, 1996, **435**, 575-582.
[42] Y. Zhang, R. Burzynski, S. Ghosal and M.K. Casstevens, *Adv. Mater.*, 1996, **8**(2), 111-125.
[43] W.E. Moerner, A. Grunnet-Jepsen and C.L. Thompson, *Ann. Review of Mater. Sci.*, 1997, **27**, 585-623.
[44] B. Kippelen, K. Meerholz, B. Volodin and N. Peyghambarian, *Optical Mater.*, 1995, **4**(2-3), 354-357.

9 Composite materials for magnetic applications

9.1 Background on magnetic behavior

9.1.1 Magnetic moment

An electron is associated with a spin, which results in a magnetic moment. The magnetic moment of an electron is 9.27×10^{-24} A.m^2, or 1 *Bohr magneton* (β). A filled orbital contains two electrons of opposite spin, so the net magnetic moment of a filled orbital is zero.

Example 1: O (oxygen atom)

Atomic oxygen has an electronic configuration of

$$1s^2\ 2s^2\ 2p^4$$

The 1s orbital is filled; so is the 2s orbital. The 2p subshell contains two unpaired electrons. Therefore, the magnetic moment of one oxygen atom is 2 (9.27×10^{-24} A.m^2), or 2 Bohr magnetons, or $\beta = 2$.

2p | ↑↓ | ↑ | ↑ |

Example 2: O^{2-} (oxide ion)

An O^{2-} ion has an electronic configuration of $1s^2\ 2s^2\ 2p^6$. The 1s, 2s, as well as all three 2p orbitals, are full, so the magnetic moment of an O^{2-} ion is zero.

Example 3: O$_2$ (oxygen molecule)

An O$_2$ molecule consists of two atoms covalently bonded by the sharing of two pairs of electrons (i.e., a double bond). Each oxygen atom in O$_2$ has two unpaired electrons. The 2p$_y$ electron cloud of one atom overlaps end to end with the 2p$_y$ cloud of the other atom, forming a σ bond (Figure 9.1(a)). The 2p$_z$ electron cloud of one atom overlaps side to side with the 2p$_z$ electron cloud of the other atom, forming a π bond (Figure 9.1(b)). The combination of a σ bond and a π bond is a double bond, i.e. O=O. The electron sharing is such that the two electrons being

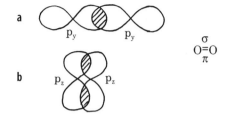

Figure 9.11 (a) Overlap of $2p_y$ electron clouds of adjacent atoms end to end to form a σ bond. (b) Overlap of $2p_z$ electron clouds of adjacent atoms side to side to form a π bond.

shared have opposite spins. Therefore, an O_2 molecule has a magnetic moment of zero.

2p [↑↓|↑|↑] $P_x P_y P_z$

Example 4: Fe (iron atom)

Iron (Fe) solid consists of Fe atoms, each having an electronic configuration of … $3d^6 4s^2$. The 4s orbital is full, but the 3d subshell has four unpaired electrons, so $\beta = 4$, or the magnetic moment per Fe atom is 4 (9.27×10^{-24} A.m²).

3d [↑↓|↑|↑|↑|↑]

Example 5: Co (cobalt atom)

Cobalt (Co) solid consists of Co atoms, each having an electronic configuration of … $3d^7 4s^2$. The 4s orbital is full, but the 3d subshell has three unpaired electrons, so $\beta = 3$.

3d [↑↓|↑↓|↑|↑|↑]

Example 6: Ni (nickel atom)

Ni is a solid with electronic configuration … $3d^8 4s^2$, so $\beta = 2$.

3d [↑↓|↑↓|↑↓|↑|↑]

Example 7: Cu (copper atom)

Cu is a solid with electronic configuration … $3d^{10} 4s^1$. The 3d subshell is full, but the 4s orbital has one unpaired electron, so $\beta = 1$.

Example 8: Zn (zinc atom)

Zn is a solid with electronic configuration … $3d^{10} 4s^2$. All orbitals are full, so $\beta = 0$.

9.1.2 Ferromagnetic behavior

Since the magnetic moment of an electron is in a certain direction, the magnetic moment of an atom or ion is also in a certain direction. Atoms or ions that are neighboring or close by in a solid can communicate with each other so that, even in the absence of an applied magnetic field, atoms or ions within a limited volume have all their magnetic moments oriented with respect to one another. This communication is a form of quantum-mechanical coupling called the *exchange interaction*. A material that exhibits this behavior in such a way that the communication causes the magnetic moments to be oriented in the same direction is called a *ferromagnetic material*. A region in which all atoms have magnetic moments in the same direction is called a *ferromagnetic domain*. Because there is a tendency for the magnetic moment to be along a certain crystallographic direction, a domain is usually smaller than a grain in a polycrystalline material. In other words, a grain can contain more than one domain.

When no external magnetic field has been applied to a ferromagnetic material (i.e., when the magnetic field strength $H = 0$; the unit for H is A/m), the magnetic moments of various domains cancel one another, so the magnetic induction (also called the magnetic flux density) B is zero. In the presence of an external magnetic field ($H > 0$), the domains with magnetic moments in the same direction as H grow while those with magnetic moments in other directions shrink. This process involves the movement of the domain boundaries and results in $B > 0$.

The slope of the plot of B versus H varies with H and is known as the *permeability* (μ), as shown in Figure 9.2. Hence,

$$B = \mu H.$$

Alternatively, this equation can be written as

$$B = \mu_o (H + M),$$

where μ_o = permeability of free space (vacuum), and M = *magnetization* ≡ magnetic moment per unit volume.

The quantity $\mu_o H$ is the magnetic induction in vacuum (no material), so $\mu_o M$ is the extra magnetic induction present due to having a material rather than vacuum as the medium.

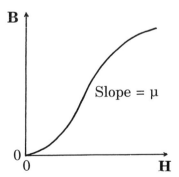

Figure 9.2 Variation of magnetic induction B with magnetic field strength H during first application of H (virgin curve). The slope of the curve is the permeability μ.

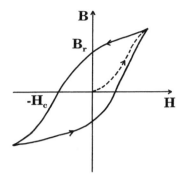

Figure 9.3 Variation of B with H during first application of H (dashed curve) and during subsequent cycling of H (solid curve).

The unit for the magnetic moment is A.m², so the unit for the magnetic moment per unit volume (i.e., the magnetization) is $A.m^2/m^3 = A.m^{-1}$, which is the same as the unit for H.

The *relative permeability* (μ_r) of a material is defined as

$$\mu_r = \frac{\mu}{\mu_o}.$$

For a ferromagnetic material, $\mu_r \gg 1$ (up to 10^6). For example, μ_r is 5000 for iron (99.95 Fe), 10^5 for Permalloy (79Ni–21Fe), and 10^6 for Supermalloy (79Ni–15Fe–5Mo). μ_o is a universal constant, with

$$\mu_o = 4\pi \times 10^{-7} \text{ H/m},$$

where H (henry) = wb/A and wb means weber. The unit for B is wb/m².

The *saturation magnetization* is reached when the magnetic moments of all domains are in the same direction. Subsequent decrease of H back to zero does not remove the magnetization totally – a *remanent magnetization* (or *remanence*) remains at $H = 0$ (Figure 9.3). In order to bring the magnetization back to zero, H must be applied in the reverse direction, i.e., $H < 0$. At $H = -H_c$, the magnetization is zero; H_c is called the *coercive field* (or *coercivity*). The coercivity is 4 kA/m for carbon steel (0.9C–1Mn), 44 kA/m for Cunife (20Fe–20Ni–60Cu), 123 kA/m for Alnico V (50Fe–14Ni–25Co–8Al–3Cu), 600 kA/m for samarium cobalt ($SmCo_5$) and 900 kA/m for $Nd_2Fe_{14}B$. Cycling of H between positive and negative values results in a hysteresis loop.

Examples of ferromagnetic materials are Fe, Co, Ni and their alloys.

9.1.3 Paramagnetic behavior

A *paramagnetic material* has its atoms or ions separately having non-zero magnetic moments, but the atoms fail to communicate sufficiently (say, because the atoms are relatively far apart) so that the directions of the magnetic moments of different atoms (even adjacent atoms) are independent of one another in the absence of an applied magnetic field. There are no domains. $B = 0$ when $H = 0$. When $H > 0$, more atoms have their magnetic moments in the direction of H, so $B > 0$, but there are still no domains. As always,

Composite materials for magnetic applications

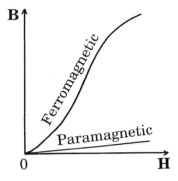

Figure 9.4 Variation of B with H during first application of H for (**a**) a ferromagnetic material and (**b**) a paramagnetic material.

$$B = \mu H,$$

but $\mu_r = \dfrac{\mu}{\mu_o} > 1$, up to 1.01. μ_r of a paramagnetic material is much less than that of a ferromagnetic material (Figure 9.4). Since thermal energy opposes alignment of the magnetic moments, the higher the temperature, the lower is μ_r.

A ferromagnetic material usually becomes paramagnetic upon heating past a certain temperature, which is also known as the *Curie temperature*. This is a reversible phase transition which involves a change in the magnetic structure (ordering of the directions of the magnetic moments of different atoms or ions) rather than a change in the crystal structure. As the temperature increases, thermal agitation is more severe and the communication (exchange interaction) between atoms is less effective. At the Curie temperature, the communication vanishes and the material becomes paramagnetic. The Curie temperature is ~768°C for carbon steel, 410°C for Cunife, 887°C for Alnico V, 727°C for $SmCo_5$ and 347°C for $Nd_2Fe_{14}B$.

9.1.4 Ferrimagnetic behavior

In a *ferrimagnetic material*, the atoms (or ions) having non-zero magnetic moments (referred to as magnetic atoms or ions) within a ferrimagnetic domain communicate with one another via a form of quantum-mechanical coupling called *superexchange interaction* (as the magnetic atoms or ions are often separated by non-magnetic ones), so that the magnetic moment of every magnetic atom or ion in the domain is in one of two antiparallel (opposite) directions and the vector sum of the magnetic moments of all these atoms or ions in the domain is non-zero. Just as a ferromagnetic material, a ferrimagnetic material has B = 0 before any H is applied and cycling H results in a hysteresis loop in the plot of B versus H.

Ferrimagnetic oxides such as Fe_3O_4 and $NiFe_2O_4$ are known as ferrites, which are unrelated to the ferrite (α) in the Fe–C phase diagram. In particular, Fe_3O_4 is called magnetite (or lodestone). Ferrites are examples of ceramic magnets (rather than metallic magnets).

Fe_3O_4 exhibits the inverse spinel structure. In this crystal structure, the O^{2-} ions occupy the corners and face centers of a cube, the Fe^{2+} ions occupy one-quarter of the octahedral interstitial sites (formed by the anion structure), and the Fe^{3+} ions occupy one-quarter of the octahedral interstitial sites and one-eighth of the tetrahedral interstitial sites. For every four O^{2-} ions, there must be one Fe^{2+} ion and two Fe^{3+} ions in order to have electrical neutrality and be consistent with the chemical formula Fe_3O_4.

Let us count the number of each type of ion per cube.

$$O^{2-} \quad 8\left(\frac{1}{8}\right) \quad + \quad 6\left(\frac{1}{2}\right) = 4$$
$$\text{Corners} \quad \text{Face centers}$$

$$Fe^{2+} \quad 4\left(\frac{1}{4}\right) = 1$$

There are four octahedral interstitial sites per cube.

$$Fe^{3+} \quad 4\left(\frac{1}{4}\right) \quad + \quad 8\left(\frac{1}{8}\right) = 2$$
$$\text{Octahedral} \quad \text{Tetrahedral}$$

There are eight tetrahedral interstitial sites per cube.

Since a fraction of certain sites are occupied in a cube, different cubes can have different arrangements of ions even though the ion counts are the same for all the cubes. Because of this, a cube is not a true unit cell. It turns out that the true unit cell consists of eight cubes (Figure 9.5). The lattice constant of the true unit cell is 8.37 Å. In a true unit cell, there are 4 (8) = 32 O^{2-} ions, 2 (8) = 16 Fe^{3+} ions and 1 (8) = 8 Fe^{2+} ions. The arrangement of the ions in a true unit cell is quite complex (Figure 9.6).

The O^{2-} ions have zero magnetic moments. Half of the Fe^{3+} ions (i.e., the Fe^{3+} ions occupying octahedral interstitial sites) and all of the Fe^{2+} ions have their magnetic moments in the same direction, while the remaining Fe^{3+} ions (i.e., those occupying tetrahedral interstitial sites) have their magnetic moments in the opposite direction. In other words, all cations occupying octahedral sites have

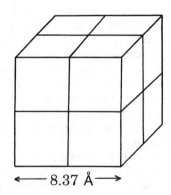

Figure 9.5 Eight cubes making up a true unit cell of dimension 8.37 Å for Fe_3O_4.

Composite materials for magnetic applications

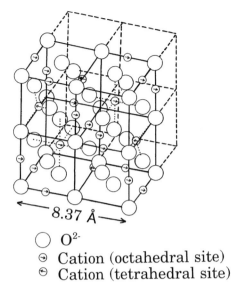

○ O^{2-}
⊖ Cation (octahedral site)
⊙ Cation (tetrahedral site)

Figure 9.6 Arrangement of ions in a true unit cell (Figure 9.6) of Fe_3O_4, which is ferrimagnetic. Only the front half of the unit cell is shown. The magnetic moment (directions shown by arrows) of the cations in octahedral (sixfold) sites and the cations in tetrahedral (fourfold) sites are opposite in direction. The large circles are O^{2-} ions, which have no magnetic moment.

their magnetic moments in the same direction, while all cations occupying tetrahedral sites have their magnetic moments in the opposite direction.

The magnetic moment of Fe ($3d^6\,4s^2$) is such that $\beta = 4$; that of Fe^{2+} ($3d^6$) is such that $\beta = 4$; that of Fe^{3+} ($3d^5$) is such that $\beta = 5$.

Let us calculate the magnetic moment per unit cell.

Interstitial site	Spin	Fe^{2+}	Fe^{3+}	β	Magnetic moment ($A.m^2$)
Octahedral	↑	8		+32	+32 (9.27×10^{-24})
Octahedral	↑		8	+40	+40 (9.27×10^{-24})
Tetrahedral	↓		8	−40	−40 (9.27×10^{-24})
				+32	+32 (9.27×10^{-24})

The saturation magnetization (i.e., magnetic moment per unit volume at saturation)

$$= \frac{\text{magnetic moment per unit cell}}{\text{volume of a unit cell}}$$

$$= \frac{32 \left(9.27 \times 10^{-24}\right) A.m^2}{\left(8.37 \times 10^{-10} m\right)^3}$$

$$= 5 \times 10^5 \, A.m^{-1}.$$

$NiFe_2O_4$ (called nickel ferrite) consists of Ni^{2+} (instead of Fe^{2+}), Fe^{3+} and O^{2-} ions. For Ni^{2+}, $\beta = 2$. Repeating the above calculation for $NiFe_2O_4$ shows that the magnetic moment per unit cell is +16 ($9.27 \times 10^{-24} \, A.m^2$).

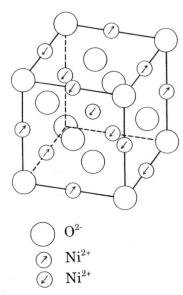

Figure 9.7 Arrangement of ions in a unit cell of NiO, which is antiferromagnetic. The small circles are Ni^{2+} ions (with magnetic moment in the direction shown by the arrows); the large circles are O^{2-} ions, which have no magnetic moment.

9.1.5 Antiferromagnetic behavior

In an *antiferromagnetic material*, each atom (or ion) has a non-zero magnetic moment and the atoms within an antiferromagnetic domain communicate with one another so that the magnetic moment of every atom in the domain is in one of two antiparallel (opposite) directions and the vector sum of the magnetic moments of all the atoms in the domain is zero.

An example of an antiferromagnetic material is NiO, which consists of Ni^{2+} and O^{2-} ions arranged in the rock salt structure (Figure 9.7). The O^{2-} ions have no magnetic moment, but the Ni^{2+} ions ($\beta = 2$) do. However, half of the Ni^{2+} ions have their magnetic moments one way, while the remaining Ni^{2+} ions have their magnetic moments in the opposite direction. Thus, the material has no magnetization.

9.1.6 Hard and soft magnets

Precipitates, grain boundaries, dislocations, etc., tend to anchor domain boundaries, thereby causing a large remanent magnetization and a large H_c (i.e., a wide hysteresis loop). A hard (permanent) magnet is characterized by a wide hysteresis loop, whereas a soft magnet is characterized by a narrow hysteresis loop (Figure 9.8). Note that magnetic hardness is totally different from mechanical hardness.

The product of B and H has the unit J/m^3 (i.e., energy per unit volume). The power of a hard magnet is defined as the maximum product of B and H in the demagnetization portion of the hysteresis loop. The $(BH)_{max}$ is 1.6 kJ/m³ for

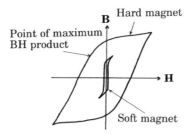

Figure 9.8 The B–H loop for a hard magnet and a soft magnet.

carbon steel (0.9C–1Mn), 12 kJ/m³ for Cunife (20Fe–20Ni–60Cu), 36 kJ/m³ for Alnico V (50Fe–14Ni–25Co–8Al–3Cu), 140 kJ/m³ for $SmCo_5$ and 220 kJ/m³ for $Nd_2Fe_{14}B$.

9.1.7 Diamagnetic behavior

A *diamagnetic material* is one with relative permeability $\mu_r < 1$ (about 0.99995) (Figure 9.9) because the electrons respond to the applied magnetic field by setting up a slight opposing field. This behavior can be a consequence of the electron shells of the atom or ion being full so that the electron spins cancel and there is no net magnetic moment. As a result, the magnetic field does not interact with the spins, but rather interacts with the electron orbit (i.e., the shape of the electron cloud) by distorting it.

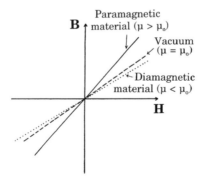

Figure 9.9 Variation of B with H during first application of H for a paramagnetic material (solid curve), vacuum (dashed curve) and a diamagnetic material (dotted curve).

9.1.8 Magnetostriction

Magnetostriction [1–9] refers to the change in dimension upon magnetization of a ferromagnetic material. It provides a mechanism of actuation, but not a mechanism of sensing, since the converse effect does not occur. The dimensional change is reversible and is due to the rotation of the magnetic dipoles (spins) and the resulting change in bond length between atoms (a phenomenon known as *spin–orbit coupling*, i.e., coupling between the electron spin and the electron

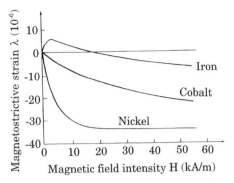

Figure 9.10 Variation of magnetostrictive strain with the magnetic field H for iron, cobalt and nickel.

orbit). The strain can be expansion (*positive magnetostriction*) or contraction (*negative magnetostriction*). It is typically small, of the order of 10^{-6} in magnitude. This strain is denoted by λ and is called the *magnetostrictive constant*. Figure 9.10 shows the strain versus magnetic field H for iron, cobalt and nickel. The strain is negative for all these three materials, except for iron at a low magnetic field. A consequence of magnetostriction is that domain boundaries may be torn at the edges of the triangular domains (called *closure domains*, since they allow the loop of magnetic flux path to complete within the solid, rather than in the air around the solid, so as to lower the energy), as illustrated in Figure 9.11. Since the dimensional change increases with domain size, the problem of Figure 9.11(a) and (b) can be alleviated by having smaller domains (Figure 9.11(c)).

The low magnetostrictive strains for the transition metals (elements with 3d subshell, e.g., Fe, Co and Ni, Figure 9.10) are not suitable for practical application in actuation. The magnetostrictive strains are larger for the rare earth elements (elements with 4f subshell, e.g., samarium (Sm), terbium (Tb) and dysprosium (Dy)), because of their large spin–orbit coupling and the highly anisotropic

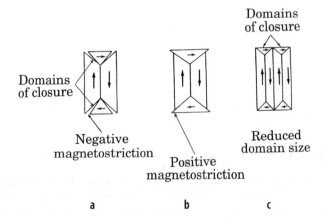

Figure 9.11 Boundaries of closure domains being torn due to (**a**) negative magnetostriction and (**b**) positive magnetostriction. The problem in (**a**) or (**b**) is alleviated by reducing the domain size, as shown in (**c**).

localized nature of the 4f electronic charge distribution. In particular, Tb and Dy are hexagonal and the large magnetostrictive strains are in the basal plane. On the other hand, the localized nature of the 4f electronic charge distribution causes the exchange interaction between atoms to be weak, thereby making the Curie temperature low (<240 K, below room temperature – not practical) compared to those of the transition metals, which have stronger exchange interaction. This shortcoming of the rare earth elements can be alleviated by alloying a rare earth element with a transition metal to form an intermetallic compound. In this way, the Curie temperature is increased. Such compounds have formula AB_2, where A is a rare earth element and B is a transition metal, e.g., $SmFe_2$ (samfenol), $TbFe_2$ (terfenol) and $DyFe_2$. They are cubic in crystal structure. In general, compounds with formula AB_2 (where A and B are any elements) are called *Laves phases* (e.g., $MgCu_2$, $MgZn_2$ and $MgNi_2$). In $TbFe_2$, saturation strain is 3.6×10^{-3} at room temperature and above, and the Curie temperature is 431°C. The phenomenon is known as *giant magnetostriction*.

Magnetostriction involves rotation of magnetic dipoles, i.e., domain rotation. The ease of rotation depends on the crystallographic orientation of the dipole. *Easy directions* are directions that are easy (requiring little energy) to rotate into or out of. *Hard directions* are directions that are difficult (requiring large energy) to rotate into or out of. The energy required to rotate the magnetization out of the easy direction is called the *magnetocrystalline anisotropy constant* (K), which is also called the *magnetocrystalline anisotropy energy*. It is equal in magnitude to the energy required to magnetize in the easy direction. This parameter is usually positive. A negative value of this parameter can occur when there are many equivalent easy axes at an angle of 90° from the easy axis – a situation known as *easy-plane anisotropy*.

For example, the spins of the iron atoms prefer to be in the [100] direction, so the easy direction for iron is [100] and the hard direction is [111]; a small magnetic field is enough to cause magnetic saturation in the easy direction, but a

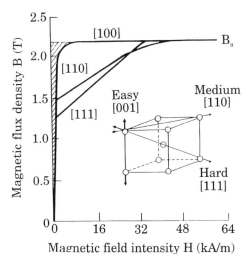

Figure 9.12 Variation of B with H during first application of H along the [100], [110] and [111] directions of iron. The [100] direction is the easy direction for iron. The shaded area is the magnetocrystalline anisotropy energy for the [100] direction.

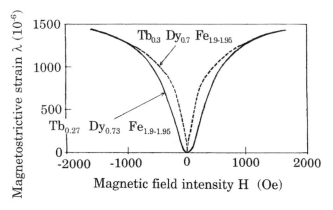

Figure 9.13 Variation of the magnetostrictve strain with the applied magnetic field **H** for Terfenol-D.

larger field is necessary in the hard direction. The [110] direction is intermediate. Since the product of B and H is energy per unit volume, the shaded area in Figure 9.12 is the magnetocrystalline anisotropy energy per unit volume for iron. For $TbFe_2$ and $SmFe_2$, the easy direction is [111].

A large magnetocrystalline anisotropy energy inhibits domain rotation, particularly at low temperatures. Furthermore, it causes the domain wall motion to be irreversible, thereby resulting in hysteresis. To decrease the magnetocrystalline anisotropy energy, $TbFe_2$ and $DyFe_2$ (both highly magnetostrictive, with λ positive) are combined, because the magnetocrystalline anisotropy constant K is negative for $TbFe_2$ and positive for $DyFe_2$. Magnetostriction/anisotropy ratio maximization (anisotropy compensation, $K \cong 0$) occurs in $Tb_xDy_{1-x}Fe_2$, where $x = 0.27$, at room temperature. It differs at other temperatures. This ternary alloy is called Terfenol-D, because Tb, Fe and Dy are in the alloy and the Naval Ordnance Laboratory (NOL), now Naval Surface Warfare Center, investigated this material first. This alloy exhibits the largest room temperature magnetostrictive strain of any commercially available material. Figure 9.13 shows the variation of magnetostrictive strain with the applied magnetic field for Terfenol-D. Note that the strain is positive for both positive and negative values of the magnetic field.

9.1.9 Magnetoresistance

Magnetoresistance [10–23] refers to the phenomenon in which electrical resistivity changes with magnetic field. Mathematically, it is defined as the resistivity at a non-zero magnetic field minus that at zero magnetic field. This phenomenon is useful for magnetic field sensing, which is important for magnetic recording (e.g., read heads), data storage (e.g., random access memory) and manufacturing control. Magnetoresistance is said to be positive if resistance increases with magnetic field, and is said to be negative if resistance decreases with magnetic field. The magnetic field may be parallel or perpendicular to the current used to measure resistivity. The magnetoresistance in the two directions may be different, even different in sign. Magnetoresistance tends to be small for a non-magnetic (i.e., diamagnetic, paramagnetic or antiferromagnetic) material, but it tends to be large for a magnetic material (e.g., a ferromagnetic or ferrimagnetic material).

The fractional change in resistivity due to the application of a magnetic field (also known as the magnetoresistance ratio) is typically 2–3% for ferromagnetic materials. However, by using a multilayer consisting of magnetic thin films (e.g., Fe films) that are separated by non-magnetic (and electrically conducting) thin films (e.g., Cr films) such that the adjacent (but separated) magnetic thin films exhibit antiferromagnetic coupling, the fractional increase in resistivity in the direction perpendicular to the multilayer due to a magnetic field can be much higher (5–40%). This large magnetoresistance is known as *giant magnetoresistance* (GMR). This large effect is due to the scattering of electrons at the interfaces between the layers of the multilayer due to the antiparallel spins of the adjacent magnetic layers. This scattering is known as *spin scattering*.

Giant magnetoresistance is usually positive. However, a giant negative magnetoresistance occurs in polymer–matrix composites containing iron oxide (ferrite) nanoparticles (~10 nm) [24].

An even larger effect than GMR is *colossal magnetoresistance* (CMR), which is associated with a magnetoresistance ratio as high as 100,000%. CMR is exhibited by certain oxides such as $(La,A)MnO_3$ (A = Ca, Sr or Ba). The crystal structure is perovskite but orthorhombic (with the lattice parameters $a \cong b \neq c$). Intrinsic to such oxides is ferromagnetic ordering in the a–b plane of the unit cell and antiferromagnetic ordering in the c-axis. The ferromagnetically ordered Mn–O layers of the a–b plane are separated by a non-magnetic La(A)–O layer.

9.2 Metal–matrix composites for magnetic applications

Because metals are among the most common ferromagnetic materials, metal–matrix composites are quite common among magnetic composite materials. In general, these composites can be classified into three groups, namely (i) composites with a magnetic (ferromagnetic or ferrimagnetic) metal matrix and a magnetic filler, (ii) composites with a magnetic metal matrix and a non-magnetic (antiferromagnetic, paramagnetic or diamagnetic) filler and (iii) composites with a non-magnetic metal matrix and a magnetic filler. In any group, ferromagnetic metals are dominant among the magnetic metals.

Composites in group (i) include those containing an α-Fe (a soft magnetic material) and $R_2Fe_{14}B$ (a hard magnetic material, with R = a rare earth element such as Nd) [25,26]. The soft magnetic grains cause a large spontaneous magnetization, while the hard magnetic grains induce a large coercive field [26]. Other examples of composites in group (i) are those containing an nickel–iron alloy and nickel zinc ferrite [27], those containing cobalt and EuS [28] and those containing Sb and MnSb [29].

Composites in group (ii) include those containing a ferromagnetic metal (e.g., Co, Ni and Fe) and an antiferromagnetic phase (e.g., CoO, NiO and FeS) [30–32]. The large antiferromagnetic–ferromagnetic interface causes enhancement of the room temperature coercivity of the ferromagnetic phase [32].

Composites in group (iii) include those with aluminum [33], Zn–22Al [34,35], Cu–Zn–Al [36], $Nb_{0.33}Cr_{0.67}$ [37], silver [38–40], copper [40] and other metals [41] as matrices. The fillers include iron fibers [33], barium ferrite powder [34], strontium powder [35], Fe–Cr flakes [36], SmCo$_5$ [37], iron nitride [38], iron oxide [39], CoFe [40] and ferrite [41]. In spite of the non-magnetic character of the matrix, these composites are attractive in the electrical conductivity and formability resulting from the metal matrix. The formability is particularly high for composites with a superplastic metal matrix, such as Zn–22Al [34].

9.3 Polymer–matrix composites for magnetic applications

Although polymers are usually not magnetic, their processability makes the fabrication of polymer–matrix composites relatively simple. These composites contain a non-magnetic polymer matrix and a magnetic filler, which is commonly in powder form. They provide a monolithic and low-cost form of magnetic material. Examples of fillers are ferrite [42–44], Fe$_2$O$_3$ [45,46], iron [47], nickel [48] and cobalt–nickel [49]. Both conventional and specialized magnetic applications benefit from magnetic polymer–matrix composites. The specialized applications are described below.

Owing to the interaction of a magnetic material with electromagnetic radiation, magnetic polymer–matrix composites are used for electromagnetic interference (EMI) shielding [50]. Composites with elastomeric matrices (e.g., silicone) are attractive for magnetostrictive actuation [51–57], heat-shrink applications [57], magnetoresistive switching [54,55] and piezoinductive current generation [53,58]. Some of these applications (e.g., magnetoresistance and piezoinduction) benefit from a certain degree of electrical conductivity, so the conductivity of the ferromagnetic particles and, less commonly, that of the polymer matrix, are sometimes exploited. Examples of conducting polymer matrices are polypyrrole [59] and polyaniline [60–63]. Nevertheless, non-conducting polymer matrices, such as polyethylene [64], wax [65] and others [66–71], are most common, because of their low cost, wide availability and processability.

For reducing the energy loss due to eddy current flow in a magnetic core of a current transformer, low electrical conductivity is desired for the core material. To attain a low conductivity, polymer–matrix composites with a non-conducting matrix are used [72,73]. Because of the requirement for AC operation of the transformer, the filler is a soft magnetic powder.

Ferromagnetic and electrically conductive particles are used as a filler in composites which require anisotropic conduction, because the particles can be aligned during composite fabrication by using a magnetic field [74].

When the ferromagnetic particles are asymmetric, the particles can be oriented by the flow of the polymer during injection molding. The orientation results in enhanced magnetic permeability when the direction of orientation and the magnetic field are parallel to one another [75]. Akin to asymmetric particles are flakes, particularly nanoflakes (submicron thickness and aspect ratio 10–100 [76]), which are potentially useful for microwave effective media [77].

Ferromagnetic particle polymer–matrix composites in the form of films formed by spin coating on glass or semiconductor substrates are potentially useful as magneto-optical media for optical devices and integrated optics [78]. A related type of composite is a polymer (e.g., silicone) substrate implanted with iron and cobalt ions [79].

9.4 Ceramic–matrix composites for magnetic applications

For transformer cores and related applications, polymer–matrix composites are attractive, owing to the low conductivity of most polymers. In contrast, metal–matrix composites have high conductivity. However, polymer–matrix composites cannot withstand high temperatures. Therefore, ceramic–matrix composites, which typically have low conductivity and high temperature resistance, are desirable. However, ceramic–matrix composites suffer from the high cost of fabrication.

The most common ceramic matrix is aluminum oxide (alumina, or Al_2O_3), which can be obtained from alumina gels (AlOOH) by calcination and reduction [80]. The ferromagnetic fillers used in alumina–matrix composites include iron particles [80–82], nickel particles [83,84] and cobalt particles [85]. The metal particles can be obtained by reduction of the corresponding metal oxide particles, e.g., Fe from Fe_2O_3 [81], and Ni from NiO [83,84], or reduction of the corresponding metal salts, e.g., $Ni(NO_3)_2$, to the metal oxides, e.g., NiO, prior to further reduction to the metal, e.g., Ni [83,84].

One of the next most common ceramic matrices is zirconium oxide (zirconia, or ZrO_2) [85], which is commonly doped with Y_2O_3 (yttria) [86–88]. The fillers used include ferrimagnetic ferrite particles such as barium hexaferrite ($BaFe_{12}O_{19}$) [86–88] and ferromagnetic nickel particles [85]. Other ceramic matrices used include aluminum nitride, boron nitride and silicon nitride [89].

Ceramic–matrix composites in the form of films can be prepared by reactive sputtering of appropriate alloys or compounds (such as nickel aluminide) to form a metal nitride (such as nickel nitride), followed by heat treatment in a vacuum to obtain the metal from the metal nitride. The heat treatment can be performed locally by means of a focused laser beam in order to generate microscopic features that are useful for magnetic data storage [89]. Instead of reactive sputtering, composite films can be deposited by vacuum codeposition, e.g., codeposition of Fe and MgO to form an MgO–matrix Fe particle composite film [90,91].

Materials that absorb microwaves are needed for avoiding the detection of aircraft, missiles and ships by radar. This is known as *Stealth* technology or *low observability*. The interaction of a magnetic material with electromagnetic radiation provides a mechanism for absorption. Therefore, aluminosilicate ($Al_2O_3.2SiO_2$) and related low-cost ceramic matrices are used to form magnetic structural composites [92].

Glass–matrix composites containing magnetic particles are attractive for magneto-optical applications [93]. Examples are a sodium borosilicate glass matrix composite containing ferrimagnetic garnet particles [93], a silica (SiO_2) matrix composite containing Fe_3O_4 particles [94] and an aluminum borate matrix composite containing Fe particles [95].

In the hyperthermia treatment of cancers, bone-repairing materials in the form of bioactive glass–ceramics containing ferrimagnetic particles are useful as thermoseeds [96].

Lead zirconate titanate (PZT) is both ferroelectric and ferromagnetic, so its use, especially in a ferrite matrix, presents electric dipoles and magnetic dipoles to the electromagnetic radiation, thereby attaining high electromagnetic absorption ability [97,98].

A superconductor is characterized by the ability to exclude magnetic flux. This is known as the Meissner effect. A composite with a ferromagnetic matrix and a superconductor particle filler is potentially useful for novel magnetic devices, because the ferromagnetic matrix affects the flux exclusion of the superconductor filler [99].

9.5 Multilayers for magnetic applications

A *multilayer* is a composite in the form of a stack of two or more thin films which are deposited successively (one on top of another) on a substrate. When there are only two layers, the multilayer is known as a *bilayer*. Each layer is typically less than 2000 Å in thickness and is deposited by sputtering, vacuum evaporation or other vapor methods.

A common configuration involves the layers being alternately magnetic (e.g., ferromagnetic) and non-magnetic (e.g., paramagnetic), so that nearest magnetic layers are separated by non-magnetic layers [100,101]. The separation allows the nearest magnetic layers to be magnetized in opposite directions in the plane of the layers. This magnetization configuration causes the magnetoresistance to be high in a direction perpendicular to the layers. The multilayer is magnetically anisotropic; it is magnetic in the plane of the layers and non-magnetic in the direction perpendicular to the layers. Applications include magnetic data storage and magnetoresistive switching. For the latter application, both magnetic and non-magnetic layers must be electrically conducting.

Another configuration involves the layers being alternately a hard magnetic material (e.g., CoPt) and a soft magnetic material (e.g., Co) [102–104]. This configuration allows tailoring of the coercivity and other aspects of the magnetic hysteresis loop.

Yet another configuration involves magnetic (e.g., ferromagnetic) layers separated by electrically insulating layers, which are usually non-magnetic [105]. The purpose is to suppress energy loss due to the eddy current induced by a magnetic field.

A rather specialized configuration involves superconductor layers (e.g., niobium) separated by ferromagnetic layers (e.g., iron), so as to decouple the nearest superconductor layers for the purpose of tailoring the superconducting properties [106].

9.6 Magnetic composites for non-destructive evaluation

Structural composites are used in strategic applications such as aircraft and helicopter blades. Non-destructive monitoring of the structural health of the structural materials is needed in order to prevent hazards.

A method of non-destructive monitoring involves adding ferromagnetic particles to a structural composite during composite fabrication [107,108]. This is known as the *particle tagging* technique. It allows non-destructive evaluation to be performed, as a crack in the composite will cause distortion of the magnetic flux lines.

A related method for monitoring stress and defects involves adding magnetostrictive particles to a structural composite [107,109]. This technique is known as *magnetotagging*. It is also known as the particle tagging technique. The stress-monitoring ability is based on the ability of the magnetostrictive particles to exhibit a change in magnetic field when they are mechanically stressed. The defect-monitoring ability is based on the distortion of the magnetic flux lines (i.e., magnetic permeability variation) by the defects, as in the case of tagging using ferromagnetic particles.

9.7 Summary

Composites for magnetic application include metal–matrix, polymer–matrix and ceramic–matrix composites, in addition to multilayers. Magnetic and non-magnetic components are used in the composites in order to tailor the magnetic and electrical properties, and to improve processability. In addition, magnetic particles are added to structural composites for the purpose of non-destructive evaluation.

Example problems

1. How many Bohr magnetons of magnetic moment are associated with (a) a samarium (Sm) atom, (b) a terbium (Tb) atom?

Solution

(a) The electronic configuration of Sm is … $4f^6 5d^0 6s^2$.

   4f | ↑ | ↑ | ↑ | ↑ | ↑ | ↑ |

   There are six unpaired electrons, so $\beta = 6$.

(b) The electronic configuration of Tb is … $4f^9 5d^0 6s^2$.

   4f | ↑↓ | ↑↓ | ↑ | ↑ | ↑ | ↑ | ↑ |

   There are five unpaired electrons, so $\beta = 5$.

2. What is the magnetic induction B for a magnetic field strength H of 80 kA/m in (a) a vacuum, (b) a material with relative permeability $\mu_r = 5000$? (c) What is the magnetization M in (b)?

Solution

(a) In a vacuum:

$$B = \mu_o H$$
$$= (4\pi \times 10^{-7} \text{ H/m})(80 \text{ kA/m})$$
$$= 1.0 \times 10^{-4} \text{ H.kA/m}^2$$
$$= 1.0 \times 10^{-1} \text{ H.A/m}^2.$$

Since $H = wb/A$,

$$H \cdot A/m^2 = \frac{wb}{A} \cdot \frac{A}{m^2} = wb/m^2.$$

Hence,

$$B = 1.0 \times 10^{-1} \text{ wb/m}^2.$$

(b) In a material with $\mu_r = 5000$:

$$B = \mu H$$
$$= \mu_o \mu_r H$$
$$= (4\pi \times 10^{-7} \text{ H/m}) (5000) (80 \text{ kA/m})$$
$$= 500 \text{ wb/m}^2.$$

(c) $$B = \mu_o (H + M) = \mu_o H + \mu_o M.$$

Hence,

$$M = \frac{B - \mu_o H}{\mu_o}$$

$$= \frac{B}{\mu_o} - H$$

$$= \frac{500 \text{ wb/m}^2}{4\pi \times 10^{-7} \text{ H/m}} - 80 \text{ kA/m}$$

$$= 4 \times 10^8 \text{ wb/H} \cdot \text{m} - 8 \times 10^4 \text{ A/m}.$$

Since $H = wb/A$,

$$\frac{wb}{H \cdot m} = \frac{wb}{(wb/A) \cdot m} = A/m$$

Hence,

$$M = 4 \times 10^8 \text{ A/m} - 8 \times 10^4 \text{ A/m}$$
$$= 4 \times 10^8 \text{ A/m}.$$

Note that $\mu_o H$ is negligible, so that $B \sim \mu_o M$. This is because μ_r is large.

3. What is the magnetic moment in 1 cm³ of ferrite (Fe_3O_4) that has been fully magnetized?

Solution

The saturation magnetization of ferrite is 5×10^5 A·m^{-1}.

$$\text{magnetic moment} = (\text{saturation magnetization}) (\text{volume})$$
$$= (5 \times 10^5 \text{ A} \cdot \text{m}^{-1}) (1 \times 10^{-6} \text{ m}^3)$$
$$= 0.5 \text{ A} \cdot \text{m}^2.$$

4. Magnesium oxide (MgO) consists of Mg^{2+} and O^{2-} ions. It exhibits the rock salt (NaCl) crystal structure, just like NiO (Figure 9.7). Estimate the magnetic

moment per unit cell of MgO by considering the contribution due to the spin of the electrons. Explain your answer.

Solution

Both Mg^{2+} and O^{2-} ions have full shell electronic configurations, so there are no unpaired electrons for either ion. Hence, the magnetic moment is estimated to be zero.

5. Metallic iron is ferromagnetic and is body-centered cubic (BCC) in crystal structure, with lattice parameter (length of an edge of the cubic unit cell) 2.8665 Å. Estimate the maximum magnetic moment per unit volume (i.e., saturation magnetization) by considering the contribution due to the spin of the electrons.

Solution

From Example 4 in Section 9.1, the magnetic moment per Fe atom is 4 (9.27 × 10^{-24} A.m²). There are two atoms per BCC unit cell. Thus,

$$\text{magnetic moment per unit volume}$$
$$= \frac{\text{magnetic moment per unit cell}}{\text{volume of a unit cell}}$$
$$= \frac{(2)(4)(9.27 \times 10^{-24} \text{A.m}^2)}{(2.8665 \times 10^{-10} \text{m})^3}$$
$$= 3.15 \times 10^6 \text{ A.m}^{-1}.$$

Review questions

1. What is meant by a Bohr magneton?
2. Give an example of a ferromagnetic material.
3. Give an example of a ferrimagnetic material.
4. Give an example of an antiferromagnetic material.
5. What is the difference between a hard magnetic material and a soft magnetic material?
6. What is an advantage for a magnetic material to be electrically non-conductive?
7. Why are a hard magnetic material and a soft magnetic material used together in a composite material?
8. Name an application for a multilayer involving alternately magnetic and non-magnetic layers.
9. Why are ferromagnetic particles added to a structural composite?

References

[1] F.T. Calkins and A.B. Flatau, *Proc. – National Conference on Noise Control Engineering*, Inst. of Noise Control Engineering, Poughkeepsie, NY, 1997, 2, 373–382.

[2] F. Claeyssen, N. Lhermet, R. Le Letty and P. Bouchilloux, *J. of Alloys & Compounds*, 1997, **258**(1–2), 61–73.
[3] M.R.J. Gibbs, R. Watts, W. Karl, A.L. Powell and R.B. Yates, *Sensors & Actuators A – Physical*, 1997, **59**(1–3), 229–235.
[4] J.R. Cullen, *Scripta Metallurgica et Materialia*, 1995, **33**(10–11), 1849–1867.
[5] A.E. Clark, *J. of Intelligent Mater. Syst. & Struct.*, 1993, **4**(1), 70–75.
[6] N.C. Koon, C.M. Williams and B.N. Das, *J. of Magnetism & Magnetic Mater.*, 1991, **100**(1–3), 173–185.
[7] F. Bucholtz, D.M. Dagenais, K.P. Koo and S. Vohra, *Proc. SPIE – International Society for Optical Engineering*, Int. Soc. for Optical Engineering, Bellingham, WA, 1990, **1367**, 226–235.
[8] M.R.J. Gibbs, *J. of Magnetism & Magnetic Mater.*, 1990, **83**(1–3), 329–333.
[9] E. du Tremolet de Lacheisserie, K. Mackay, J. Betz and J.C. Peuzin, *J. of Alloys & Compounds*, 1998, **275-277**, 685–691.
[10] M.A. Subramanian, A.P. Ramirez and G.H. Kwei, *Solid State Ionics*, 1998, **108**(1–4), 185–191.
[11] J.P. Heremans, *Magnetic Ultrathin Films, Multilayers and Surfaces*, Materials Research Society Symposium Proceedings, MRS, Warrendale, PA, 1997, **475**, 63–74.
[12] P.B. Allen, *Solid State Communications*, 1997, **102**(2–3), 127–134.
[13] P.M. Levy and S. Zhang, *J. of Magnetism & Magnetic Mater.*, 1996, **164**(3), 284–292.
[14] W. Schwarzacher and D.S. Lashmore, *IEEE Transactions on Magnetics*, 1996, **32**(4), pt 2, 3133–3153.
[15] A. Barthelemy, V. Cros, J.L. Duvail, A. Fert, R. Morel, F. Parent, F. Petroff and L.B. Steren, *Nanostructured Mater.*, 1995, **6**(1–4), 217–226.
[16] R. von Helmolt, J. Wecker, K. Samwer and K. Baerner, *J. of Magnetism & Magnetic Mater.*, 1995, **151**(3), 411–416.
[17] P.M. Levy and S. Zhang, *J. of Magnetism & Magnetic Mater.*, 1995, **151**(3), 315–323.
[18] A. Fert, P. Grunberg, A. Barthelemy, F. Petroff and W. Zinn, *J. of Magnetism & Magnetic Mater.*, 1995, **140-144**(pt 1), 1–8.
[19] B. Dieny, *J. of Magnetism & Magnetic Mater.*, 1994, **136**(3), 335–359.
[20] W.P. Pratt Jr, S-F. Lee, P. Holody, Q. Yang, R. Loloee, J. Bass and P.A. Schroeder, *J. of Magnetism & Magnetic Mater.*, 1993, **126**(1–3), 406–409.
[21] J. Heremans, *J. of Phys. D – Appl. Phys.*, 1993, **26**(8), 1149–1168.
[22] S. Bednarek, *Mater. Sci. & Eng. B – Solid State Mater. for Adv. Tech.*, 1998, (3), 196–201.
[23] A.P. Ramirez, *J. of Phys. – Condensed Matter*, 1997, **9**(39), 8171–8199.
[24] D. Yu. Godovsky, A.V. Varfolomeev, G.D. Efremova, V.M. Cherepanov, G.A. Kapustin, A.V. Volkov and M.A. Moskvina, *Adv. Mater. Optics Electronics*, 1999, **9**(3), 87–93.
[25] R. Grossinger, H. Hauser, M. Dahlgren and J. Fidler, *Physica B: Condensed Matter* 2000, **275**(1–3), 248–252.
[26] R. Fischer, T. Schrefl, H. Kronmueller and J. Fidler, *J. Magnetism & Magnetic Mater.*, 1996, **153**(1–2), 35–49.
[27] N. Hiratsuka, K. Saito, H. Kobayashi and T. Mitamura, *J. Japan Soc. Powder & Powder Metallurgy*, 1993, **40**(10), 998–1001.
[28] J. Tang, C.E. O'Connor and L. Feng, *J. Alloys & Compounds*, 1998, **275-277**, 606–610.
[29] Y. Pan and G. Sun, *Scripta Materialia*, 1999, **41**(8), 803–807.
[30] D.S. Geoghegan, P.G. McCormick and R. Street, *Mater. Sci. Forum*, 1995, **179-181**, 629–634.
[31] J. Sort, J. Nogues, X. Amils, S. Surinach, J.S. Munoz and M.D. Baro, *Mater. Res. Soc. Symp. Proc.*, 2000, **581**, 641–646.
[32] J. Sort, J. Nogues, X. Amils, S. Surinach, J.S. Munoz and M.D. Baro, *Mater. Sci. Forum*, 2000, **343**, 812–818 (2000).
[33] M. Abe and S. Oie, *R&D, Res. & Development* (Kobe Steel, Ltd), 1987, **37**(3), 34–36.
[34] K. Okimoto, T. Satoh and N. Horiishi, *J. Japan Soc. Powder & Powder Metallurgy*, 1988, **35**(2), 47–52.
[35] K. Okimoto, T. Satoh, H. Matsuyama and M. Oka, *J. Japan Soc. Powder & Powder Metallurgy*, 1991, **38**(5), 673–680.
[36] K. Shin, C.R. Wong and S.H. Whang, *Mater. Sci. & Eng. A – Struct. Mater.: Properties, Microstructure & Processing*, 1993, (1), 35–43.
[37] R.L. Schalek, D.L. Leslie-Pelecky, J. Knight, D.J. Sellmyer and S.C. Axtell, *IEEE Transactions on Magnetics*, 1995, **31**(6), pt 2, 3772–3774.
[38] T.A. Yamamoto, K. Nishimaki, T. Harabe, K. Shiomi, T. Nakagawa and M. Katsura, *Nanostructured Mater.*, 1999, **12**(1), 523–526.
[39] T. Nakayama, T.A. Yamamoto, Y.-H. Choa and K. Niihara, *J. Mater. Sci.*, 2000, **35**(15), 3857–3861.
[40] V. Franco, X. Batlle and A. Labarta, *J. Magnetism & Magnetic Mater.*, 2000, **210**(1–3), 295–301.

[41] M. Sugimoto, *4th Int. Conf. on Ferrites*, Part 1, Advances in Ceramics, American Ceramic Society, Columbus, OH, 1984, **15**, 5-10.
[42] Y. Iijima, Y. Houjou and R. Sato, *IEEE Int. Symp. Electronic Compatibility*, IEEE, Piscataway, NJ, 2000, **2**, 547-549.
[43] S. Labbe and P.-Y. Bertin, *J. Magnetism & Magnetic Mater.*, 1999, **206**(1), 93-105.
[44] T. Tsutaoka, T. Kasagi, K. Hatakeyama, *J. European Ceramic Soc.*, 1999, **19**(6-7), 1531-1535.
[45] R.F. Ziolo, E.P. Giannelis, B.A. Weinstein, M.P. O'Horo, B.N. Ganguly, V. Mehrotra, M.W. Russell and D.R. Huffman, *Science*, 1992, **257**(5067), 219-223.
[46] F.G. Jones, T.R. Shrout, S.-J. Jang and M.T. Lanagan, *1990 IEEE 7th Int. Symp. on Applications of Ferroelectrics*, IEEE Service Center, Piscataway, NJ, 1990, pp. 455-458.
[47] M. M. Khvorov, T.N. Amelichkina, E.P. Zhelibo and V.E. Vember, *Soviet Powder Metallurgy & Metal Ceramics*, 1991, **30**(1), 80-84.
[48] I.M. Papisov, Y.S. Yablokov and A.I. Prokofev, *Vysokomolekularnye Soedineniya Seriya A*, **36**(2), 352-354 (1994).
[49] G. Viau, f. Ravel, O. Acher, F. Fievet-Vincent and F. Fievet, *J. Magnetism & Magnetic Mater.*, 1995, **140-144**(pt 1), 377-378.
[50] T.I. Fil', L.S. Radkevich, M.A. Pashkov and A.M. Savitskij, *Poroshkovaya Metallurgiya*, 1993, (9-10), 81-84.
[51] H.S. Gokturk, T.J. Fiske and D.M. Kalyon, *IEEE Transactions on Magnetics*, 1993, **29**(6), pt 3, 4170-4176.
[52] S. Bednarek, *Appl. Phys. A: Mater. Sci. & Processing*, 1998, **66**(6), 643-650.
[53] S. Bednarek, *J. Magnetism & Magnetic Mater.*, 1998, **188**(1-2), 71-84.
[54] S. Bednarek, *Mater. Sci. & Eng. B - Solid State Mater. for Adv. Tech.*, 1998, (3), 201-209.
[55] S. Bednarek, *Mater. Sci. & Eng. B - Solid State Mater. for Adv. Tech.*, 1999, **63**(3), 228-233.
[56] S. Bednarek, *Appl. Phys. A - Mater. Sci. & Processing*, 1999, **68**(1), 63-67.
[57] S. Bednarek, S. Bednarek, *Mater. Sci. & Eng. B - Solid State Mater. for Adv. Tech.*, 2000, **77**(1), 120-127.
[58] S. Bednarek, *Appl. Phys. A - Mater. Sci. & Processing*, 1997, **163**(1), 181-193.
[59] J. Liu and M. Wan, *J. Polym. Sci. Part A - Polym. Chem.*, 2000, **38**(15), 2734-2739.
[60] M. Wan, W. Zhou and J. Li, *Synthetic Metals*, 1996, **78**(1), 27-31.
[61] M. Wan and J. Fan, *J. Polym. Sci. Part A - Polym. Chem.*, 1998, **36**(15), 2749-2755.
[62] M. Wan and J. Liu, *J. Polym. Sci. Part A - Polym. Chem.*, 1998, **36**(15), 2799-2805.
[63] M.X. Wan and J.C. Li, *Synthetic Metals*, 1999, **101**(1), 844-845.
[64] T.J. Fiske, H.S. Gokturk and D.M. Kalyon, *J. Mater. Sci.*, 1997, 32(20), 5551-5560.
[65] P. Lederer and C. Brewit-Taylor, *IEE Proc. A - Sci. Measurement & Tech.*, 2000, **147**(4), 209-211.
[66] S.V. Mikhajlin and Y.A. Lapshin, *Elektrotekhnika*, 1996, (11), 16-18.
[67] J. Slama, R. Vicen, P. Krivosik, A. Gruskova and R. Dosoudil, *J. Magnetism & Magnetic Mater.*, 1999, **196**, 359-361.
[68] S.A. Sosnina, I.A. Polunina and A.A. Baranenko, *Colloid J. Russian Academy of Sci.*, 1999, **61**(6), 776-780.
[69] T. Yogo, T. Nakamura, W. Sakamoto and S.-I. Hirano, *J. Mater. Res.*, 1999, **14**(7), 1855-1860.
[70] R. Mathur, D.R. Sharma, S.R. Vadera, S.R. Gupta, B.B. Sharma and N. Kumar, *Nanostructured Mater.*, 1999, **11**(5), 677-686.
[71] M. Ata, M. Machida, H. Watanabe and J. Seto, *Japanese J. Appl. Phys. Part 1 - Regular Papers Short Notes & Review Papers*, 1865-1871.
[72] B. Weglinski, *Reviews on Powder Metallurgy & Physical Ceramics*, 1990, **4**(2), 79-154.
[73] P. Sankaran, V.J. Kumar and V. Jayashankar, *IEEE Transactions on Magnetics*, 1990, **26**(6), 3086-3088.
[74] B.Z. Janos and N.W. Hagood, *Proc. SPIE - Int. Soc. Optical Eng.*, 1999, **3675**, 10-21.
[75] T. Fiske, H.S. Gokturk, R. Yazici and D.M. Kalyon, *Polym. Eng. & Sci.*, 1997, **37**(5), 826-837.
[76] R.M. Walser and W. Kang, *IEEE Transactions on Magnetics*, 1998, **34**(4) pt 1, 1144-1146.
[77] O. Acher, P. Le Gourrierec, G. Perrin, P. Baclet and O. Roblin, *IEEE Transactions on Microwave Theory & Techniques*, 1996, **44**(5), 674-684.
[78] K. Baba, F. Takase and M. Miyagi, *Optics Communications*, 1997, **137**(1-3), 35-38.
[79] B.Z. Rameev, B. Aktas, R.I. Khaibullin, V.A. Zhikharev, Y.N. Osin and I.B. Khaibullin, *Vacuum*, 2000, **58**(2), 551-560.
[80] M. Verelst, K.R. Kannan, G.N. Subbanna, C.N.R. Rao, C. Laurent and A. Rousset, *J. Mater. Res.*, 1992, 7(11), 3072-3079.
[81] A. Marchand, X. Devaux, B. Barbara, P. Mollard, A. Rousset and M. Brieu, *J. Mater. Sci.*, 1993, **28**(8), 2217-2226.
[82] M. Verelst, K.R. Kannan, G.N. Subbanna, C.N.R. Rao, M. Brieu and A. Rousset, *Mater. Res. Bull.*,

1993, **28**(4), 293-303.
[83] T. Sekino, T. Nakajima and K. Niihara, *Mater. Lett.*, 1996, **29**(1-3), 165-169.
[84] T. Sekino, T. Nakajima, S. Ueda and K. Niihara, *J. Amer. Ceramic Soc.*, 1997, **80**(5), 1139-1148.
[85] T. Sekino, S. Etoh, Y.-H. Choa and K. Niihara, *Proc. 1997 MRS Fall Symposium on Surface-Controlled Nanoscale Materials for High-Added-Value Applications*, Materials Research Society, Warrendale, PA, 1998, **501**, 289-294.
[86] T. Kojima, W. Sakamoto, T. Yogo, T. Fujii and S. Hirano, *Nippon Seramikkusu Kyokai Gakujutsu Ronbunshi - J. Ceramic Soc. Japan*, 1999, **107**(1249), 796-800.
[87] Y. Suzuki, M. Awano, N. Kondo, T. Ohji, *J. Amer. Ceramic Soc.*, 1999, **82**(9), 2557-2559.
[88] Y. Suzuki, M. Awano, N. Kondo, T. Ohji, *J. Japan Soc. Powder & Powder Metallurgy*, 2000, **47**(2), 208-212.
[89] L. Maya, M. Paranthaman, J.R. Thompson, T. Thundat and R.J. Stevenson, *Proc. 1996 MRS Fall Symp. on Nanophase and Nanocomposite Materials II*, Materials Research Society, Pittsburgh, PA, 1997, **457**, 213-218.
[90] N. Tanaka, F. Yoshizaki, K. Katzuda and K. Mihama, *Acta Metallurgica et Materialia*, 1992, **40**(Suppl.), S275-S280.
[91] S. Matsuo, T. Matsuura, I. Nishida and N. Tanaka, *Japanese J. Appl. Phys. Part 1 - Regular Papers Short Notes & Review Papers*, 1994, **33**(7A), 3907-3912.
[92] V. Vendange, E. Flavin and P. Colomban, *J. Mater. Sci. Lett.*, 1996, **15**(2), 137-141.
[93] A. Yasumori, T. Katsuyama, Y. Kameshima and K. Okada, *J. Sol-Gel Sci. & Tech.*, 2000, **19**(1-3), 237-242.
[94] M.R. Ayers, X.Y. Song and A.J. Hunt, *J. Mater. Sci.*, 1996, **31**(23), 6251-6257.
[95] S. Guggilla and A. Manthiram, *Mater. Sci. & Eng. B - Solid State Mater. Adv. Tech.*, 1996, (2-3), 191-197.
[96] T. Kokubo, *Mater. Sci. Forum*, 1999, **293**, 65-82.
[97] W. Huang, J. Chen, M. Tu, Y. Liu, J. Mao, S. Deng, Q. Chen and W. Jiao, *Gongneng Cailiao/J. Functional Mater.*, 1996, **27**(5), 431-433.
[98] J. Chen, W. Huang, M. Tu, Y. Liu and J. Mao, *Fuhe Cailiao Xuebao/Acta Materiae Compositeae Sinica*, 1997, **14**(3), 41-44.
[99] R.B. Flippen, *Solid State Communications*, 1992, **81**(1), 105-107.
[100] L.A. Chebotkevich, Y.D. Vorob'ev, E.V. Il'in, A.I. Kuznetsov and I.M. Slabszennikova, *Fizika Metallov i Metallovedenie*, 1994, **77**(5), 73-76.
[101] S. Kikkawa, *Mater. Trans. Jim.*, 1996, **37**(3), 420-425.
[102] S. Li, C. Wang, W. Bao, P. Wang and R. Ma, *J. Rare Earths*, 1998, **16**(1), 25-31.
[103] J. Kim, K. Barmak, L.H. Lewis, D.C. Crew and D.O. Welch, *Mater. Res. Soc. Symp. Proc.*, 1999, **562**, 327-332.
[104] J. Kim, K. Barmak, L.H. Lewis, D.C. Crew and D.O. Welch, *Mater. Res. Soc. Symp. Proc.*, 1999, **577**, 353-358.
[105] J.L. Wallace, *6th IEEE Pulsed Power Conf. Dig. Tech. Papers*, IEEE Service Center, Piscataway, NJ, 1987, p. 17-20.
[106] A. Orito, A. Fukushima, S. Katsumoto and Y. Iye, *Solid-State Electronics*, 1998, **42**(7-8), 1481-1488.
[107] V. Giurgiutiu, Z. Chen, F. Lalande, C.A. Rogers, R. Quattrone and J. Berman, *J. Intelligent Mater. Sys. & Struct.*, 1996, **7**(6), 623-634.
[108] W.G. Clark, Jr, *Int. SAMPE Tech. Conf.*, SAMPE, Covina, CA, 1991, **23**, 233-246.
[109] G.N. Weisensel, *Mater. Tech.*, 1995, **10**(7-8), 142-144.

10 Composite materials for electrochemical applications

10.1 Background on electrochemical behavior

Electrochemical behavior pertains to chemical processes brought about by the movement of charged species (ions and electrons) under the influence of an electric field. It occurs when there are ions that can move in a medium (liquid or solid) due to an electric field (i.e., a voltage gradient). The positive ions (cations) move toward the negative end of the voltage gradient, while the negative ions (anions) move toward the positive end of the voltage gradient (Figure 10.1). The medium containing the movable ions is called the *electrolyte*. The electrical conductors in contact with the electrolyte for the purpose of applying the electric field are called electrodes. The electrode at the positive end of the voltage gradient is called the *positive electrode* or *anode*. The electrode at the negative end of the voltage gradient is called the *negative electrode* or *cathode*. The electrodes must be sufficiently inert so that they do not react with the electrolyte. In order to apply the electric field, the electrodes are connected to a DC power supply (or a battery) such that the cathode is connected to the negative end of the power supply (so that the cathode becomes negative) and the anode is connected to the positive end of the battery (so that the anode becomes positive). In this way, electrons flow in the electrical leads (the outer circuit) from the anode to the cathode, while ions flow in the electrolyte. The electron flow in the outer circuit and the ion flow in the electrolyte constitute a loop of charge flow. Note that the anions flow from cathode to anode in the electrolyte, while the electrons flow from anode to cathode in the outer circuit and both anions and electrons are negatively charged.

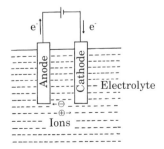

Figure 10.1 Flow of electrons in the outer circuit from anode to cathode and flow of cations toward cathode and of anions toward anode in the electrolyte in response to applied voltage, which is positive at the anode.

At the anode, a chemical reaction occurs that gives away one or more electrons. In general, a reaction involving the loss of one or more electrons is known as an *oxidation reaction*. The electron lost through this reaction supplies the anode with the electron to release to the outer circuit. An example of an oxidation (anodic) reaction is

$$Cu \rightarrow Cu^{2+} + 2e^-, \qquad (10.1)$$

where Cu is the anode, which is oxidized to Cu^{2+}, thereby becoming corroded and releasing two electrons to the outer circuit. The Cu^{2+} ions formed remain in the electrolyte, while the electrons go through the anode (which must be electrically conducting) to the outer circuit.

At the cathode, a chemical reaction occurs that takes one or more electrons from the cathode. In general, a reaction involving the gain of one or more electrons is known as a *reduction reaction*. The electron gained through this reaction is supplied to the cathode (which must be electrically conducting) by the outer circuit. Always, oxidation occurs at the anode and reduction occurs at the cathode.

The product of the reduction reaction at the cathode may be a species coated on the cathode, as in the electroplating of copper, where the reduction (cathodic) reaction is

$$Cu^{2+} + 2e^- \rightarrow Cu, \qquad (10.2)$$

where Cu^{2+} is the cation in the electrolyte and Cu is the product of the reduction reaction and is the species electroplated on the cathode. The product of the reduction reaction may be OH^- ions, as in the case when oxygen and water are present (a very common situation), i.e.,

$$O_2 + 2 H_2O + 4e^- \rightarrow 4 (OH^-). \qquad (10.3)$$

The product of the reaction may be H_2 gas, as in the case when H^+ (in an acid) ions are present, i.e.,

$$2H^+ + 2e^- \rightarrow H_2 \uparrow. \qquad (10.4)$$

The product of the reaction may be H_2 gas together with OH^- ions, as in the case when H_2O is present without O_2 (an anerobic condition), i.e.,

$$2H_2O + 2e^- \rightarrow H_2 \uparrow + 2(OH)^-. \qquad (10.5)$$

The product of the reaction may be water, as in the case when O_2 and H^+ (an acid) are present, i.e.,

$$O_2 + 4H^+ + 4e^- \rightarrow 2H_2O. \qquad (10.6)$$

Which of the abovementioned five cathodic reactions takes place depends on the chemical species present. The most common of the five reactions is (10.3), as oxygen and water are present in most environments. The variety of cathodic reactions is much greater than that of anodic reactions.

The anodic and cathodic reactions together make up the overall reaction. For example, the anodic reaction (10.1) and cathodic reaction (10.3) together become

$$\begin{array}{r} 2Cu \rightarrow 2Cu^{2+} + 4e^- \\ \underline{O_2 + 2H_2O + 4e^- \rightarrow 4(OH^-)} \\ 2Cu + O_2 + 2H_2O \rightarrow 2Cu^{2+} + 4(OH^-) \end{array}$$

A Cu^{2+} ion and two OH^- ions formed by the overall reaction react to form $Cu(OH)_2$, which resides in the electrolyte.

Both anode and cathode must be electrically conducting, but they do not have to participate in the anodic or cathodic reaction. The anodic reaction (10.1) involves the anode (Cu) as the reactant, but the cathodic reactions (10.2), (10.3), (10.4), (10.5) and (10.6) do not involve the cathode. Thus, the cathode is usually a platinum wire or other electrical conductor that does not react with the electrolyte.

The electrolyte must allow ionic movement in it. It is usually a liquid, although it can be a solid. For the cathodic reaction (10.2) to occur, the electrolyte must contain Cu^{2+} ions (e.g., a copper sulfate or $CuSO_4$ solution). For the cathodic reaction (10.4) to occur, the electrolyte must contain H^+ ions (e.g., a sulfuric acid or H_2SO_4 solution). The anodic reaction occurs at the interface between the anode and the electrolyte; the cathodic reaction occurs at the interface between the cathode and the electrolyte. The ions that are either reactants (e.g., H^+ in reaction (10.4)) or reaction products (e.g., OH^- in reaction (10.3)) must move to or from the electrode/electrolyte interface.

Whether an electrode ends up being the anode or the cathode depends on the relative propensity for oxidation and reduction at each electrode. The electrode at which oxidation has higher propensity than at the other electrode becomes the anode and the other electrode becomes the cathode. The propensity depends on the types and concentrations of chemical species present.

Different species have different propensities to be oxidized. For example, iron (Fe → Fe^{2+} + 2e^-) has more propensity for oxidation than copper (Cu → Cu^{2+} + 2e^-), when the Fe^{2+} and Cu^{2+} ions in the electrolyte are at the same concentration, so when Fe and Cu are the electrodes and the Fe^{2+} and Cu^{2+} ion concentrations are the same in the electrolyte, Fe will become the anode and Cu will become the cathode (Figure 10.2). Fe and Cu constitute a *galvanic couple* and this set-up constitutes a *galvanic cell* or a *composition cell*. The anodic reaction will be Fe → Fe^{2+} + 2e^-, while the cathodic reaction will be (10.2), (10.3), (10.4), (10.5) or

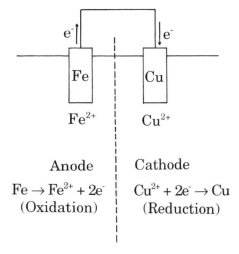

Figure 10.2 Composition cell with Fe as the anode and Cu as the cathode. Note that electrons and ions flow without the need for an applied voltage between anode and cathode.

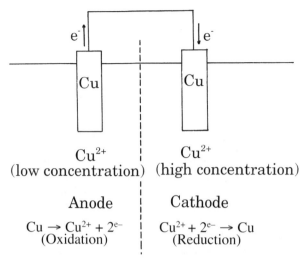

Figure 10.3 Concentration cell with Cu electrodes, a low Cu^{2+} concentration at the anode side and a high Cu^{2+} concentration at the cathode side.

(10.6), depending on the relative propensity for these competing reduction reactions.

The concentrations of the species also affect the propensity. In general, high concentrations of the reactants increase the propensity for a reaction, and high concentrations of the reaction products decrease the propensity for a reaction. This is known as *Le Chatelier's principle*. Thus, a high concentration of Cu^{2+} in the electrolyte increases the propensity for reaction (10.2). When copper forms both electrodes, but the Cu^{2+} concentrations are different in the vicinities of the two electrodes, then the electrode with the higher Cu^{2+} ion concentration will become the cathode (i.e., high propensity for reaction (10.2), which is the cathodic reaction) and the electrode with the lower Cu^{2+} ion concentration will become the anode (i.e., high propensity for reaction (10.1), which is the anodic reaction)

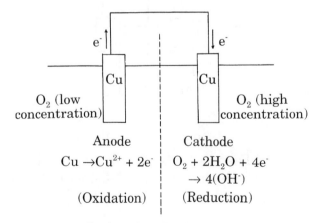

Figure 10.4 Oxygen concentration cell with Cu electrodes, a low O_2 concentration at the anode side and a high O_2 concentration at the cathode side.

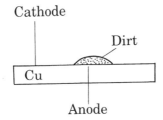

Figure 10.5 A piece of copper partly covered by dirt is an oxygen concentration cell.

(Figure 10.3). This set-up involving a difference in concentration constitutes a *concentration cell*. Similarly, when copper forms both electrodes, but the O_2 concentrations are different in the vicinities of the two electrodes (Figure 10.4), then the electrode with the higher O_2 concentration will become the cathode (i.e., high propensity for reaction (10.3), which is the cathodic reaction) and the electrode with the lower O_2 concentration will become the anode (i.e., low propensity for reaction (10.3), forcing the electrode to become the anode, with reaction (10.1) being the anodic reaction). A difference in oxygen concentration is commonly encountered, since oxygen comes from air and access to air can be varied. Therefore, a concentration cell involving a difference in O_2 concentration is particularly common and is called an *oxygen concentration cell*. For example, a piece of copper with dirt covering a part of it (Figure 10.5) is an oxygen concentration cell, because the part of the copper covered by dirt has less access to air and becomes the anode (with anodic reaction (10.1)), thus forcing the uncovered part of the copper to be the cathode (with cathodic reaction (10.3)). Hence, the covered part gets corroded. Similarly, a piece of metal (an electrical conductor) having a crack in it (Figure 10.6) is an oxygen concentration cell, because the part of the metal at the crack has less access to air and becomes the anode (with anodic reaction (10.1)), thus forcing the part of the metal that is exposed to open air to be the cathode.

When the two electrodes are open circuited at the outer circuit, a voltage (called the *open-circuit voltage*) exists between the two electrodes. This voltage describes the difference in the propensity for oxidation between the two electrodes, so it is considered the driving force for the cathodic and anodic reactions that occur when the electrodes are short circuited at the outer circuit. The greater is the difference in oxidation propensity, the higher is the open-circuit voltage. This is the same principle behind batteries, which provide a voltage as the output. Because the voltage pertains to the difference between a pair of electrodes, it is customary to describe the voltage of any electrode relative to the same electrode. In this way, each electrode is associated with a voltage and a scale is established

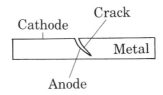

Figure 10.6 A piece of metal with a crack is an oxygen concentration cell.

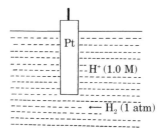

Figure 10.7 The standard hydrogen reference half-cell.

for ranking various electrodes in terms of the oxidation propensity. The electrode that serves as the reference involves the oxidation reaction

$$H_2 \rightarrow 2H^+ + 2e^-$$

such that the H_2 gas is at a pressure of 1 atm, the H^+ ion concentration in the electrolyte is 1.0 M (1 M = 1 molar = 1 mole/liter) and the temperature is 25°C. In this reference, the electrode does not participate in the reaction, so it only needs to be an electrical conductor. Therefore, platinum is used. The platinum electrode is immersed in the electrolyte and H_2 gas is bubbled through the electrolyte, as illustrated in Figure 10.7. The electrode in Figure 10.7 is called the *standard hydrogen reference half-cell* (half of a cell). Relative to this reference half-cell, copper has a lower oxidation propensity (when the Cu^{2+} ion concentration is 1 M and the temperature is 25°C), such that the open circuit voltage is 0.340 V and the positive end of the voltage is at the cathode, which is copper (Figure 10.8). That the positive end of the open circuit voltage is at the cathode is because the anode has a higher oxidation propensity than the cathode and it "wants" to release electrons to the outer circuit, which cannot accept the electrons owing to the open circuit situation. The *standard electrode potential* for copper is thus +0.340 V. A ranking of the electrodes in terms of the standard electrode potential is shown in Table 10.1, which is known as the *standard electromotive force (emf) series*. Note that this ranking is for the case in which the ion concentration is 1 M and the temperature is 25°C. If the ion concentration or the temperature is changed, the ranking can be different. The higher the electrode is in Table 10.1, the more

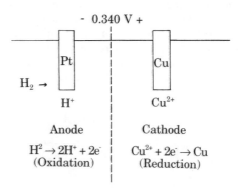

Figure 10.8 A cell with copper as the cathode and hydrogen half-cell as the anode under open circuit condition. The open circuit voltage is 0.340 V, positive at the cathode.

Table 10.1 Standard electromotive force series

	Electrode reaction	Standard electrode potential (V)
Increasingly inert (cathodic) ↑	$Au^{3+} + 3e^- \rightarrow Au$	+ 1.420
	$O_2 + 4H^+ + 4e^- \rightarrow 2H_2O$	+ 1.229
	$Pt^{2+} + 2e^- \rightarrow Pt$	~ + 1.2
	$Ag^+ + e^- \rightarrow Ag$	+ 0.800
	$Fe^{3+} + e^- \rightarrow Fe^{2+}$	+ 0.771
	$O_2 + 2H_2O + 4e^- \rightarrow 4(OH^-)$	+ 0.401
	$Cu^{2+} + 2e^- \rightarrow Cu$	+ 0.340
	$2H^+ + 2e^- \rightarrow H_2$	0.000
	$Pb^{2+} + 2e^- \rightarrow Pb$	− 0.126
	$Sn^{2+} + 2e^- \rightarrow Sn$	− 0.136
	$Ni^{2+} + 2e^- \rightarrow Ni$	− 0.250
	$Co^{2+} + 2e^- \rightarrow Co$	− 0.277
	$Cd^{2+} + 2e^- \rightarrow Cd$	− 0.403
	$Fe^{2+} + 2e^- \rightarrow Fe$	− 0.440
Increasingly active (anodic) ↓	$Cr^{3+} + 3e^- \rightarrow Cr$	− 0.744
	$Zn^{2+} + 2e^- \rightarrow Zn$	− 0.763
	$Al^{3+} + 3e^- \rightarrow Al$	− 1.662
	$Mg^{2+} + 2e^- \rightarrow Mg$	− 2.363
	$Na^+ + e^- \rightarrow Na$	− 2.714
	$K^+ + e^- \rightarrow K$	− 2.924

cathodic it is. The lower the electrode is in Table 10.1, the more anodic it is. For example, from Table 10.1, Fe is cathodic relative to Zn, and the open-circuit voltage between the Fe and Zn electrodes (called the *overall cell potential*) is

$$[(-0.440) - (-0.763)] \text{ V} = 0.323 \text{ V},$$

such that the positive end of the voltage is at the Fe electrode (Figure 10.9). Note from Figure 10.9 that the electrolyte in the vicinity of the Fe electrode (i.e., the right compartment of the electrolyte) has Fe^{2+} ions at 1.0 M, while the electrolyte in the vicinity of the Zn electrode (i.e., the left compartment of the electrolyte) has Zn^{2+} ions at 1.0 M. The two compartments are separated by a membrane, which allows ions to flow through (otherwise it would be open circuited within the electrolyte) but provides enough hindrance to the mixing of the electrolytes in the two compartments. The anode reaction in Figure 10.9 is

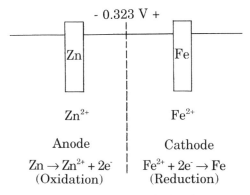

Figure 10.9 A cell with zinc as the anode and iron as the cathode under open circuit condition. The open circuit voltage is 0.323 V, positive at the cathode.

$$Zn \rightarrow Zn^{2+} + 2e^-.$$

The cathode reaction is

$$Fe^{2+} + 2e^- \rightarrow Fe.$$

The overall reaction is

$$Zn + Fe^{2+} \rightarrow Zn^{2+} + Fe.$$

The cell of Figure 10.9 is commonly written as

$$Zn \,|\, Zn^{2+} \,|\, Fe^{2+} \,|\, Fe, \qquad (10.7)$$

where Zn and Fe^{2+} are the reactants, Zn^{2+} and Fe are the reaction products, and the vertical lines denote phase boundaries.

10.2 Background on batteries

A *battery* is a number of galvanic cells connected in series, so that it gives a potential equal to the sum of the potentials given by the cells. In the connection, the anode of one cell is connected to the cathode of the adjacent cell. A battery allows the conversion of chemical energy to electrical energy. It is a source of direct current (DC).

A cell of the *dry cell* flashlight battery (the acid version) (Figure 10.10) consists of an outer zinc cylinder (the anode) and an inner carbon rod (the current collector), in addition to a moist paste of solid MnO_2 (the cathode), NH_4Cl acid (the electrolyte) and carbon (an electrically conductive additive for attaining a composite cathode that is conductive, since MnO_2 itself is not conductive). Carbon is used as the current collector because of its electrical conductivity and chemical inertness. The anode reaction is

$$Zn \rightarrow Zn^{2+} + 2e^-. \qquad (10.8)$$

The cathode reaction is

$$2NH_4^+ + 2MnO_2 + 2e^- \rightarrow Mn_2O_3 + 2NH_3 + H_2O. \qquad (10.9)$$

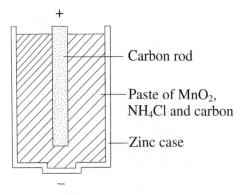

Figure 10.10 Dry cell.

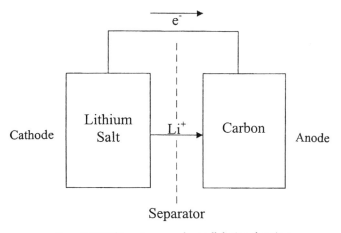

Figure 10.11 Lithium ion secondary cell during charging.

The anode reaction causes Zn^{2+} ions to leave the anode, thereby leaving electrons at the anode, which becomes negative in charge and serves as the negative terminal. The Zn^{2+} ions are attracted to the Cl^- ions from NH_4Cl. The cathode reaction involves NH_4^+ and MnO_2 pulling electrons from the carbon rod, which becomes positive in charge. The carbon rod does not participate in either anode or cathode reaction, so it is neither anode nor cathode, but is a current collector (the positive terminal). The reactions occur until the potential difference between the terminals is 1.5 V, at which no further reaction occurs because the potential difference across the terminals opposes further pulling of electrons from the carbon rod and further exit of Zn^{2+} ions from the anode.

Figure 10.10 shows the open-circuit state of the cell. When the cell is connected in a circuit, electrons exit from negative terminal to the outer circuit and arrive at the positive terminal. This allows further anodic and cathodic reactions to occur, thereby maintaining the cell voltage of 1.5 V.

Another important type of battery is the *nickel–cadmium battery*. Cadmium metal is the anode and nickel dioxide (NiO_2) is the cathode. In practice, the nickel dioxide is in the form of a coating on a nickel mesh. The electrolyte is water, in which both electrodes are immersed. The anode reaction is

$$Cd + 2OH^- \rightarrow Cd(OH)_2 + 2e^-. \tag{10.10}$$

The cathode reaction is

$$NiO_2 + 2H_2O + 2e^- \rightarrow Ni(OH)_2 + 2OH^-. \tag{10.11}$$

The reaction products $Cd(OH)_2$ and $Ni(OH)_2$ adhere to the respective electrodes. As a result, the cell can be recharged by running the reactions in the opposite direction.

A rechargeable cell is also called a *secondary cell*. One that is not rechargeable is called a *primary cell*. The state-of-the-art secondary cell is *the lithium-ion cell*. The cathode of this cell is a lithium salt (such as $LiMn_2O_4$, $LiCoO_2$ or $LiNiO_2$) in which lithium ions can be intercalated. (Intercalation refers to a chemical reaction in which a reactant goes into a host.) The anode is carbon, which can also be intercalated by lithium ions. During charging, lithium ions enter the anode (Figure 10.11). The reaction during charging at the carbon is

$$yC + xLi^+ + xe^- \to Li_xC_y. \tag{10.12}$$

During discharging, lithium ions enter the cathode. The overall reaction during charging for the case of $LiCoO_2$ being the cathode is

$$LiCoO_2 + yC \to Li_{1-x}CoO_2 + Li_xC_y. \tag{10.13}$$

The anode and cathode of a lithium-ion cell are separated by a porous polymer (e.g., polyethylene) separator. The electrodes and separator are immersed in an electrolyte, which can be an organic liquid (e.g., propylene carbonate) in which a lithium salt (e.g., $LiPF_6$) is dissolved.

10.3 Background on fuel cells

A *fuel cell* is a galvanic cell in which the reactants for the anodic and cathodic reactions are continuously being supplied. In the most common type of fuel cell, the *hydrogen–oxygen fuel cell*, hydrogen gas is the reactant for the anodic reaction, oxygen is the reactant for the cathodic reaction and H_2O is the overall reaction product. The anodic reaction is

$$2H_2 + 4OH^- \to 4H_2O + 4e^-. \tag{10.14}$$

The cathodic reaction is

$$4e^- + O_2 + 2H_2O \to 4OH^-. \tag{10.15}$$

The overall reaction, as obtained by summing Equations (10.14) and (10.15), is

$$2H_2 + O_2 \to 2H_2O. \tag{10.16}$$

Figure 10.12 shows a schematic of the hydrogen–oxygen fuel cell. Hydrogen is the anode. Oxygen is the cathode. The reactions need a catalyst (typically platinum), which is applied on a porous carbon that serves as the current collector. There are two current collectors: one for the anode side and one for the cathode side. The cathodic reaction produces OH^- ions, which move from the cathode side to the anode side, where they combine with hydrogen to provide the anodic reaction. Electrons produced by the anodic reaction are collected by the current collector at the anode side. They move through the outer circuit and enter the current collector at the cathode side, where the electrons are consumed to provide the cathodic reaction. The product of the overall reaction is H_2O (steam) which exits from the bottom of the cell. Between the two current collectors is the electrolyte, which can be phosphoric acid, potassium carbonate or a solid that allows ionic conduction.

The outermost plates that sandwich the rest of the components of a fuel cell are called *bipolar plates*. A bipolar plate serves as an electrical connection between single cell units that make up a stack. Moreover, it serves to supply reactant gases to the current collectors via a flow field in its pattern of surface ridges. Requirements for the bipolar plate material are low gas permeability, corrosion resistance, low weight, high strength and manufacturability. Graphite is conventionally used for bipolar plates, but carbon–carbon composites are receiving attention for this application, because of their high strength and low density.

Composite materials for electrochemical applications

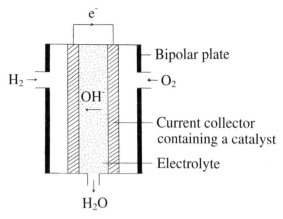

Figure 10.12 Hydrogen–oxygen fuel cell.

Fuel cells are important for electric vehicles. They are economically attractive owing to the low cost of the reactants.

10.4 Background on electric double-layer capacitors

Capacitors are used for electrical energy storage. Those based on dielectric materials are addressed in Section 7.8. This section addresses those based on electrochemistry.

An *electric double-layer capacitor* (also called an *electrochemical capacitor*, a *supercapacitor* or an *ultracapacitor*) consists of two electrodes immersed in an electrolyte and electrically connected to a power supply (Figure 10.13). The power supply causes one electrode to be positively charged, thus attracting the anions in the electrolyte. It causes the other electrode to be negatively charged, thus attracting the cations in the electrolyte. This results in a layer of negative ions on the positive electrode and a layer of positive ions on the negative electrode. This means the occurrence of polarization in the electrolyte. By using electrodes with a high specific surface area (e.g., activated carbon and, more exotically, single-walled carbon nanotubes [1]), the capacitance can be substantial. However, this type of capacitor cannot operate at high AC frequencies, owing to the sluggishness of ion movement in the electrolyte.

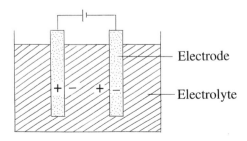

Figure 10.13 Electric double-layer capacitor.

Electric double-layer capacitors are advantageous compared to secondary batteries for energy storage in their long cycle life (>100 000 cycles), simple principle and construction, short charging time, safety and high power density (10 times that of a secondary battery), although they suffer in their lower energy density [1]. (Note that power is energy per second.) As the stored energy is given by $CV^2/2$, where C is the capacitance and V is the voltage, a high capacitance is valuable for increasing the energy density.

10.5 Composite materials for electrodes and current collectors

Composite materials are used for electrodes, including anodes and cathodes, as well as current collectors. Some of these materials are designed to enhance electrochemical behavior, such as capacity and rechargeability in the case of a battery. Some (such as those containing a conductive additive) are designed to render the electrode more conductive (electrically and/or thermally), as some electrode materials are by themselves not conductive. Yet some other composites (such as those containing fibers or a polymer binder) are designed to enhance the processability, handleability, chemical stability and electrolyte absorptivity.

10.5.1 Composites for improved electrochemical behavior

Composites for improved electrochemical behavior have been designed for batteries, fuel cells and electric double-layer capacitors. The design is aimed at increasing the capacity for conducting electrochemical reactions and for accommodating reaction products, and improving the electrode–electrolyte interface, in the case of batteries. It is also aimed at extending the service lifetime of electrodes and current collectors in the case of batteries and fuel cells, and increasing the specific surface area of the electrodes in the case of electric double-layer capacitors.

In the area of lithium-ion cells, composites for improved electrochemical behavior include cathodes based on a conducting polymer (e.g., polypyrrole) and an electrochemically active material (e.g., $LiMn_2O_4$) [2,3], anodes based on tin or $SnSb_x$, preferably containing $Li_{2.6}Co_{0.4}N$ as an additive [4–6], anodes involving an inactive matrix (e.g., $SnFe_3C$) and an active filler (Sn_2Fe) [7], electrodes based on vanadium pentoxide (V_2O_5) [8], cathodes based on manganese dioxide and containing nickel [9], nanometer-size $Na_{0.7}MnO_2$ particles [10] or carbon [11], anodes based on carbon and containing silver [12], tin [13], montmorillonite [14], silicon [15,16] or nickel [17], and electrodes containing Ta_2O_5 and TiO_2 [18].

In the area of fuel cells, composites for improved electrochemical behavior [19] include current collectors based on yttria-stabilized zirconia, preferably containing nickel [20] or lathanum strontium manganese oxide ($La_{0.9}Sr_{0.1}MnO_3$) [20–24].

In the area of electric double-layer capacitors, composite electrodes include those based on ruthenium oxide (RuO_2) [25–27], preferably together with carbon [28–30]. Both ruthenium oxide and carbon are available in the form of aerogels, which have a high specific surface area.

10.5.2 Composites for improved conductivity

Electrical conductivity is essential for electrodes and current collectors. Thermal conductivity is also desired, especially for batteries that operate at a high rate (i.e., a high current), owing to the heat generated by the battery operation. Composites for improved conductivity contain conductive additives, which can be in particle and fibrous forms.

Due to the fact that some electrochemical electrode materials are not conductive electrically, a conductive additive is added to the electrochemically active species in forming the electrode [31–36]. An example is manganese dioxide (MnO_2), which is not conducting and serves as the cathode of various electrochemical cells, including lithium cells. Owing to the chemically inert and electrically conductive nature of carbons, carbons are used as the conductive additive. Among the carbons, carbon black is particularly common for this purpose.

Carbon black is more effective than submicron-diameter carbon filament without graphitization as a conductive additive for the MnO_2 cathode [37], even though a carbon black compact without MnO_2 exhibits higher resistivity than a carbon filament compact without MnO_2 [38]. Thus, a low resistivity for a carbon compact does not imply a low resistivity for an MnO_2-carbon composite. This is because the resistivity of an MnO_2-carbon composite depends on the dispersion and connectivity of the carbon in the midst of MnO_2 particles. In spite of the large aspect ratio of the carbon filament, carbon black is superior as a conductive additive, because of the spreadability of carbon black between the MnO_2 particles [37].

There are many examples of conductive additives for electrodes and current collectors. Carbon is used for MnO_2 [37,39,40], nickel and $LaMnO_3$ are conductive additives that are used for yttria-stabilized zirconia [41–43], nickel is used for samaria-doped ceria and RuO_2 [44–47], V_2O_5 [48] and polymers [49,50], conducting polymers are used for $LiMn_2O_4$ [51,52], V_2O_5 [53–55], molybdenum-doped ruthenium selenide [55] and carbon fluoride [57,58], and silver is used for $Bi_{1.5}Y_{0.5}O_3$ [59].

The effectiveness of a conductive additive increases with its conductivity and aspect ratio. In addition it is enhanced by the ability of the conductive additive to spread itself upon being squeezed between the matrix particles during composite fabrication. For example, carbon black has spreadability, but carbon fibers do not. As a result, carbon black is effective in spite of its low aspect ratio.

Other than electrochemically active materials and the conductive additive, the electrode contains a binder, which is commonly a thermoplastic polymer, such as polytetrafluoroethylene (PTFE, or Teflon) and polyvinylidenefluoride (PVDF) in the form of particles. The binder serves to bind the ingredients together to form a shaped object (such as a disk) that can be handled. The type and amount of binder affect not only the bindability, but also the distribution of the carbon additive, thereby affecting the conductivity of the composite.

10.5.3 Composites for improved processability, handleability, chemical stability and electrolyte absorptivity

Owing to the typically discontinuous nature (usually particulate) of the active components in electrodes, binders or matrices are used in composite electrodes for improving processability and handleability, and in some cases chemical stability as well. Most binders are polymers such as PTFE [60,61], polypropylene [62], polyester [63], polyvinyl chloride (PVC) [64] and polyethylene [65,66]. More specialized binders or matrices are electrically conductive polymers (e.g., polypyrrole) [67] and TiN [68]. Electrically conductive polymers can function as both binder and conducting matrix [69].

A problem with binders is the reactivity with the electrolyte or other components in the electrochemical cell. This problem can be alleviated by using carbon filaments to help serve the binding function [70]. Furthermore, the filaments tend to enhance electrolyte absorptivity and the rate of electrolyte absorption [70].

10.5.4 Carbon composites

Carbon is one of the most important materials for current collectors, owing to its electrical conductivity and chemical inertness [70]. Graphite, as made by mixing coke particles with pitch (a binder) and subsequently carbonizing the pitch, is commonly used as current collectors in industrial electrochemical processes, such as the reduction of minerals (e.g., iron oxide and aluminum oxide) to metals (e.g., iron and aluminum). Such artificial graphite can be considered a carbon–carbon composite, although it is not usually referred to as such.

Relevant carbon composites include polymer-matrix composites [71–73], particularly those containing carbon black. Carbon composites in the form of polymer-matrix and carbon-matrix composites are used for the bipolar plates of fuel cells [74–76].

A mixture of different types of carbon is often used for the anode of the lithium-ion cell. An example is a mixture of hard carbon (i.e., non-graphitizable carbon, which is attractive for its high capacity) and mesocarbon microbeads (MCMB, which is made from pitch and is attractive for its low irreversible capacity) [77]. Another example is a composite in the form of graphite encased in coke, which is disordered and thus protects the graphite core from exfoliation and improves the diffusion rate of lithium ions [78]. Other carbon composites for the anode of the lithium-ion cell include carbon–clay composites [79] and carbon fiber carbon–matrix composites [80].

Carbon fiber carbon–matrix composites are used for the bipolar plates of fuel cells. However, they are not as conductive electrically as graphite, which is the conventional material for this application. A method to increase the conductivity of a carbon–carbon composite involves the addition of electroconductive carbon black particles to the carbon matrix during fabrication of the composite [81].

10.6 Composite materials for electrolytes

Electrolytes are ionic conductors. In contrast, electrodes are electronic conductors. Electrolytes are commonly in liquid form. For safety, solid electrolytes are preferred, although ionic diffusivity tends to be lower in solids than liquids. Another problem with solid electrolytes is associated with the relatively poor contact between the solid electrolyte and the electrode, compared to the contact between a liquid electrolyte and an electrode. As electron transfer occurs across the electrode–electrolyte interface, an intimate contact is necessary. A solid electrolyte may also serve as a separator, i.e., a membrane sandwiched by two electrodes or two current collectors.

Solid electrolytes include ceramics, polymers and ceramic–polymer composites. Polymer electrolytes include dry solid systems, polymer gels and polymer composites [82]. The dry solid systems are relatively poor in ionic conductivity, but they are safer owing to the absence of organic solvents.

Polymer-based electrolytes include conducting polymer blends (e.g., polypyrrole/polysulfide [83], polypyrrole/polyetherimide [84] and polyaniline/polyaniline-sulfonic acid [85]), composites based on perfluorinated polymers (e.g., Nafion) [86–89], polyethylene oxide (PEO) [90–96], sulfonated polyetheretherketone (PEEK) [97], poly(acrylonitrile-co-methylmethacrylate) [98], polyethylene glycol [99] and PVDF [100]. A polymer may be doped with a lithium salt for enhancing the ionic conductivity [90,94,95,98,99,101,102]. It may also contain a ceramic filler (e.g., $LiAlO_2$, $LiBF_4$, $LiClO_4$ and zeolite) for enhancing ionic conductivity [102–106], a ceramic filler (e.g., SiO_2 and Al_2O_3) for enhancing the thermal stability [107–114], or a ceramic filler (e.g., MgO, CaO and Si_3N_4) for improving the electrode–electrolyte interfacial stability [115].

Ceramic-based solid electrolytes include those based on zirconia [116–119], lanthanum strontium manganate [120,121], lanthanum gallate [122,123], lithium titanium phosphate [124], lithium phosphate [125], lithium sulfate [126], lithium sodium sulfate [127], lithium hectorite clay [128], ceria [129–134], glass [135–137], sodium chloride [138], sodium carbonate [139], alkaline and alkaline earth fluorides [140–143], hydrofluoride [144], silver sulfate [145] and gadolinium titanium oxide [146]. The composites may contain ionic salts (e.g., $LiNO_3$ [124]) or ceramics (e.g., zirconia [120,126]) for enhancing ionic conductivity, ceramics (e.g., Al_2O_3 and MgO) for enhancing the toughness [116], and polymers for enhancing mechanical flexibility [135].

10.7 Composite materials for multiple functions

A composite material may by tailored with concentration gradients of the components, so that it is an ionic conductor at one extreme of the gradient for use as an electrolyte, an electronic conductor at the other extreme for use as an electrode, and a mixed conductor in between. This is known as *functionally gradient electrode–electrolyte structure*, which is used to improve the electrode–electrolyte interface. An example is the system ZrO_2–In_2O_3, in which ZrO_2 is the ionic conductor and In_2O_3 is the electronic conductor [117].

A composite may contain both the current collector (e.g., carbon and conducting polymers) and the catalyst (e.g., platinum) needed for the cell reaction [147–156]. In this way, the composite serves two functions.

A composite may be a separator film (e.g., polypropylene) with an electrochemically active electrode material (e.g., lithium) deposited on it [157]. In this way, the composite serves two functions.

10.8 Summary

Composite materials are used in batteries, fuel cells and electric double-layer capacitors as electrodes, current collectors and electrolytes, in addition to serving multiple functions (e.g., electrode + electrolyte, current collector + catalyst). Components in the composites include electronic conductors, ionic conductors, components for enhancing processability, handleability and stability, and catalysts. Carbon, polymers, ceramics and metals are all involved.

Review questions

1. Describe the components in a standard hydrogen reference half-cell.
2. Give an example of an oxygen concentration cell.
3. Why is an MnO_2 cathode containing carbon black as an electrically conductive additive more conducting electrically than that containing submicron-diameter carbon filaments as a conductive additive?
4. What is the main function of the carbon material used in the lithium-ion secondary cell?

References

[1] K.H. An, W.S. Kim, Y.S. Park, J.-M. Moon, D.J. Bae, S.C. Lim, Y.S. Lee and Y.H. Lee, *Adv. Funct. Mater.*, 2001, **11**(5), 387–392.
[2] A. Du Pasquier, F. Orsini, A.S. Gozdz and J.-M. Tarascon, *J. Power Sources*, 1999, **81**, 607–611.
[3] N. Oyama and O. Hatozaki, *Molecular Crystals and Liquid Crystals Science and Technology Section A – Molecular Crystals and Liquid Crystals*, 2000, **349**, 329–334.
[4] J. Yang, Y. Takeda, N. Imanishi, T. Ichikawa and O. Yamamoto, *Solid State Ionics*, 2000, **135**(1–4), 175–180.
[5] J. Yang, Y. Takeda, Q. Li, N. Imanishi and O. Yamamoto, *J. Power Sources*, 2000, **90**(1), 64–69.
[6] M. Winter and J.O. Besenhard, *Electrochimica Acta*, 1999, **45**(1), 31–50.
[7] O. Mao, R.L. Turner, I.A. Courtney, B.D. Fredericksen, M.I. Buckett, L.J. Krause and J.R. Dahn, *Electrochemical and Solid-State Letters*, 1999, **2**(1), 3–5.
[8] M.J. Parent, S. Passerini, B.B. Owens and W.H. Smyrl, *J. Electrochem. Soc.*, 1999, **146**(4), 1346–1350.
[9] P. Lavela, L. Sanchez, J.L. Tirado, S. Bach and J.P. Pereira-Ramos, *J. Power Sources*, 1999, **84**(1), 75–79.
[10] J. Kim and A. Manthiram, *Electrochemical and Solid-State Letters*, 1998, **1**(5), 207–209.
[11] P.G. Bruce, A.R. Armstrong and H. Huang, *J. Power Sources*, 1997, **68**(1), pt 1, 19–23.
[12] K. Nishimura, H. Honbo, S. Takeuchi, T. Horiba, M. Oda, M. Koseki, Y. Maranaka, Y. Kozono and H. Miyadera, *J. Power Sources*, 1997, **68**(2), pt 2, 436–439.
[13] E. Dayalan, *Mater. Res. Soc. Symp. Proc.*, 1999, **548**, 57–63.
[14] Y.-H. Lee, W.-K. Chang, C.-H. Fang, Y.-F. Huang and A.A. Wang, *Mater. Chem. and Phys.*, 1998, **53**(3), 243–246.
[15] C.S. Wang, G.T. Wu, X.B. Zhang, Z.F. Qi and W.Z. Li, *J. Electrochem. Soc.*, 1998, **145**(8), 2751–2758.
[16] H. Li, X. Huang, L. Chen, Z. Wu and Y. Liang, *Electrochemical and Solid-State Letters*, 1999, **2**(11), 547–549.
[17] P. Yu, J.A. Ritter, R.E. White and B.N. Popov, *J. Electrochem. Soc.*, 2000, **147**(4), 1280–1285.
[18] Z.-W. Fu and Q.-Z. Qin, *J. Electrochem. Soc.*, 2000, **147**(6), 2371–2374.
[19] S. Sunde, *J. Electroceramics*, 2000, **5**(2), 153–182.
[20] M. Mogensen, S. Primdahl, M.J. Jorgensen and C. Bagger, *J. Electroceramics*, 2000, **5**(2), 141–152.
[21] J.H. Choi, J.H. Jang and S.M. Oh, *Electrochimica Acta*, 2001, **46**(6), 867–874.
[22] J.-D. Kim, G.-D. Kim, J.-W. Moon, H.-W. Lee, K.-T. Lee and C.-F. Kim, *Solid State Ionics*, 2000, **133**(1), 67–77.
[23] K. Barthel and S. Rambert, *Mater. Sci. Forum*, 1999, **308–311**, 800–805.

[24] S. Wang, Y. Jiang, Y. Zhang, J. Yan and W. Li, *J. Electrochem. Soc.*, 1998, **145**(6), 1932–1939.
[25] J.W. Long, K.E. Swider, C.I. Merzbacher and D.R. Rolison, *Langmuir*, 1999, **15**(3), 780–785.
[26] M. Musiani, F. Furlanetto and R. Bertoncello, *J. Electroanalytical Chem.*, 1999, **465**(2), 160–167.
[27] M. Ramani, B.S. Haran, R.E. White, B.N. Popov and L. Arsov, *J. Power Sources*, 2001, **93**(1–2), 209–214.
[28] J.M. Miller, B. Dunn, T.D. Tran and R.W. Pekala, *J. Electrochem. Soc.*, 1997, **144**(12), L309–L311.
[29] J.P. Zheng, *Electrochemical and Solid-State Letters*, 1999, **2**(8), 359–361.
[30] J.M. Miller and B. Dunn, *Langmuir*, 1999, **15**(3), 799–806.
[31] K. Kinoshita, *Carbon*, Wiley, New York, 1988, pp. 405.
[32] J. Lahaye, M.J. Wetterwal and J. Messiet, *J. Appl. Electrochem.*, 1984, **14**, 545.
[33] M. Dohzono, H. Katsuki and M. Egashira, *J. Electrochem. Soc.*, 1989, **137**, 1255.
[34] D. Kohler, J. Zabasajja, A. Krishnagopalan and B. Tatarchuk, *J. Electrochem. Soc.*, 1990, **137**, 136.
[35] J. Lahaye, M.J. Wetterwal and J. Messier, *J. Appl. Electrochem.*, 1984, **14**, 117.
[36] J.L. Weininger and C.R. Morelock, *J. Electrochem. Soc.*, 1975, **122**, 116.
[37] C.A. Frysz, X. Shui and D.D.L. Chung, *J. Power Sources*, 1996, **58**(1), 41–54.
[38] C.A. Frysz, X. Shui and D.D.L. Chung, *J. Power Sources*, 1996, **58**(1), 55–56.
[39] K. Terashita, T. Hashimoto and K. Miyanami, *J. Japan Soc. Powder and Powder Metallurgy*, 2000, **47**(1), 91–96.
[40] K. Terashita, T. Hashimoto and K. Miyanami, *J. Japan Soc. Powder and Powder Metallurgy*, 1998, **45**(9), 877–881.
[41] A.V. Virkar, J. Chen, C.W. Tanner and J.-W. Kim, *Solid State Ionics*, 2000, **131**(1), 189–198.
[42] T. Nishikawa, A. Takano, S. Furukawa, S. Honda and H. Awaji, *Zairyo/J. Soc. Mater. Sci., Japan*, 2000, **49**(6), 606–610.
[43] M. Marinsek, K. Zupan and J. Macek, *J. Power Sources*, 2000, **86**(1), 383–389.
[44] S. Ohara, R. Maric, X. Zhang, K. Mukai, T. Fukui, H. Yoshida, T. Inagaki and K. Miura, *J. Power Sources*, 2000, **86**(1), 455–458.
[45] R. Maric, S. Ohara, T. Fukui, T. Inagaki and J.-I. Fujita, *Electrochemical and Solid-State Letters*, 1998, **1**(5), 201–203.
[46] X. Zhang, S. Ohara, R. Maric, K. Mukai, T. Fukui, H. Yoshida, M. Nishimura, T. Inagaki and K. Miura, *J. Power Sources*, 1999, **83**(1), 170–177.
[47] A.C. Tavares and S. Trasatti, *Electrochimica Acta*, 2000, **45**(25–26), 4195–4202.
[48] F. Coustier, J.-M. Lee, S. Passerini and W.H. Smyrl, *Solid State Ionics*, pp. 279–291.
[49] S. Kuwabata, N. Tsumura, S.-I. Goda, C.R. Martin and H. Yoneyamal, *J. Electrochem. Soc.*, 1998, **145**(5), 1415–1420.
[50] N. Oyama, T. Tatsuma and T. Sotomura, *J. Power Sources*, 1997, **68**(1), pt 1, 135–138.
[51] S. Kuwabata, S. Masui and H. Yoneyama, *Electrochimica Acta*, 1999, **44**(25), 4593–4600.
[52] J.O. Kim, M.-K. Song, S.W. Lee, B.W. Cho, K.S. Yun and H.-W. Rhee, *Molecular Crystals and Liquid Crystals Science and Technology Section A – Molecular Crystals & Liquid Crystals*, 2000, **349**, 287–290.
[53] G.J.F. Demets, F.J. Anaissi and H.E. Toma, *Electrochimica Acta*, 2000, **46**(4), 547–554.
[54] F. Leroux, G. Goward, W.P. Power and L.F. Nazar, *J. Electrochem. Soc.*, 1997, **144**(11), 3886–3895.
[55] S. Kuwabata, T. Idzu, C.R. Martin and H. Yoneyama, *J. Electrochem. Soc.*, 1998, **145**(8), 2707–2710.
[56] N. Alonso-Vante, S. Cattarin and M. Musiani, *J. Electroanalytical Chem.*, 2000, **481**(2), 200–207.
[57] E. Frackowiak, *Molecular Crystals and Liquid Crystals Science and Technology Section A – Molecular Crystals and Liquid Crystals*, 1998, **310**, 403–408.
[58] E. Frackowiak and K. Jurewicz, *Adv. Mater. for Optics and Electronics*, 1998, **8**(6), 303–308.
[59] Z. Wu and M. Liu, *J. Amer. Ceramic Soc.*, 1998, **81**(5), 1215–1220.
[60] G. Oritz, M.C. Gonzalez, A.J. Reviejo and J.M. Pingarron, *Analytical Chem.*, 1997, **69**(17), 3521–3526.
[61] E. Andrukaitis, *J. Power Sources*, 1997, **68**(2), pt 2, 656–659.
[62] R. Santhanam and M. Noel, *J. Power Sources*, 1998, **76**(2), 147–152.
[63] M.M.D. Jimenez, M.P. Elizalde, M. Gonzalez and R. Silva, *Electrochimica Acta*, 2000, **45**(25–26), 4187–4193.
[64] M. Davila, M.P. Elizalde, M. Gonzalez, M.A. Perez and R. Silva, *Electrochimica Acta*, 1998, **44**(8–9), 1307–1316.
[65] J. Navarro-Laboulais, J. Trijueque, J.J. Garcia-Jareno, D. Benito and F. Vicente, *J. Electroanalytical Chem.*, 1998, **444**(2), 173–186.
[66] J. Navarro-Laboulais, J. Trijueque, J.J. Garcia-Jareno, D. Benito and F. Vicente, *J. Electroanalytical Chem.*, 1998, **443**(1), 41–48.
[67] H.N. Cong, K.E. Abbassi and P. Chartier, *Electrochemical and Solid-State Letters*, 2000, **3**(4), 192–195.

[68] I.-S. Kim, P.N. Kumta and G.E. Blomgren, *Electrochemical and Solid-State Letters*, 2000, **3**(11), 493-496.
[69] S. Kuwabata, S. Masui, H. Tomiyori and H. Yoneyama, *Electrochimica Acta*, 2000, **46**(1), 91-97.
[70] C.A. Frysz, X. Shui and D.D.L. Chung, *J. Power Sources*, 1996, **58**(1), 55-66.
[71] V. Haddadi-Asl and M.S. Rabbani, *Iranian Polym. J.* (English edn), 1998, **7**(2), 95-101.
[72] H. Maleki, J.R. Selman, R.B. Dinwiddie and H. Wang, *J. Power Sources*, 2001, **94**(1), 26-35.
[73] V. Suryanarayanan and M. Noel, *J. Power Sources*, 2001, **94**(1), 137-141.
[74] D. Busick and M. Wilson, *Mater. Res. Soc. Symp. Proc.*, 2000, **575**, 247-251.
[75] K. Ledjeff-Hey, T. Kalk, F. Mahlendorf, O. Niemzig, A. Trautmann and J. Roes, *J. Power Sources*, 2000, **86**(1), 166-172.
[76] T.M. Besmann, J.W. Klett, J.J. Henry Jr and E. Lara-Curzio, *J. Electrochem. Soc.*, 2000, **147**(11), 4083-4086.
[77] E. Buiel and J.R. Dahn, *Electrochimica Acta*, 1999, **45**(1), 121-130.
[78] W. Qiu, G. Zhang, S. Lu and Q. Liu, *Solid State Ionics*, 1999, **121**(1), 73-77.
[79] G. Sandi, K.A. Carrado, R.E. Winans, C.S. Johnson and R. Csencsits, *J. Electrochem. Soc.*, 1999, **146**(10), 3644-3648.
[80] M. Saito, K. Yamaguchi, K. Sekine and T. Takamura, *Solid State Ionics*, 2000, **135**(1-4), 199-207.
[81] S.K. Ryu, T.S. Hwang, S.G. Lee, S.A. Lee, C.S. Kim and D.H. Jeong, *Carbon 2001*, American Carbon Society, Poster P4.07.
[82] F.B. Dias, L. Plomp and J.B.J. Veldhuis, *J. Power Sources*, 2000, **88**(2), 169-191.
[83] F. Yan, G. Xue and X. Wan, *Synthetic Metals*, 1999, **107**(1), 35-38.
[84] D.-I. Kang, W.-J. Cho and C.-S. Ha, *Molecular Crystals and Liquid Crystals Science and Technology Section A - Molecular Crystals and Liquid Crystals*, 1997, **294-295**, 237-240.
[85] A. Kitani, K. Satoguchi, K. Iwai and S. Ito, *Synthetic Metals*, 1999, **102**(1-3), pt 2, 1171-1172.
[86] E.K.W. Lai, P.D. Beattie, F.P. Orfino, E. Simon and S. Holdcroft, *Electrochimica Acta*, 1999, **44**(15), 2559-2569.
[87] C. Boyer, S. Gamburzev, O. Velev, S. Srinivasan and A.J. Appleby, *Electrochimica Acta*, 1998, **43**(24), 3703-3709.
[88] J.-C. Lin, H.R. Kunz, M.B. Cutlip and J.M. Fenton, *Hazardous and Industrial Wastes - Proc. Mid-Atlantic Industrial Waste Conf.*, 1999, pp. 656-662.
[89] M. Doyle, S.K. Choi and G. Proulx, *J. Electrochem. Soc.*, 2000, **147**(1), 34-37.
[90] H.Y. Sun, Y. Takeda, N. Imanishi, O. Yamamoto and H.-J. Sohn, *J. Electrochem. Soc.*, 2000, **147**(7), 2462-2467.
[91] M. Mastragostino, F. Soavi and A. Zanelli, *Mater. Res. Soc. Symp. Proc.*, 1999, **548**, 359-365.
[92] G.B. Appetecchi, F. Croce, G. Dautzenberg and B. Scrosati, *Proc. 1998 MRS Fall Symp. on Materials for Electrochemical Energy Storage and Conversion II - Batteries, Capacitors and Fuel Cells*, Materials Research Society, Warrendale, PA, 1998, **496**, 511-516.
[93] E. Quartarone, P. Mustarelli and A. Magistris, *Solid State Ionics*, 1998, **110**(1-2), 1-14.
[94] K.M. Abraham, V.R. Koch and T.J. Blakley, *J. Electrochem. Soc.*, 2000, **147**(4), 1251-1256.
[95] P.P. Soo, B. Huang, Y.-I. Jang, Y.-M. Chiang, D.R. Sadoway and A.M. Mayes, *J. Electrochem. Soc.*, 1999, **146**(1), 32-37.
[96] G.B. Appetecchi, F. Croce, L. Persi, F. Ronci and B. Scrosati, *Electrochimica Acta*, 2000, **45**(8), 1481-1490.
[97] S.M.J. Zaidi, S.D. Mikhailenko, G.P. Robertson, M.D. Guiver and S. Kaliaguine, *J. Membrane Sci.*, 2000, **173**(1), 17-34.
[98] K.-H. Lee, Y.-G. Lee, J.-K. Park and D.-Y. Seung, *Solid State Ionics*, 2000, **133**(3), 257-263.
[99] T.-C. Wen and W.-C. Chen, *J. Power Sources*, 2001, **92**(1-2), 139-148.
[100] D.A. Boysen, C.R.I. Chisholm, S.M. Haile and S.R. Narayanan, *J. Electrochem. Soc.*, 2000, **147**(10), 3610-3613.
[101] M. Mastragostino, F. Soavi and A. Zanelli, *J. Power Sources*, 1999, **81**, 729-733.
[102] Z. Wen, T. Itoh, M. Ikeda, N. Hirata, M. Kubo and O. Yamamoto, *J. Power Sources*, 2000, **90**(1), 20-26.
[103] E. Morales and J.L. Acosta, *Solid State Ionics*, 1998, **111**(1-2), 109-115.
[104] Z. Poltarzewski, W. Wieczorek, J. Przyluski and V. Antonucci, *Solid State Ionics*, 1999, **119**(1), 301-304.
[105] E. Quartarone, C. Tomasi, P. Mustarelli and A. Magistris, *Electrochimica Acta*, 1998, **43**(10-11), 1321-1325.
[106] H.-W. Park, D.-W. Shin, S.-H. Kang, M.-H. Bae, K.-J. Lim and B.-H. Kim, *Proc. 1997 5th IEEE Int. Conf. on Properties and Applications of Dielectric Materials*, IEEE, Piscataway, NJ, 1997, **1**, 600-602.
[107] I. Honma, S. Hirakawa, K. Yamada and J.M. Bae, *Solid State Ionics*, 1999, **118**(1-2), 29-36.
[108] P. Staiti, M. Minutoli and S. Hocevar, *J. Power Sources*, 2000, **90**(2), 231-235.

[109] A.S. Arico, P. Creti, P.L. Antonucci and V. Antonucci, *Electrochemical and Solid-State Letters*, 1998, **1**(2), 66-68.
[110] P.L. Antonucci, A.S. Arico, P. Creti, E. Ramunni and V. Antonucci, *Solid State Ionics*, 1999, **125**(1), 431-437.
[111] J. Fan, S.R. Raghavan, X.-Y. Yu, S.A. Khan, P.S. Fedkiw, J. Hou, G.L. Baker, *Solid State Ionics*, 1998, **111**(1-2), 117-123.
[112] I. Honma, S. Hirakawa and J.M. Bae, *Mater. Res. Soc. Symp. Proc.*, 1999, **548**, 377-382.
[113] H.J. Walls, J. Zhou, J.A. Yerian, P.S. Fedkiw, S.A. Khan, M.K. Stowe and G.L. Baker, *J. Power Sources*, 2000, **89**(2), 156-162.
[114] E. Strauss, D. Golodnitsky, G. Ardel and E. Peled, *Electrochimica Acta*, 1998, **43**(10-11), 1315-1320.
[115] B. Kumar and L.G. Scanlon, *J. Electroceramics*, 2000, **5**(2), 127-139.
[116] A. Yuzaki, A. Kishimoto and Y. Nakamura, *Solid State Ionics*, 1998, **109**(3-4), 273-277.
[117] K. Sasaki and L.J. Gauckler, *Key Eng. Mater.*, 1999, **169**, 197-200.
[118] J.-H. Lee, B.-K. Kim, K.-Y. Lee, H.-I. Kim and K.-W. Han, *Sensors and Actuators B - Chemical*, 1999, **59**(1), 9-15.
[119] G. Dotelli, F. Casartelli, I.N. Sora, C. Schmid and C.M. Mari, *Mater. Res. Soc. Symp. Proc.*, 2000, **575**, 331-336.
[120] X. Zhang, S. Ohara, K. Mukai, M. Ogawa and T. Fukui, *Electrochemistry*, 2000, **68**(1), 11-16.
[121] E.P. Murray, T. Tsai and S.A. Barnett, *Solid State Ionics*, 1998, **110**(3-4), 235-243.
[122] R. Maric, S. Ohara, T. Fukui, H. Yoshida, M. Nishimura, T. Inagaki and K. Miura, *J. Electrochem. Soc.*, 1999, **146**(6), 2006-2010.
[123] S.M. Choi, K.T. Lee, S. Kim, M.C. Chun and H.L. Lee, *Solid State Ionics*, 2000, **131**(3), 221-228.
[124] Y. Kobayashi, M. Tabuchi and O. Nakamura, *J. Power Sources*, 1997, **68**(2), pt 2, 407-411.
[125] B. Kumar and L.G. Scanlon, *Proc. 1997 32nd Intersociety Energy Conversion Engineering Conf. on Aerospace Power Systems and Technologies*, IEEE, Piscataway, NJ, 1997, **1**, 19-24.
[126] B. Zhua, *Solid State Ionics*, 1999, **119**(1), 305-310.
[127] S. Tao, B. Zhu, D. Peng and G. Meng, *Mater. Res. Bull.*, 1999, **34**(10-11), 1651-1659.
[128] M.W. Riley, P.S. Fedkiw and S.A. Khan, *Mater. Res. Soc. Symp. Proc.*, 2000, **575**, 137-142.
[129] J.-I. Hamagami, Y. Inda, T. Umegaki and K. Yamashita, *Solid State Ionics*, 1998, **113-115**, 235-239.
[130] Y. Mishima, H. Mitsuyasu, M. Ohtaki and K. Eguchi, *J. Electrochem. Soc.*, 1998, **145**(3), 1004-1007.
[131] R. Chiba, F. Yoshimura and J. Yamaki, *Proc. 1998 MRS Fall Symp. on Materials for Electrochemical Energy Storage and Conversion II - Batteries, Capacitors and Fuel Cells*, MRS, Warrendale, PA, 1998, **496**, 185-191.
[132] A. Tsoga, A. Naoumidis, A. Gupta and D. Stoever, *Mater. Sci. Forum*, 1999, **308-311**, 794-799.
[133] B. Zhu, I. Albinsson, B.-E. Bengt and G. Meng, *Solid State Ionics*, 1999, **125**(1), 439-446.
[134] B. Zhu, *J. Power Sources*, 2001, **93**(1-2), 82-86.
[135] J. Cho, G. Kim and H. Lim, *Proc. 1998 13th Annual Battery Conf. on Applications and Advances*, IEEE, Piscataway, NJ, 1998, pp. 381-385.
[136] J. Cho, G. Kim, H. Lim and M. Liu, *J. Electrochem. Soc.*, 1998, **145**(6), 1949-1952.
[137] H. Aono and Y. Sadaoka, *J. Electrochem. Soc.*, 2000, **147**(11), 4363-4367.
[138] S. Tao and G. Meng, *J. Mater. Sci. Lett.*, 1999, **18**(1), 81-84.
[139] K. Singh, P. Ambekar and S.S. Bhoga, *Solid State Ionics*, 1999, **122**(1-4), 191-196.
[140] B. Zhu, I. Albinsson and B.-E. Mellander, *Solid State Ionics*, 2000, **135**(1-4), 503-512.
[141] B. Zhu, *J. Mater. Sci. Lett.*, 1999, **18**(13), 1039-1041.
[142] G. Rog, M.M. Bucko, A. Kielski and A. Kozlowska-Rog, *Ceramics Int.*, 1999, **25**(7), 623-630.
[143] X. Sun, H.S. Lee, X.Q. Yang and J. McBreen, *J. Electrochem. Soc.*, 1999, **146**(10), 3655-3659.
[144] B. Zhu, *Int. J. Energy Res.*, 2000, **24**(1), 39-49.
[145] K. Singh, J. Randhawa, P. Khadakkar and S.S. Bhoga, *Solid State Ionics*, 1999, **126**(1), 47-53.
[146] J.J. Sprague, O. Porat and H.L. Tuller, *Proc. 1997 MRS Fall Symp. on Electrically Based Microstructural Characterization II*, MRS, Warrendale, PA, 1998, **500**, 321-326.
[147] K. Scott, W.M. Taama and P. Argyropoulos, *J. Power Sources*, 1999, **79**(1), 43-59.
[148] K. Scott, W.M. Taama and P. Argyropoulos, *J. Appl. Electrochem.*, 1998, **28**(12), 1389-1397.
[149] R. Coppola, L. Giorgi, A. Lapp and M. Magnani, *Physica B: Condensed Matter*, 2000, **276**, 839-840.
[150] S. Wang and Y. Zhao, *Taiyangneng Xuebao/Acta Energiae Solaris Sinica*, 2000, **21**(3), 229-233.
[151] A.S. Arico, P. Creti, P.L. Antonucci, J. Cho, H. Kim and V. Antonucci, *Electrochimica Acta*, 1998, **43**(24), 3719-3729.
[152] W.-Y. Tu, W.-J. Liu, C.-S. Cha and B.-L. Wu, *Electrochimica Acta*, 1998, **43**(24), 3731-3739.

[153] Y. Kiros and S. Schwartz, *J. Power Sources*, 2000, **87**(1), 101–105.
[154] K. Asaka and K. Oguro, *J. Electroanalytical Chem.*, 2000, **480**(1–2), 186–198.
[155] Z. Qi, M.C. Lefebvre and P.G. Pickup, *J. Electroanalytical Chem.*, 1997, pp. 9–14.
[156] M. Grzeszczuk and P. Poks, *Electrochimica Acta*, 2000, **45**(25–26), 4171–4177.
[157] S. Zeng and P.R. Moses, *J. Power Sources*, 2000, **90**(1), 39–44.

11 Composite materials for biomedical applications

11.1 Background on biomedical materials and applications

The repair or reconstruction of damaged or diseased parts of the musculoskeletal system requires various *implants* [1,2]. In particular, implants are used for the internal fixation of fractured bones and joints (such as hips, knees, fingers and jaws). They can also serve as prosthetic heart valves, intervertebral disk spacers, teeth, tendons and ligaments.

The materials for implants [3,4] must be biologically compatible (i.e., *biocompatible*), corrosion resistant and mechanically strong and tough. More exactly, biocompatibility refers to the ability of a material to provide an appropriate host response (i.e., no or a tolerable adverse reaction of a living system to the presence of the material) in a specific application. An adverse reaction may be due to the toxicity of the material, for example.

In the case of bone replacement, the materials should have an elastic modulus that matches that of the bone, in order to maintain the load distribution in the skeletal system for the purpose of avoiding the natural weakening of the bones that encounter abnormally low stresses, i.e., *stress shielding*. The modulus can be tailored by composite engineering.

The blood and tissue response to implants is of concern. In the case of prosthetic heart valves, the material should be *thromboresistant*, i.e., resistant to thrombosis, which is the local coagulation of blood in the heart, forming an obstruction to circulation.

Carbon, such as *pyrolytic carbon* (i.e., carbon made by the thermal decomposition of carbonaceous molecules such as pitch), turns out to be quite thromboresistant, in addition to being biocompatible. Titanium is a metal which is biocompatible. However, most metals (even gold) are not biocompatible, although they are mechanically strong and tough. Thus, metal implants may require removal in a second operation. To alleviate this problem, orthopedic implants are often coated with a biocompatible material.

Since the mineral phase of bone comprises calcium salts (such as apatite, which is a form of calcium phosphate), coatings are commonly composites which contain apatite or hydroxyapatite, which is *bioactive*, i.e., having the ability to promote biological processes such as tissue growth. For the purpose of tissue growth, porous materials or *scaffolds* are used to allow room for tissue growth. Therefore, some implant materials are coated with a porous coating for fixation by bone ingrowth. This is an example of *non-cemented fixation*. In contrast, *cemented fixation* involves attachment of the prosthesis to the host bone by an adhesive, such as self-curing acrylic [5].

Certain components of an implant (e.g., sutures [6]) may be needed only temporarily, as their need diminishes over time due to healing or tissue growth. Thus, *resorbable* (i.e., able to be reabsorbed) materials (e.g., tricalcium phosphate and some polymers [7]) are needed.

Biodegradable polymers (such as enzymatically degradable polymers [8]) are used to carry biologically active compounds, such as medical drugs. They are also used in plastic waste management of packaging materials [9]. The possible toxicity of the degradation products of a degradable implant is a concern, so naturally occurring metabolites (such as amino acids, lactic acid and glycolic acid) are used as the monomers for numerous biodegradable polymers [10]. Polylactic acid (PLA), an aliphatic polyester [11], degrades by hydrolysis and can be fashioned into porous scaffolds for tissue growth in orthopedics [12].

Polymers in the form of breathable adhesives are needed for breathable wound dressings, which resemble a temporary synthetic artificial skin [13]. Related applications are skin substitutes and bandages [14]. Examples of breathable materials are porous polymers in the form of hydrogels [15], which are water-swollen polymer networks [16].

Polymers in fiber form are important for suture threads, gauzes and vessel tubes for prosthetics of arteries, veins and the intestine. They include (i) polymers which undergo fast biodegradation in the living body (from 2–3 weeks to 3–4 months) for use in sutures, for example, (ii) moderately degradable polymers with a service life ranging from 2–3 weeks to 2–3 years, such that the biodegradation products are excreted from the body, and (iii) polymers which retain their initial properties (strength, shape, etc.) after being placed in the body for a long time (decades), for use in vessel, gauze and woven prostheses [17].

Glass–ionomer cements and polymer–matrix composites (in competition with traditional materials such as silver amalgam) are increasingly used for restorative dentistry [18,19], such as the restoration of decayed teeth. Glass–ionomers are adhesives that consist of interpenetrating networks of inorganic and organic components forming a matrix in which particles of unreacted glass are embedded [20]. The polymer–matrix composites are commonly adhesives filled with fine ceramic (e.g., SiO_2) particles or short fibers. The fillers serve as reinforcements, in addition to reducing the coefficient of thermal expansion (CTE) of the composite. The matching of the CTE of a tooth repair material and that of the tooth itself is desirable, as teeth encounter temperature excursions (say, due to hot coffee) and even a tiny gap between the repair material and the tooth can cause bacteria growth. Teeth repaired using adhesives are stronger and tend to last longer than those repaired using silver amalgam fillings. This is due to the need to remove less healthy tissue during the repair and also less leakage (better bond).

Dental adhesives are used not only for tooth repair. They are also used for securing brackets to the teeth in orthodontic treatment and for installing artificial teeth (known as jacket crowns) [14]. Because a natural tooth has dentine underneath the enamel coating, which may be removed during the repair of a decayed tooth, adhesives designed to bond to dentine are needed. Dentine has an inorganic part (apatite) and an organic part (collagen) [21], so the adhesive needs to bond to both parts.

Polymers are important in pharmaceutics. Biodegradable polymers are used as drug carriers (e.g., controlled drug-release agents [22]) and for drug conjunction [23]. Polymers that are pH sensitive are used in self-regulated drug delivery systems, which provide therapeutics in response to physiological requirements [24].

Polymers are also used as encapsulant materials applied to implanted device (e.g., pacemaker) electronics and leads. Examples are polyurethanes, silicones, epoxies and polyimides. However, their limited stability in an ionic liquid medium is of concern [25]. Among these materials, polyurethanes are particularly attractive, owing to their elastomeric character and blood compatibility [26–28].

11.2 Polymer–matrix composites for biomedical applications

Polymer–matrix composites for biomedical applications contain particulate and fibrous fillers, such as hydroxyapatite, apatite, calcium phosphate, carbon fibers, glass fibers, polymer fibers, bioactive glass particles and silica particles. The choice of filler is governed by the requirements of biocompatibility, bioactivity, mechanical properties, and processability into the size and shape required for the application.

The choice of polymer matrix is important, as it impacts the biocompatibility, toxicity, resorbability, biodegradability, breathability, long-term durability, bondability and *leachability*. Leaching (not mentioned in Section 11.1) is the reverse phenomenon of sorption; it involves small molecules diffusing outward from the bulk to the surface of the material and then entering the physiological milieu [29]. Leaching is not desirable. Not only is the choice of polymer matrix important, but also the surface treatment of the polymer, as it affects the surface chemistry, surface energy and surface morphology [30].

The detailed chemistry, as affected by the choice of monomer used in making the polymer, affects the toxicity and other properties. For example, polymethyl methacrylate (PMMA) is used in dental cement and bone cement [31]. The use of the butyl methacrylate monomer instead of the methyl methacrylate monomer results in less toxicity [32]. The use of the 2-hydroxyethyl methacrylate monomer allows the formation of a hydrogel, which is used as a membrane material [33].

Consideration should be given to the biocompatibility of the degradation products, in addition to that of the material itself, as the degradation products enter the physiological milieu. Thus polymers of the types present in the human body are desirable. An example is a copolymer of polylactic acid and polyglycine [34].

Polymer–matrix fiber composites are increasingly used in orthopedic implants [35], owing to the desirable mechanical properties imparted by the fibers. Although continuous fibers are more effective than short fibers as a reinforcement, short fibers are often used, owing to the requirement of processing the composite (say by injection molding) into an object of a complex shape and a small size. The object size tends to be particularly small for dental applications, so particles are most commonly used in dental composites.

Pylons (artificial legs) are relatively large objects, so they involve continuous-fiber composites made by filament winding [36]. Continuous fibers are valuable not only for attaining high strength and modulus in the composite, but they also allow tailoring of the bending stiffness of the composite object (e.g., an artificial foot) through control of fiber orientation [37,38].

Hip prostheses involve hip stems, which are highly loaded. Continuous fiber composites of appropriate modulus are used for the stems [39–41]. The composites can be in the form of laminates [42], or of a continuous fiber polymer–matrix wrapping on a metallic core material (e.g., stainless steel), so that the resulting macroscopic composite exhibits the desirable combination of modulus and fatigue performance [43].

Short-fiber polymer–matrix composites are used as a bearing material in hip joints involved in total hip replacement [44–46]. For this application, the wear behavior (which depends on the counterface material) is critical. A softer and more graphitic carbon fiber (e.g., pitch-based carbon fiber) is preferred to a harder and more abrasive filler, and a ceramic (e.g., zirconia) counterface is preferred to a metal counterface in reducing the wear rate [46].

Continuous-fiber polymer–matrix composites in the form of wires are used as orthodontic arch wires. For application, the wire needs to resist sliding against the orthodontic brackets through friction. The composite arch wire has a higher friction coefficient than stainless steel, but a lower coefficient than titanium [47].

Continuous-fiber polymer–matrix composites made by pultrusion and exhibiting high flexural strength are used for bone replacement [48]. These composites (usually involving carbon fibers) are tailored in terms of the modulus, which needs to match that of certain bones. In some applications, the composites need to be resorbable, so resorbable fibers made of calcium phosphate (CaP), polyglycolic acid (PGA) or chitin (the principal constituent of the hard covering of insects and crustaceans) are used [49]. In contrast, carbon fibers are not resorbable. In self-fixation, the composite needs to swell; a polymer that provides the required swelling characteristic is required, while the fibers provide the required structural properties [50].

Continuous-fiber polymer–matrix composites are used for spine fusion cages, which are structural systems for stabilization after full or partial vertebrectomy (removal of the vertebrae) [51]. The composite needs to have a modulus close to that of the cortical bone.

Bioactive glass in particle [52–54] and fiber [55–58] forms are used as fillers in polymer–matrix composites to provide bioactivity, as needed for bone reconstruction. In contrast, carbon fibers, although biocompatible, are not bioactive. In addition, glass particles [59,60] and fibers [61–66] are used as reinforcements; the applications include orthodontic wires [62,63], prosthetic dentistry [65] and dental appliances [66]. Furthermore, bioactive, biodegradable and resorbable composites are made by using bioactive glass in combination with biodegradable and resorbable polymers such as polylactic acid (PLA) [67–71] and starch [72].

Not only carbon fibers and glass fibers are used in biomedical composites. Polymers fibers are also used, partly because of their ductility. Polyethylene fibers are used in a rubber matrix to provide a soft composite material for heart valve prosthesis [73]. Woven polyethylene fibers are used to reinforce a polyethylene–matrix hydroxyapatite bone-substitute composite [74]. Kevlar, Nylon 66 and polyethylene terephthalate (PET) fibers are also used for the reinforcement of orthopedic and dental implants [75]. Moreover, helically wound PET fibers are used to reinforce polymer hydrogels in order to mimic the architecture of collagen fibers in soft natural issue, as needed for ligament prostheses [76]. Such soft composite materials are also used for implants in the form of tendons and intervertebral disks [77].

Owing to their biocompatibility, titanium fibers [78] and meshes [79] are used as fillers in polymers. A particularly important filler is hydroxyapatite (HAp, $Ca_{10}(PO_4)_6(OH)_2$), which is the main mineral component of calcified tissues. It mimics natural bone mineral and has good bone-bonding properties [80], in addition to serving as a reinforcement [80,81]. Hydroxyapatite is commonly in the form of particles, which are used in a variety of polymer matrices, including polyetheretherketone (PEEK) [82], polyhydroxybutyrate (PHB) [83], a network polymer involving chitosan and gelatin [84], polyethylene [85,86] and polypropylene [85] for bone implants.

Polylactic acid [87–94] and starch [95–98] are used as matrices for hydroxyapatite to provide biodegradable and resorbable composites as scaffolds in bone tissue engineering. Both the hydroxyapatite and the polymer are resorbable, although the former tends to take longer (e.g., 10 months) to resorb [99].

Hydroxyapatite is also bioactive, so it is used to provide bioactive composites for tissue replacement [100]. Akin to hydroxyapatite are other forms of bioactive calcium phosphate [101–111].

Porous hydroxyapatite with pore size exceeding 100 μm is valuable for bone ingrowth in dentistry [112]. Reticulated ceramics, such as aluminum oxide (Al_2O_3, which is bio-inert) with pore size ranging from 150 to 1700 μm and infiltrated with a biodegradable polymer, are used to facilitate natural tissue regeneration and provide mechanical support during healing in hard tissue replacement [113].

In the area of restorative dentistry such as tooth filling, polymer–matrix composites with fine ceramic particles (e.g., silica and $BaSiO_2$) as the filler and a methacrylate-based polymer matrix are common [114–119]. Methacrylate-based polymers are attractive for their mechanical properties, processability and low water diffusivity. Methacrylate resin is also used in bioactive bone cement that includes a glass–ceramic powder (e.g., $MgO–CaO–SiO_2–P_2O_5–CaF_2$) as the bioactive filler [120].

Owing to the toughness and strength of bones and shells, layered organic/inorganic nanocomposites that mimic bone and shell structures have been the dream of materials chemists. An example of such a material is a silica/polymer thin-film nanocomposite [121].

11.3 Ceramic–matrix composites for biomedical applications

Ceramic–matrix composites for biomedical applications include those with hydroxyapatite (and other types of calcium phosphate), bioactive glass and bio-inert ceramics (e.g., Al_2O_3) as matrices.

Sintered hydroxyapatite and related calcium phosphates suffer from their low strength [122], so they are reinforced with titania [123], $CaO–P_2O_5$-based glass [124], fibers (e.g., Al_2O_3, stainless steel, titanium and carbon) [125,126] or related materials. In addition, glass and titanium are used in combination in hydroxyapatite in order to obtain bioactivity (due to hydroxyapatite), biochemical stability (due to the glass) and mechanical strength (due to titanium) [127]. Moreover, glass and titanium can be graded in their spatial distribution to attain a functionally gradient material for bioactive artificial joints and dental implants [127,128].

Bioactive glass has the capacity to bond to living soft and hard tissues [129]. For enhancing the strength and toughness, bioactive glass (e.g., modified soda-lime–silica glass) is reinforced with metal fibers [130,131] or particles [132].

Alumina is attractive for its biocompatibility and good mechanical properties except toughness, so it is reinforced with ceramic particles such as TiN [133].

11.4 Carbon–matrix composites for biomedical applications

Carbon–carbon composites with a carbon matrix (typically made by pyrolysis of pitch) and continuous carbon fibers as the reinforcement are very important for orthopedic prosthetics, owing to the biocompatibility of carbon and the superior mechanical properties of the composites [134]. However, the shear strength between a carbon–carbon composite implant and bone is lower than that between a hydroxyapatite implant and bone, and is higher than that between surgical steel and bone [135]. On the other hand, carbon–carbon composites can be tailored in terms of modulus by control of the fiber orientation [136], and in terms of porosity by control of the composite fabrication condition (such as the carbonization condition for making the carbon matrix) [137]. Furthermore, functionally graded materials using a hybrid matrix consisting of silicon carbide and carbon in a graded composition profile can be made in order to form a graded interface between the bone and the prosthesis [138]. The bioactivity of a carbon–carbon composite can be improved by coating with calcium phosphate [139] or pyrolytic carbon [140].

Carbon–carbon composites are not only used for bone replacement [141] and scaffolds [142], but also for screws [136]. Although carbon screws are not as good as steel screws in terms of holding capacity, their biocompatibility is valuable for applications such as osteosynthesis.

11.5 Metal–matrix composites for biomedical applications

Metals are attractive owing to their strength, toughness and processability. Metal–matrix composites for biomedical applications are mainly titanium–matrix composites, because of the biocompatibility of titanium (e.g., Ti-6Al-4V). In order to provide bioactivity, hydroxyapatite [143–146] and glass–ceramic [147] are used as fillers. In order to enhance the lubricity, which is important for joints, graphite particles [148,149] are used as a filler. Titanium carbide, formed during sintering of titanium and graphite particles, enhances wear resistance [148,149].

Metal–matrix composites containing glass or metal oxide filler particles are used as cements in dentistry. The glass and metal oxide serve as binders [150].

11.6 Summary

Polymer-matrix, ceramic-matrix, carbon-matrix and metal-matrix composites, when appropriately designed in terms of the modulus, biocompatibility, biodegradability, resorbability, bioactivity, lubricity, wear resistance and processability, all find biomedical applications, particularly in orthopedics, prosthetics, bone and tissue engineering, and restorative dentistry. Important components in biomedical composites include hydroxyapatite, bioactive glass, biodegradable polymers, carbon and titanium.

Review questions

1. Carbon–carbon composites are used for hip replacement. Why?
2. Define "biocompatibility".
3. Define "stress shielding".
4. What is the function of a scaffold?
5. Give an example of a biodegradable polymer.
6. What is a glass–ionomer?
7. Name a pharmaceutical application of biodegradable polymers.
8. What is hydroxyapatite?
9. Give an example of a metal which is biocompatible.
10. What filler is used to enhance the lubricity of a metal–matrix composite?

References

[1] L.L. Hench, *J. Amer. Ceramic Soc.*, 1998, **81**(7), 1705–1728.
[2] K. de Groot, H.B. Wen, Y. Liu, P. Layrolle and F. Barrere, *Mater. Res. Soc. Symp. Proc.*, 2000, **599**, 109–116.
[3] P.P. Lutton and B. Ben-Nissan, *Mater. Tech.*, 1997, **12**(2), 59–63.
[4] G.O. Hofmann, *Clinical Mater.*, 1992, **10**(1–2), 75–80.
[5] K.B. Kwarteng, *SAMPE Quarterly – Society for the Advancement of Material and Process Engineering*, 1988, **20**(1), 28–32.
[6] I. Capperauld, *Clinical Mater.*, 1989, **4**(1), 3–12.
[7] R. Po, *J. Macromolecular Sci. – Rev. in Macromolecular Chem. and Phys.*, 1994, (4), 607–662.
[8] G. Pkhakadze, M. Grigorieva, I. Gladir and V. Momot, *J. Mater. Sci. – Mater. in Medicine*, 1996, **7**(5), 265–267.
[9] W. Amass, A. Amass and B. Tighe, *Polym. Int.*, 1998, **47**(2), 89–144.
[10] S.I. Ertel and J. Kohn, *Proc. Amer. Chem. Soc. Spring Meeting on Polym. Mater. Sci. and Eng.*, ACS, Books and Journals Division, Washington, DC, **66**, 224–225.
[11] S. Jacobsen, P. Degee, H.G. Fritz, P. Dubois and R. Jerome, *Polym. Eng. and Sci.*, 1999, **39**(7), 1311–1319.
[12] C.M. Agrawal, K.A. Athanasiou and J.D. Heckman, *Mater. Sci. Forum*, 1997, **250**, 115–128.
[13] M. Szycher and S.J. Lee, *J. Biomater. Appl.*, 1992, **7**(2), 142–213.
[14] T.S. Kryzhanovskaya and N.A. Lavrov, *Plasticheskie Massy*, 1995, (2), 44–47.
[15] S. Woerly, *Mater. Sci. Forum*, 1997, **250**, 53–68.
[16] A.S. Hoffman, *Polym. Preprints, Division of Polym. Chem.*, American Chemical Society, Books and Journals Division, Washington, DC, 1989, **31**(1), 220–221.
[17] L.G. Privalova and G.E. Zaikov, *Polym. – Plastics Tech. & Eng.*, 1990, **29**(5–6), 455–520.
[18] S. Deb, *Proc. Inst. Mech. Eng., Part H – J. Eng. Medicine*, 1998, **212**(H6), 453–464.
[19] J.W. Nicholson, *Int. J. Adhesion and Adhesives*, 1998, **18**(4), 229–236.
[20] J.W. Nicholson, *Biomater.*, 1998, **19**(6), 485–494.
[21] S.A.M. Ali and D.F. Wiliams, *Clinical Mater.*, 1993, **14**(3), 243–254.
[22] J.H. Braybrook and L.D. Hall, *Progress in Polym. Sci.*, 1990, **15**(5), 715–734.
[23] A.J. Domb, A. Bentolila and D. Teomin, *Acta Polymerica*, 1998, **49**(10–11), 526–533.
[24] J. Kost and R. Langer, *Trends in Biotechnology*, 1992, **10**(4), 127–131.
[25] M.F. Nichols, *Critical Reviews in Biomedical Engineering*, 1994, (1), 39–67.
[26] L.A. Poole-Warren, D.J. Martin, K. Schindhelm and G.F. Meijs, *Mater. Forum*, 1997, **21**, 241–256.
[27] M. Szycher and A.M. Reed, *Proc. 1996 54th Annual Tech. Conf.*, Society of Plastics Engineers, Brookfield, CT, 1996, **3**, 2758–2766.
[28] H.J. Griesser, *Polym. Degradation and Stability*, 1991, **33**(3), 329–354.
[29] S.D. Bruck, *Medical Progress through Technology*, 1990, **16**(3), 131–143.
[30] B.D. Ratner, A.B. Johnston and T.J. Lenk, *J. Biomedical Mater. Res.*, 1987, **21**(A1), 59–89.

[31] S. Saha and S. Pal, *J. Biomedical Mater. Res.*, 1984, **18**(4), 435–462.
[32] P.A. Revell, M. Braden and M.A.R. Freeman, *Biomater.*, 1998, **19**(17), 1579–1586.
[33] N.A. Peppas and H.J. Moynihan, *Hydrogels in Med. and Pharm.*, 1986, **2**, 49–64.
[34] J.M. Schakenrad and P.J. Dijkstra, *Clinical Mater.*, 1991, **7**(3), 253–269.
[35] D. Taylor and B. McCormack, *Key Eng. Mater.*, 1989, **32**, 35–44.
[36] D.A. Taylor, *Int. SAMPE Symp. and Exhib.*, 2000, **45**(I), 422–426.
[37] M. Jenkins, *Mater. World*, 2000, **8**(9), 11.
[38] B. Lehmann and W. Michaeli, *Proc. 1997 42nd Int. SAMPE Symp. and Exhib. on Evolving Technologies for the Competitive Edge*, SAMPE, Covina, CA, 1997, **42**(1), 571–581.
[39] C. Kaddick, S. Stur and E. Hipp, *Medical Eng. and Phys.*, 1997, **19**(5), 431–439.
[40] T.P. Vail, R.R. Glisson, T.D. Koukoubis and F. Guilak, *J. Biomechanics*, 1998, **31**(7), 619–628.
[41] J.A. Simoes, A.T. Marques and G. Jeronimidis, *Compos. Sci. and Tech.*, 2000, **60**(4), 559–567.
[42] S. Srinivasan, J.R. de Andrade, S.B. Biggers Jr and R.A. Latour Jr, *Compos. Struct.*, 1999, **45**(3), 163–170.
[43] H.S. Hedia, D.C. Barton, J. Fisher and A. Ibrahim, *Proc. 1996 5th World Biomaterials Congress, Society for Biomaterials*, St Louis Park, MN, 1996, **2**, 500.
[44] V.K. Polineni, A. Wang, A. Essner, R. Lin, A. Chopra, C. Stark and J.H. Dumbleton, *ASTM Special Tech. Pub.*, 1999, (1346), 266–273.
[45] A. Wang, R. Lin, C. Stark and J.H. Dumbleton, *Wear*, 1999, **225**(II), 724–727.
[46] A. Wang, R. Lin, V.K. Polineni, A. Essner, C. Stark and J.H. Dumbleton, *Tribology Int.*, 1998, **31**(11), 661–667.
[47] S.W. Zlufall, K.C. Kennedy and R.P. Kusy, *J. Mater. Sci. - Mater. in Medicine*, 1998, **9**(11), 611–620.
[48] J. Kettunen, E.A. Makela, H. Miettinen, T. Nevalainen, M. Heikkila, T. Pohjonen, P. Tormala and P. Rokkanen, *Biomater.*, 1998, **19**(14), 1219–1228.
[49] M.A. Slivka, C.C. Chu, I.A. Adisaputro, *J. Biomedical Mater. Res.*, 1997, **36**(4), 469–477.
[50] A. Abusafieh, S. Siegler and S.R. Kalidindi, *Proc. 1996 ASME Int. Mechanical Eng. Congress and Exposition on Adv. Mater.: Development, Characterization Processing, and Mechanical Behavior*, American Society of Mechanical Engineers, Materials Division, New York, 1996, **74**, 71–72.
[51] C.R. McMillin and J.W. Brantigan, *Proc. 1993 38th Int. SAMPE Symp. and Exhib. on Adv. Mater.: Performance Through Technology Insertion International*, SAMPE, Covina, CA, 1993, **38**(1), 582–590.
[52] J.K. West, A.B. Brennan, A.E. Clark and L.L. Hench, *J. Biomedical Mater. Res.*, 1997, **36**(2), 209–215.
[53] J.K. West, A.B. Brennan, A.E. Clark, M. Zamora and L.L. Hench, *J. Biomedical Mater. Res.*, 1998, **41**(1), 8–17.
[54] A.J. Aho, T. Tirri, J. Seppala, J. Rich, N. Strandberg, T. Jaakkola, T. Narhi and J. Kukkonen, *Key Eng. Mater.*, 2001, **192-195**, 685–687.
[55] M. Marcolongo, P. Ducheyne and W.C. LaCourse, *J. Biomedical Mater. Res.*, 1997, **37**(3), 440–448.
[56] M. Marcolongo, P. Ducheyne, J. Garino and E. Schepers, *J. Biomedical Mater. Res.*, 1998, **39**(1), 161–170.
[57] M. Marcolongo, P. Ducheyne, E. Schepers and J. Garino, *Proc. 1996 5th World Biomaterials Congress, Transactions of the Annual Meeting of the Society for Biomaterials in conjunction with the International Biomaterials Symposium*, Society for Biomaterials, St Louis Park, MN, 1996, **2**, 444.
[58] E. Pirhonen, G. Grandi and P. Tormala, *Key Eng. Mater.*, 2001, **192-195**, 725–728.
[59] C.I. Vallo, *J. Biomedical Mater. Res.*, 2000, **53**(6), 717–727.
[60] V.K. Krishnan, P.P. Lizymol and S.P. Nair, *J. Appl. Polym. Sci.*, 1999, **74**(3), 735–746.
[61] V.M. Miettinen, K.K. Narva and P.K. Vallittu, *Biomater.*, 1999, **20**(13), 1187–1194.
[62] F. Watari, S. Yamagata, T. Imai, S. Nakamura and M. Kobayashi, *J. Mater. Sci.*, 1998, **33**(23), 5661–5664.
[63] T. Imai, F. Watari, S. Yamagata, M. Kobayashi, K. Nagayama, Y. Toyoizumi and S. Nakamura, *Biomater.*, 1998, **19**(23), 2195–2200.
[64] T.W. Lin, A.A. Corvelli, C.G. Frondoza, J.C. Roberts and D.S. Hungerford, *J. Biomedical Mater. Res.*, 1997, **36**(2), 137–144.
[65] J. Tanner, P.K. Vallittu and E. Soderling, *J. Biomedical Mater. Res.*, 2000, **49**(2), 250–256.
[66] A.C. Karmaker, A.T. DiBenedetto and A.J. Goldberg, *J. Biomater. Appl.*, 1997, **11**(3), 318–328.
[67] Q.-Q. Qiu, P. Ducheyne and P.S. Ayyaswamy, *J. Biomedical Mater. Res.*, 2000, **52**(1), 66–76.
[68] N.C. Leatherbury, G.G. Niederauer, K. Kieswetter and M.A. Slivka, *Proc. 1998 17th Southern Biomedical Eng. Conf.*, IEEE, Piscataway, NJ, 1998, 68 pp.
[69] M. Kellomaki, H. Niiranen, K. Puumanen, N. Ashammakhi, T. Waris and P. Tormala, *Biomater.*, 2000, **21**(24), 2495–2505.

[70] Q. Qiu, P. Decheyne and P.S. Ayyaswamy, *Key Eng. Mater.*, 2001, **192–195**, 467–470.
[71] C.V. Ragel and M. Vallet-Regi, *J. Biomedical Mater. Res.*, 2000, **51**(3), 424–429.
[72] I.B. Leonor, R.A. Sousa, A.M. Cunha, Z.P. Zhong, D. Greenspan and R.L. Reis, *Key Eng. Mater.*, 2001, **192–195**, 705–708.
[73] G. Cacciola, G.W. Peters, P.J. Schreurs and J.D. Janssen, *Proc. 1997 Bioengineering Conf.*, American Society of Mechanical Engineers, Bioengineering Division, New York, 1997, 435–436.
[74] N.H. Ladizesky, E.M. Pirhonen, D.B. Appleyard, I.M. Ward and W. Bonfield, *Compos. Sci. and Tech.*, 1998, **58**(3–4), 419–434.
[75] H. Wan, R.L. Williams, P.J. Doherty and D.F. Williams, *J. Mater. Sci. - Mater. in Medicine*, 1997, **8**(1), 45–51.
[76] L. Ambrosio, R. De Santis, S. Iannace, P.A. Netti and L. Nicholais, *J. Biomedical Mater. Res.*, 1998, **42**(1), 6–12.
[77] L. Ambrosio, R. De Santis and L. Nicholais, *Proc. Inst. Mech. Eng., Part H - J. Eng. in Medicine*, 1998, **212**(2), 93–99.
[78] L.D.T. Topoleski, P. Ducheyne and J.M. Cuckler, *J. Biomedical Mater. Res.*, 1992, **26**(12), 1599–1617.
[79] M. Oka, Y. Chang, S. Hyon, Y. Ikada, W. Cha and T. Nakamura, *Proc. 1996 5th World Biomater. Congress*, Society for Biomaterials, St Louis Park, MN, 1996, **1**, 922.
[80] R. Zhang and P.X. Ma, *J. Biomedical Mater. Res.*, 1999, **44**(4), 446–455.
[81] W. Suchanek and M. Yoshimura, *J. Mater. Res.*, 1998, **13**(1), 94–117.
[82] M.S.A. Bakar, P. Cheang and K.A. Khor, *J. Mater. Processing Tech.*, 1999, **89–90**, 462–466.
[83] Z.B. Luklinska and W. Bonfield, *J. Mater. Sci. - Mater. in Medicine*, 1997, **8**(6), 379–383.
[84] Y.J. Yin, F. Zhao, X.F. Song, K.D. Yao, W.W. Lu and J.C. Leong, *J. Appl. Polym. Sci.*, 2000, **77**(13), 2929–2938.
[85] A. Sendemir and S. Altintas, *Proc. 1998 2nd Int. Conf. Biomedical Eng. Days*, IEEE, Piscataway, NJ, 1998, 114–117.
[86] M. Wang, S. Deb and W. Bonfield, *Mater. Lett.*, 2000, **44**(2), 119–124.
[87] A. Mcmanus, R. Siegel, R. Doremus and R. Bizios, *Annals of Biomedical Eng.*, 2000, **28**(Suppl. 1), S-15.
[88] N. Ignjatovic, V. Savic, S. Najman, M. Plavsic and D. Uskokovic, *Biomater.*, 2001, **22**(6), 571–575.
[89] N. Ignjatovic, M. Plavsic, S. Najman, V. Savic and D. Uskokovic, *Mater. Sci. Forum*, 2000, **352**, 143–150.
[90] L. Calandrelli, B. Immirzi, M. Malinconico, M.G. Volpe, A. Oliva and F.D. Ragione, *Polym.*, 2000, **41**(22), 8027–8033.
[91] K.G. Marra, P.G. Campbell, P.A. Dimilla, P.N. Kumta, M.P. Mooney J.W. Szem and L.E. Weiss, *Mater. Res. Soc. Symp. Proc.*, 1999, **550**, 155–160.
[92] K.G. Marra, J.W. Szem, P.N. Kumta, P.A. DiMilla and L.E. Weiss, *J. Biomedical Mater. Eng.*, 1999, **47**(3), 324–335.
[93] P.X. Ma, R. Zhang, G. Xiao and R. Franceschi, *J. Biomedical Mater. Res.*, 2001, **54**(2), 284–293.
[94] T. Furukawa, Y. Matsusue, T. Yasunaga, Y. Nakagawa, Y. Okada, Y. Shikinami, M. Okuno and T. Nakamura, *J. Biomedical Mater. Res.*, 2000, **50**(3), 410–419.
[95] R.L. Reis, A.M. Cunha, P.S. Allan and M.J. Bevis, *Adv. Polym. Tech.*, 1997, **16**(4), 263–277.
[96] R.L. Reis, A.M. Cunha and M.J. Bevis, *Proc. 1997 55th Special Areas Annual Tech. Conf.*, Soc. Plastics Engineers, Brookfield, CT, 1997, 2849–2853.
[97] J.F. Mano, C.M. Vaz, S.C. Mendes, R.L. Reis and A.M. Cunha, *J. Mater. Sci. - Mater. Medicine*, 1999, **10**(12), 857–862.
[98] P.B. Malafaya, F. Stappers and R.L. Reis, *Key Eng. Mater.*, 2001, **192–195**, 243–246.
[99] A. Piattelli, M. Franco, G. Ferronato, M.T. Santello, R. Martinetti and A. Scarano, *Biomater.*, 1997, **18**(8), 629–633.
[100] B. Chua and M. Wang, *Mater. Res. Soc. Symp. Proc.*, 2000, **599**, 45–50.
[101] P. Weiss, M. Lapkowski, R.Z. Legeros, J.M. Bouler, A. Jean and G. Daculsi, *J. Mater. Sci. - Mater. Medicine*, 1997, **8**(10), 621–629.
[102] O. Gauthier, J.-M. Bouler, P. Weiss, J. Bosco, G. Daculsi and E. Aguado, *J. Biomedical Mater. Res.*, 1999, **47**(1), 28–35.
[103] M. Kikuchi, J. Tanaka, Y. Koyaman and K. Takakuda, *J. Biomedical Mater. Res.*, 1999, **48**(2), 108–110.
[104] O. Gauthier, I. Khairoun, P. Weiss, J.-M. Bouler, E. Aguado and G. Daculsi, *Key Eng. Mater.*, 2001, **192–195**, 801–804.
[105] G. Daculsi, R. Rohanizadeh, P. Weiss and J.M. Bouler, *J. Biomedical Mater. Res.*, 2000, **50**(1), 1–7.
[106] M. Kikuchi, Y. Koyama, K. Takakuda, H. Miyairi and J. Tanaka, *Key Eng. Mater.*, 2001, **192–195**, 677–680.

[107] D. Skrtic, J.M. Antonucci, E.D. Eanes, F.C. Eichmiller and G.E. Schumacher, *J. Biomedical Mater. Res.*, 2000, **53**(4), 381–391.
[108] G. Grimandi, P. Weiss, F. Millot and G. Daculsi, *J. Biomedical Mater. Res.*, 1998, **39**(4), 660–666.
[109] S.J. Peter, S.T. Miller, G. Zhu, A.W. Yasko and A.G. Mikos, *J. Biomedical Mater. Res.*, 1998, **41**(1), 1–7.
[110] M.A. Slivka and C.C. Chu, *J. Biomedical Mater. Res.*, 1997, **37**(3), 353–362.
[111] K.E. Watson, K.S. Tenhuisen and P.W. Brown, *J. Mater. Sci. – Mater. Medicine*, 1999, **10**(4), 205–213.
[112] G. Carotenuto, G. Spagnuolo, L. Ambrosio and L. Nicolais, *J. Mater. Sci. Medicine*, 1999, **10**(10), 671–676.
[113] K. Thorne, P.D. Saint-Pierre, *Proc. 1997 16th Southern Biomedical Eng. Conf.*, IEEE, Piscataway, NJ, 1997, 107–110.
[114] P.P. Paul, S.F. Timmons, W.J. Machowski, *Proc. 1997 Las Vegas ACS Meeting on Polymer Preprints*, American Chemistry Society, Division of Polymer Chemistry, Washington, DC, 1997, **2**, 124.
[115] B. Sandner, S. Baudach, K.W.M. Davy, M. Braden and R.L. Clarke, *J. Mater. Sci. – Mater. Medicine*, 1997, **8**(1), 39–44.
[116] Y. Wei, D. Jin, G. Wei, D. Yang and J. Xu, *J. Appl. Polym. Sci.*, 1998, **70**(9), 1689–1699.
[117] J. Luo, R. Seghi and J. Lannutti, *Mater. Sci. and Eng. C: Biomimetic Mater., Sensors and Sys.*, 1997, **5**(1), 15–22.
[118] G. Wang, B.M. Culbertson and R.R. Seghi, *J. Macromolecular Sci. – Pure and Appl. Chem.*, 1999, **36**(3), 373–388.
[119] K.-D. Ahn, C.-M. Chung and Y.-H. Kim, *J. Appl. Polym. Sci.*, 1999, **71**(12), 2033–2037.
[120] H. Fujita, T. Nakamura, J. Tamura, M. Kobayashi, Y. Katsura, T. Kokubo and T. Kikutani, 1998, **40**(1), 145–152.
[121] A. Sellinger, P.M. Weiss, A. Nguyen, Y. Lu, R.A. Assink and C.J. Brinker, *Proc. 1998 MRS Spring Symp. on Organic/Inorganic Hybrid Mater.*, Materials Research Society Symposium, Warrendale, PA, 1998, **519**, 95–101.
[122] M. Court, J.C. Viguie and A. Royer, *17th Annual Meeting of the Society for Biomaterials in Conjunction with the 23rd International Biomaterials Symposium*, Society for Biomaterials, Algonquin, IL, 1991, **14**, 90.
[123] J. Li, *Biomater.*, 1993, **14**(3), 229–232.
[124] P.L. Silva, J.D. Santos, F.J. lMonteiro and J.C. Knowles, *Surface and Coatings Tech.*, 1998, **102**(3), 191–196.
[125] M. Knepper, B.K. Milthrope and S. Moricca, *J. Mater. Sci. – Mater. in Medicine*, 1998, **9**(10), 589–596.
[126] K. Park and T. Vasilos, *J. Mater. Sci. Lett.*, 1997, **16**(12), 985–987.
[127] S. Maruno, S. Ban, Y.-F. Wang, H. Iwata and H. Itoh, *Nippon Seramikkusu Kyokai Gakujutsu Ronbunshi – J. Ceramic Soc. Japan*, 1992, **100**(1160), 362–367.
[128] S. Maruno, H. Iwata, S. Ban and H. Itoh, *Mater. Sci. Forum*, 1999, **308–311**, 344–349.
[129] D. Day, *Amer. Ceramic Soc. Bull.*, 1995, **74**(12), 64–68.
[130] P. Ducheyne and L.L. Hench, *J. Mater. Sci.*, 1982, **17**(2), 595–606.
[131] P. Ducheyne, *Publ. by Plastics and Rubber Inst.*, London, 1985, 8, 1–8, 15.
[132] E. Verne, C.V. Brovarone and D. Milanese, *J. Biomedical Mater. Res.*, 2000, **53**(4), 408–413.
[133] S. Begin-Colin, A. Mocellin, J. Von Stebut, K. Bordji and D. Mainard, *J. Mater. Sci.*, 1998, **33**(11), 2837–2843.
[134] D. Devlin, J. Cowie, D. Carroll and D. Rivero, *Proc. 1995 ASME Int. Mech. Eng. Congress and Exposition on Advances in Bioengineering*, 1995, **31**, 31–32.
[135] M. Lewandowska-Szumiel, J. Komender, A. Gorecki and M. Kowalski, *J. Mater. Sci. – Mater. in Medicine*, 1997, **8**(8), 485–488.
[136] S. Blazewicz, J. Chlopek, A. Litak, C. Wajler and E. Staszkow, *Biomater.*, 1997, **18**(5), 437–439.
[137] D. Devlin, D. Carroll, R. Barbero, J. Klawitter, P. Strzepa and W. Ogilive, *Proc. 1997 ASME Int. Mech. Eng. Congress and Exposition on Adv. in Bioengineering*, American Society of Mechanical Engineers, Bioengineering Division, Fairfield, NJ, 1997, **36**, 265.
[138] Y. Miyamoto, *Mater. Tech.*, 1996, **11**(6), 230–236.
[139] S. Li, Z. Zheng, Q. Liu, J.R. de Wijn and K. de Groot, *J. Biomedical Mater. Res.*, 1998, **40**(4), 520–529.
[140] V. Pesakova, Z. Klezl, K. Balik and M. Adam, *J. Mater. Sci. – Mater. Medicine*, 2000, **11**(12), 793–798.
[141] M. Mohanty, T.V. Kumary, A.V. Lal and R. Sivakumar, *Bull. Mater. Sci.*, 1998, **21**(6), 439–444.
[142] M. Lewandowska-Azumiel, J. Komender and J. Chlopek, *J. Biomedical Mater. Res.*, 1999, **48**(3), 289–296.

[143] K.A. Khor, C.S. Yip and P. Cheang, *J. Thermal Spray Tech.*, 1997, **6**(1), 109–115.
[144] K.A. Khor, Z.L. Dong, C.H. Quek and P. Cheang, *Mater. Sci. and Eng. A: Struct. Mater.: Properties, Microstructure and Processing*, 2000, **281**(1), 221–228.
[145] T. Nonami, A. Kamiya, K. Naganuma and T. Kameyana, *Mater. Sci. and Eng. C: Biomimetic Mater., Sensors & Systems*, 1998, **6**(4), 281–284.
[146] K.A. Khor, C.S. Yip and P. Cheang, *J. Mater. Processing Tech.*, 1997, **71**(2), 280–287.
[147] C. Mueller-Mai, H.-J. Schmitz, V. Strunz, G. Fuhrmann, T. Fritz and U.M. Gross, *J. Biomedical Mater. Res.*, 1989, **23**(10), 1149–1168.
[148] S.H. Teoh, R. Thampuran, K.H.W. Seah and J.C.H. Goh, *J. Mater. Sci. Lett.*, 1997, **16**(8), 639–641.
[149] S.H. Teoh, R. Thampuran and W.K.H. Seah, *Wear*, 1998, **214**(2), 237–244.
[150] H.J. Prosser and A.D. Wilson, *Mater. and Design*, 1986, **7**(5), 262–266.

12 Composite materials for vibration damping

12.1 Introduction

Vibrations are undesirable for structures, owing to the need for structural stability, position control, durability (particularly durability against fatigue), performance and noise reduction. Vibrations are of concern to large structures such as aircraft, as well as small structures such as electronics.

Vibration reduction can be attained by increasing the *damping capacity* (which is expressed by the *loss tangent*, tan δ) and/or increasing the stiffness (which is expressed by the *storage modulus*). The *loss modulus* is the product of these two quantities and thus can be considered a figure of merit for the vibration reduction ability.

Damping of a structure can be attained by passive or active methods. *Passive* methods make use of the inherent ability of certain materials (whether structural or non-structural materials) to absorb the vibrational energy, thereby providing passive energy dissipation. *Active* methods make use of sensors and actuators to attain vibration sensing and activation to suppress the vibration in real time. The sensors and actuators can be piezoelectric devices [1]. This chapter focuses on materials for passive damping because of its relatively low cost and ease of implementation.

Materials for vibration damping are mainly metals [2] and polymers [3], because of their viscoelastic character. Rubber is commonly used as a vibration damping material owing to its viscoelasticity [4]. However, viscoelasticity is not the only mechanism for damping. Defects such as dislocations, phase boundaries, grain boundaries and various interfaces also contribute to damping, since defects may move slightly and surfaces may slip with respect to one another during vibration, thereby dissipating energy. Thus, the microstructure greatly affects the damping capacity of a material [5]. The damping capacity depends not only on the material, but also on the loading frequency, as the viscoelasticity as well as defect response depend on the frequency. Moreover, the damping capacity depends on the temperature.

12.2 Metals for vibration damping

Metals for vibration damping include *shape-memory alloys* (SMAs), ferromagnetic alloys and other alloys. The SMAs provide damping for the reasons explained in the following paragraphs. Ferromagnetic alloys provide damping through the magnetomechanical mechanism (i.e., movement of the magnetic domain in boundaries during vibration). Other alloys provide damping through microstructural design. The last type is most commonly used because of their low cost. However, more than one mechanism may apply to the same alloy.

The shape-memory effect refers to the ability of a material to transform to a phase having a twinned microstructure that, after subsequent plastic deformation, can return the material to its initial shape when heated. The initial phase is called *austenite*. The highly twinned phase to which austenite transforms is called *martensite*. Martensite generally has less crystallographic symmetry than austenite. However, the twinning enables plastic deformation to occur easily. Austenite begins to transform to martensite at a temperature M_s upon cooling and the transformation is completed upon further cooling to temperature M_f. Deformation is applied to the martensite. It can occur through either the growth of favorably oriented twins or deformation twinning. Upon unloading and subsequent heating, martensite transforms back to austenite through reversal of the deformation mechanisms involving twinning and the shape recovers. Martensite begins to transform to austenite upon heating at a temperature A_s and the transformation is completed upon further heating to a temperature A_f. In general, $M_f < M_s < A_s < A_f$. The shape-memory transformation is reversible but has a large hysteresis.

The *martensitic transformation* may be induced by stress rather than by temperature. Beyond a certain stress, martensite starts to form from austenite and results in elastic elongation that exceeds the elasticity of ordinary alloys by a factor of 10 or more. Upon removal of the stress, the martensite changes back to austenite and the strain (shape) returns to the value prior to the martensitic transformation. This phenomenon occurs above A_f (i.e., for austenite) and is known as *superelasticity* (or *pseudoelasticity*). It is illustrated in Figure 12.1, where the stress has a plateau during loading and another plateau during unloading. Typically, a strain of 10% can be nearly fully recovered, out of which a strain of about 8% is due to the stress-induced martensitic transformation and the rest is to conventional elasticity. As shown in Figure 12.1, a superelastic material does not follow Hooke's law, but gives a nearly constant stress (the plateau) when strained typically between 1.5% and 7%.

The large hysteresis between loading and unloading in Figure 12.1 means that a significant part of the strain energy put into the SMA is dissipated as heat. This energy dissipation provides a mechanism for vibration damping. Furthermore, the motion of the coherent interfaces between martensite and austenite occurs quite easily under small stresses, thus causing the absorption of vibrational energy. As a result, SMAs tend to have excellent damping ability [6–18].

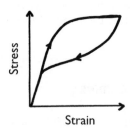

Figure 12.1 Stress–strain curve of an SMA at temperature $T > A_f$.

Alloys for vibration damping include those based on iron (e.g., cast iron, steel, Fe–Ni–Mn, Fe–Al–Si, Fe–Al, Fe–Cr, Fe–Cr–V, Fe–Mn and Fe–Mn–Co) [19–40], aluminum (e.g., Al–Ge, Al–Co, Al–Zn, Al–Cu, Al–Si, alloys 6061, 2017, 7022 and 6082) [41–48], zinc (e.g., Zn–Al) [49–52], lead [52], tin (e.g., Sn–In) [53], titanium (e.g., Ti–Al–V, Ti–Al–Sn–Zr–Mo and Ti–Al–Nb–V–Mo) [54–57], nickel (e.g., superalloys, Ni_3Al and NiAl) [58–60], zirconium (e.g., Zr–Ti–Al–Cu–Ni) [61,62], copper (e.g., Cu–Al–Zn–Cd) [63] and magnesium (e.g., Mg–Ca) [64,65]. In addition, metal–matrix composites (e.g., Al/SiC_p, $Al/graphite_p$, $Mg/carbon_f$, NiAl/AlN and Al–Cu/Al_2O_3) [66–73] and metal laminates (e.g., Fe/Cu) [74] have attractive damping ability.

Owing to the interface between reinforcement (particles, whiskers or fibers) and matrix in a composite, composite formation tends to increase the damping capacity, in addition to the well-known effect of increasing the stiffness. A high stiffness is useful for vibration reduction. However, metal–matrix composites are expensive to make and their competition with the high damping alloys is difficult. A particularly common form of composite is a laminate in which a high damping layer is sandwiched and constrained by stiff layers [75]. The shear deformation of the constrained layer provides damping, while the stiff layers allow structural use of the laminate.

12.3 Polymers for vibration damping

Due to their viscoelastic behavior, polymers (particularly thermoplastics) can provide damping [3,76,77]. Rubber is particularly well known for its damping ability. However, rubber suffers from low stiffness, which results in a rather low value of the loss modulus [78]. Other polymers used for vibration damping include polyurethane [79], a polypropylene/butyl rubber blend [80], a polyvinyl chloride/chlorinated polyethylene/epoxidized natural rubber blend [81], a polyimide/polyimide blend [82], a polysulfone/polysulfone blend [82], a nylon-6/polypropylene blend [83] and a urethane/acrylate interpenetrating polymer network [84]. In general, elastomers and other amorphous thermoplastics with a glass transition temperature below room temperature are attractive for damping. Polymer blends and interpenetrating networks are also attractive, owing to the interface between the components in the blend or network providing a mechanism for damping.

In relation to fibrous structural composites, viscoelastic polymeric interlayers between the laminae of continuous fibers are often used for damping [85–87]. However, the presence of the interlayer degrades the stiffness of the composite, particularly when the temperature is high (e.g., 50°C). The use of 0.1 μm diameter carbon filaments in place of the viscoelastic interlayer alleviates this problem and is particularly attractive when the temperature is high [88,89]. Figure 12.2 [89] shows the loss modulus of continuous carbon fiber polymer–matrix composites. The composite with the viscoelastic interlayer exhibits a higher value of the loss modulus than that with the composite with a filament interlayer at 25°C, but the reverse occurs at 50°C. Both composites are superior to the composite without interlayer. The large amount of interface between the 0.1 μm diameter filaments and the polymer matrix contributes to damping.

A related application involves sandwiching a high damping polymeric layer between steel layers in laminated steel [90].

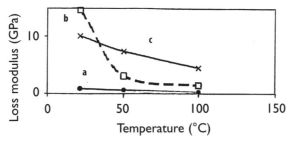

Figure 12.2 Effect of temperature on the loss modulus of continuous carbon fiber thermoplastic–matrix composite (longitudinal configuration) at 0.2 Hz: (**a**) composite without interlayer; (**b**) composite with viscoelastic interlayer; (**c**) composite with treated carbon filament interlayer.

12.4 Ceramics for vibration damping

Ceramics are not good for damping, but are high in stiffness. Nevertheless, the improvement of the damping capacity of structural ceramics is valuable for ceramic structures. The use of the structural material itself for damping reduces the need for non-structural damping materials, which tend to be limited in durability and temperature resistance, in addition to being low in stiffness.

The most widely used structural ceramic is concrete, a cement–matrix composite. The addition of silica fume (a fine particulate) as an admixture in the cement mix results in a large amount of interface and hence a significant increase in the damping capacity [91]. The addition of latex as an admixture also enhances damping, owing to the viscoelastic nature of latex [91]. The addition of sand or short 15 μm diameter carbon fibers to the mix does not help the damping [92,93], owing to the large unit size of these components and the relatively high damping associated with the inhomogeneity within cement paste.

The damping capacity of conventional ceramics [67] and of high-temperature ceramic–matrix composites (e.g., $MoSi_2/Si_3N_4$) [94] is also of interest.

12.5 Comparison among representative materials

Due to the differences in testing method and specimen configuration in the work of different researchers, quantitative comparison of the damping capacity of the numerous materials mentioned in this chapter is difficult. Nevertheless, Table 12.1 provides a comparison among representative materials (including polymers, metals, cement-based materials and metal–matrix and polymer–matrix composites), all tested in the author's laboratory by using the same method and equipment [77,78,88,91,92,95]. Among these classes of materials, polymers give the highest damping capacity (tan δ), whereas metals give the highest loss modulus. Although cement-based materials have less damping capacity than polymers, the loss modulus is comparable (except for plain mortar). The continuous carbon fiber polymer–matrix composites are worst in damping capacity, but the loss modulus is as high as those of metals if a vibration damping interlayer is used in the composite.

Table 12.1 Dynamic flexural behavior of materials at 0.2 Hz.

Material	tan δ	Storage modulus (GPa)	Loss modulus (GPa)	Ref.
Cement paste (plain)	0.016	13.7	0.22	91
Mortar (plain)	<10^{-4}	9.43	<0.001	92
Mortar with silica fume (treated) (15% by wt. of cement)	0.021	13.11	0.28	92
Aluminum, pure	0.019	51	1.0	95
Al/AlN$_p$ (58 vol.%)	0.025	120	3.0	95
Zn–Al	0.021	74	1.5	95
Zn–Al/SiC$_w$ (27 vol.%)	0.032	99	3.0	95
Carbon fiber epoxy–matrix composite (without interlayer)	0.008	101	0.8	88
Carbon fiber epoxy–matrix composite (with vibration damping interlayer)	0.017	92	1.6	88
Neoprene rubber	0.67	0.0075	0.0067	77
PTFE	0.189	1.2	0.23	78
PMMA	0.09	3.6	0.34	78
PA-66	0.04	4.4	0.19	78
Acetal	0.03	3.7	0.13	78
Epoxy	0.03	3.2	0.11	78

Neoprene rubber exhibits an outstandingly high value of tan δ, but its storage modulus is outstandingly low, so that its loss modulus is almost the lowest among the materials of Table 12.1. Among the thermoplastics polymethyl methacrylate (PMMA), PTFE, polyamide-66 (PA-66) and acetal, PMMA exhibits the highest loss modulus, while PTFE exhibits the highest loss tangent. Epoxy (a thermoset) and acetal exhibit the lowest loss tangent among the polymers listed in Table 12.1.

The loss tangent of cement paste (even the plain one, i.e., no admixture at all) is comparable to those of aluminum and carbon fiber epoxy–matrix composites, although the storage modulus is lower. Thus, cement paste itself has a high damping capacity, even without admixtures. Addition of sand to cement paste results in mortar, which exhibits a very low damping capacity. However, addition of silica fume to the mortar greatly increases the damping capacity, bringing the loss tangent back to the level of cement paste.

The carbon fiber epoxy–matrix composite without interlayer is poorer in damping than pure aluminum. However, the use of a filament interlayer in the composite increases the damping capacity, so that it is comparable to that of pure aluminum. On the other hand, the storage modulus is much higher for the composites than pure aluminum. As a result, the loss modulus is higher for the composite with interlayer than for pure aluminum.

Comparison of Al and Al/AlN$_p$ and of Zn–Al and Zn–Al/SiC$_w$ shows that composite formation increases both loss tangent and storage modulus. However, comparison of Al and Zn–Al shows that alloying also increases both quantities. Alloying is much less expensive than composite formation.

The loss modulus is lower for cement-based materials than for metal-based and polymer-based materials, owing to the low storage modulus of cement-based materials.

12.6 Summary

Materials for vibration damping include metals, polymers, ceramics and their composites. Metals and polymers tend to be better than cement for damping owing to their viscoelasticity. However, it is attractive to use a structural material (such as concrete) to provide some damping. Damping enhancement mainly involves microstructural design in the case of metals, interface design in the case of polymers, and admixture use in the case of cement.

Review questions

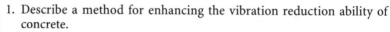

1. Describe a method for enhancing the vibration reduction ability of concrete.
2. What are the main reasons why a shape-memory alloy is good for vibration damping?
3. What are the main disadvantages of using a viscoelastic interlayer in a fiber composite for vibration reduction?
4. Why is mortar less effective for vibration damping than cement paste?
5. Why is the addition of silica fume to mortar effective for enhancing the vibration damping ability?
6. Define the loss modulus.
7. A viscoelastic material such as rubber is good for vibration damping. However, it has some disadvantages. Describe a main disadvantage.

References

[1] J.J. Hollkamp and R.W. Gordon, *Smart Mater. and Struct.*, 1996, **5**(5), 715.
[2] I.G. Ritchie and Z.-L. Pan, *Metallurgical Transactions A – Physical Metallurgy and Mater. Sci.*, 1991, **22A**(3), 607.
[3] E.M. Kerwin Jr and E.E. Ungar, *Proceedings of the ACS Division of Polymeric Materials: Science and Engineering*, ACS, Books and Journals Division, Washington, DC, 1989, Vol. 60, 816 pp.
[4] S.N. Ganeriwala, *Proceedings of SPIE – International Society for Optical Engineering*, Smart Structures and Materials 1995: Passive Damping, Society of Photo-Optical Instrumentation Engineers, Bellingham, 1995, Vol. 2445, 200 pp.
[5] R. de Batist, *J. de Physique* (Paris), Colloque C9, 1983, **44**(12), 39.
[6] H. Kawabe and K. Yoshida, *Bulletin of the Japan Soc. of Precision Eng.*, 1987, **21**(2), 132.
[7] Y. Sato and K. Tanaka, *Res Mechanica: Int. J. of Struct. Mechanics and Mater. Sci.*, 1988, **23**(4), 381.
[8] M. Kisaichi and Y. Takahashi, *Sumitomo Metals*, 1989, **41**(2), 167.
[9] D.-Y. Ju and A. Shimamoto, *J. Intelligent Mater. Systems and Struct.*, 2000, **10**(7), 514.
[10] P. Shi, D. Yang, F. Chen and H. Shen, *Dalian Ligong Daxue Xuebao/ J. of Dalian University of Tech.*, 2000, **40**(1), 83.
[11] W.D. Armstrong and P.G. Reinhall, *Proc. of SPIE – Int. Soc. for Optical Eng.*, Smart Structures and Materials 2000 – Active Materials: Behavior and Mechanics, Society of Photo-Optical Instrumentation Engineers, Bellingham, 2000, Vol. 3992, 509 pp.
[12] E.I. Rivin and L. Xu, *Proc. of the 1994 Int. Mech. Eng. Congress and Exposition*, Materials for Noise and Vibration Control, American Society of Mechanical Engineers, Noise Control and Acoustics Division, New York, 1994, Vol. 18, 35 pp.
[13] J. Van Humbeeck and Y. Liu, *Mater. Sci. Forum*, 2000, **327**, 331.
[14] R. Fosdick and Y. Ketema, *J. Intelligent Mater. Systems and Struct.*, 1999, **9**(10), 854.
[15] D.Z. Li and Z.C. Feng, *Proc. of SPIE – Int. Soc. for Optical Eng.*, Smart Structures and Materials 1997: Smart Structures and Integrated Systems, Society of Photo-Optical Instrumentation Engineers, Bellingham, 1997, Vol. 3041, 715 pp.

[16] Y. Furuya, *J. Intelligent Mater. Systems and Struct.*, 1996, **7**(3), 321.
[17] K.-H. Wu, S.K. Dalip, Y.Q. Liu and Z.J. Pu, *Proc. of SPIE – Int. Soc. for Optical Eng.*, Smart Structures and Materials 1995: Smart Materials, Society of Photo-Optical Instrumentation Engineers, Bellingham, 1995, Vol. 2441, 139 pp.
[18] R. Krumme, J. Hayes and S. Sweeney, *Proc. of SPIE – Int. Soc. for Optical Eng.*, Smart Structures and Materials 1995: Passive Damping, Society of Photo-Optical Instrumentation Engineers, Bellingham, 1995, Vol. 2445, 225 pp.
[19] A. Visnapuu, R.W. Nash and P.C. Turner, *Rep Invest US Bur Mines*, 1987, **9068**, 15 pp.
[20] Y.G. Dorofeev and N.S. Martirosyan, *Soviet Powder Metallurgy and Metal Ceramics*, 1988, **27**(2), 115.
[21] H. Ona, S. Ichikawa, T. Nakako and A. Takezoi, *J. Mater. Processing Tech.*, 1990, **23**(1), 7.
[22] S. Watanabe, S. Sato, I. Nakagami and S. Nagashima, *Tetsu to Hagane – J. Iron and Steel Inst. Japan*, 1991, **77**(2), 306.
[23] T. Yamada, T. Takamura, S. Hashizume, T. Odake, T. Omori and K. Hattori, *Nkk Tech. Rev.*, 1992, (65), 21.
[24] M. Sato, Y. Tanaka, Y. Yutori, H. Nishikawa and M. Miyahara, *SAE (Soc. of Automotive Eng.) Transactions*, 1991, **100**(Sect. 5), 309.
[25] J. Lu, X. Liu, W. Zheng, B. Wu and H. Bi, *J. Mater. Sci. and Tech.*, 1993, **9**(4), 293.
[26] K. Aoi, *J. Mech. Eng. Lab.*, 1994, **48**(2), 58.
[27] V.A. Udovenko, S.I. Tishaev and I.B. Chudakov, *Izvestiya an Sssr: Metally*, 1994, (1), 98.
[28] S.-H. Baik, *Nuclear Eng. and Design*, 2000, **198**(3), 241.
[29] J.-H. Jun, D.-K. Kong and C.-S. Choi, *Mater. Res. Bulletin*, 1998, **33**(10), 1419.
[30] J.-H. Jun, S.-H. Baik, Y.-K. Lee and C.-S. Choi, *Scripta Materialia*, 1998, **39**(1), 39.
[31] V.G. Gavriljuk, P.G. Yakovenko and K. Ullakko, *Scripta Materialia*, 1998, **38**(6), 931.
[32] W. Hermann, *Proc. of the 1998 TMS Fall Meeting*, Interstitial and Substitutional Solute Effects in Intermetallics, Minerals, Metals and Materials Society, Warrendale, 1998, 51 pp.
[33] S.-H. Baik and N.-H. Kim, *Proc. of the 1998 ASME/JSME Joint Pressure Vessels and Piping Conf.*, Seismic, Shock, and Vibration Isolation, American Society of Mechanical Engineers, Pressure Vessels and Piping Division, Fairfield, 1998, Vol. 379, 149 pp.
[34] A.I. Skvortsov and V.M. Kondratov, *Problemy Prochnosti*, 1998, (2), 2.
[35] A. Karimi, P.H. Giauque and J.L. Martin, *ASTM Special Tech. Pub.*, 1997, **1304**, 115.
[36] I. Aaltio and K. Ullakko, *Smart Mater. and Struct.*, 1997, **6**(5), 616.
[37] K.K. Jee, W.Y. Jang, S.H. Baik, M.C. Shin and C.S. Choi, *Scripta Materialia*, 1997, **37**(7), 943.
[38] A. Karimi, P.H. Giauque, J.L. Martin, G. Barbezat and A. Salito, *J. de Physique IV*, 1996, **6**(8), 779.
[39] I. Aaltio, K. Ullakko and H. Hanninen, *Proc. of SPIE – Int. Soc. for Optical Eng.*, Smart Structures and Materials 1996: Passive Damping and Isolation, Society of Photo-Optical Instrumentation Engineers, Bellingham, 1996, Vol. 2720, 378 pp.
[40] A. Karimi, P.H. Giauque and J.L. Martin, *Mater. Sci. Forum*, 1995, **179–181**, 679.
[41] J.C. Roughan and D. Hearnshaw, *IEEE Conf. Pub.*, (297), 60.
[42] M. Hinai, S. Sawaya and H. Masumoto, *Mater. Transactions Jim*, 1991, **32**(10), 957.
[43] J. Zhang, M.N. Gungor and E.J. Lavernia, *J. Mater. Sci.*, 1993, **28**(6), 1515.
[44] M. Hinai, S. Sawaya and H. Masumoto, *Mater. Transactions Jim*, 1993, **34**(4), 359.
[45] K. Kondoh, M. Hashikura and Y. Takeda, *J. Japan Soc. of Powder and Powder Metallurgy*, 1999, **46**(7), 715.
[46] C.Y. Xie, R. Schaller and C. Jaquerod, *Mater. Sci. and Eng. A: Structural Materials, Microstructure and Processing*, 1998, (1), 78.
[47] B.-C. Moon and Z.-H. Lee, *Proc. of the 1997 TMS Annual Meeting*, Synthesis/Processing of Lightweight Metallic Materials, Minerals, Metals and Materials Society, Warrendale, 1997, 127 pp.
[48] S. Sgobba, L. Parrini, H.-U. Kunzi and B. Ilschner, *Metallurgical and Mater. Transactions A – Physical Metallurgy and Mater. Sci.*, 1995, **26A**(10), 2745.
[49] I.G. Ritchie, Z.-L. Pan and F.E. Goodwin, *Metallurgical and Mater. Transactions A – Physical Metallurgy and Mater. Sci.*, 1991, **22A**(3), 617.
[50] M. Gu, Z. Chen, Z. Wang, Y. Jin, J. Huang and G. Zhang, *Scripta Metallurgica et Materialia*, 1994, **30**(10), 1321.
[51] Y. Liu, G. Yang, Y. Lu and L. Yang, *J. Mater. Processing Tech.*, 1999, **87**(1–3), 53.
[52] F.E. Goodwin, *Adv. Mater. and Processing*, 1996, **150**(4), 2 pp.
[53] A. Dooris, R.S. Lakes, B. Myers and N. Stephens, *Mechanics of Time-Dependent Mater.*, 1999, **3**(4), 305.
[54] Y.T. Lee and G. Welsch, *Mater. Sci. and Eng. A: Structural Mater.: Properties, Microstructure and Processing*, 1990, (1), 77.
[55] J.E. Grady and B.A. Lerch, *SAMPE Quarterly – Society for the Advancement of Mater. and Process

Eng., 1992, **23**(2), 11.
[56] Wolfenden, M.S. McGuff and R.U. Vaidya, *J. Adv. Mater.*, 2000, **32**(4), 60.
[57] A. Wolfenden, M.W. Canty and R.U. Vaidya, *J. Adv. Mater.*, 1996, **27**(3), 14.
[58] P. Gaduad, A. Riviere and J. Woirgard, *ASTM Special Tech. Pub.*, 1991, (1169), 447.
[59] W. Hermann and H.-G. Sockel, *ASTM Special Tech. Pub.*, 1997, (1304), 143.
[60] W. Hermann, T.V. Ort and H.G. Sockel, *J. de Physique IV*, 1996, **6**(8), 223.
[61] A. Wolfenden, L.M. Steinocher and L.Q. Xing, *J. Mater. Sci. Letters*, 2000, **19**(12), 1099.
[62] A. Wolfenden, K.A. Barrios and L.Q. Xing, *J. Mater. Sci. Letters*, 1998, **17**(13), 1095.
[63] O.G. Zotov, S.Y. Kondrat'ev, G.Y. Yaroslavskii, B.S. Chaikovskii and V.V. Matveev, *Problemy Prochnosti*, 1985, (11), 78.
[64] D.L. Hallum, *American Machinist*, 1995, **139**(5), 48.
[65] Z. Trojanova, P. Lukac, S. Kraft, W. Riehemann and B.L. Mordike, *Mater. Sci. Forum*, 1996, **210-213**(pt 2), 825.
[66] R.J. Perez, J. Zhang and E.J. Lavernia, *Scripta Metallurgica et Materialia*, 1992, **27**(9), 1111.
[67] J. Zhang, R.J. Perez and E.J. Lavernia, *J. Mater. Sci.*, 1993, **28**(9), 2395.
[68] S. Ray, V.K. Kinra and C. Zhu, *Proc. of the 1993 ASME Winter Annual Meeting*, Dynamic Characterization of Advanced Materials, American Society of Mechanical Engineers, Noise Control and Acoustics Division, New York, 1993, Vol. 16, 69 pp.
[69] A. Wolfenden, C.M. Miller and M.G. Hebsur, *J. Mater. Sci. Letters*, 1998, **17**(21), 1861.
[70] L. Parrini and R. Schaller, *Mater. Sci. Forum*, 1996, **210-213**(pt 2), 627.
[71] A. Wolfenden, M. Jackson and N. Martirosian, *J. Mater. Sci.*, 1996, **31**(7), 1815.
[72] E.J. Lavernia, R.J. Perez and J. Zhang, *Metallurgical and Mater. Transactions A – Physical Metallurgy and Mater. Sci.*, 1995, **26A**(11), 2803.
[73] E.J. Lavernia, J. Zhang and R.J. Perez, *Key Eng. Mater.*, 1995, **104-107**(pt 2), 691.
[74] N. Igata, M. Sasaki, Y. Kogo and K. Hishitani, *J. de Physique IV*, 2000, **10**(6), Pr6-113–Pr6-117.
[75] G.X. Sui, G.H. He, L.Y. Bai and B.L. Zhou, *J. Mater. Sci. Letters*, 1995, **14**(17), 1218.
[76] S.N. Ganeriwala and H.A. Hartung, *Proc. of the ACS Division of Polymeric Materials: Science and Engineering*, ACS, Books and Journals Division, Washington, DC, 1989, Vol. 60, 605 pp.
[77] X. Luo and D.D.L. Chung, *Carbon*, 2000, **38**(10), 1510.
[78] W. Fu and D.D.L. Chung, *Polymers and Polymer Composites*, 2001, **9**(6), 423–426.
[79] J.-S.G. Lin, C.H. Newton and J.A. Manson, *Proc. of the 45th Annual Tech. Conf. – Society of Plastics Engineers*, Society of Plastics Engineers, Brookfield Center, 1987, 478 pp.
[80] F.-S. Liao, T.-C. Hsu and A.C. Su, *J. Applied Polym. Sci.*, 1993, **48**(10), 1801.
[81] N. Yamada, S. Shoji, H. Sasaki, A. Nagatani, K. Yamaguchi, S. Kohjiya and A.S. Hashim, *J. Applied Polym. Sci.*, 1999, **71**(6), 855.
[82] R.P. Chartoff, J.M. Butler, R.S. Venkatachalam and D.E. Miller, *Proc. of the 41st Annual Tech. Conf.*, Society of Plastics Engineers, Brookfield Center, 1983, 360 pp.
[83] K. Sasaki, T. Okumoto and J. Koizumi, *SAE Technical Paper Series*, 1990, 7 pp.
[84] C.-J. Tung and T.-C.J. Hsu, *J. Applied Polym. Sci.*, 1992, **46**(10), 1759.
[85] J.M. Pereira, *Proc. of the 1992 ASME Winter Annual Metting*, Dynamic Characterization of Advanced Materials, American Society of Mechanical Engineers, Noise Control and Acoustics Division, New York, 1992, Vol. 14, 51 pp.
[86] T.E. Alberts and H. Xia, *J. Vibration and Acoustics – Transactions of the ASME*, 1995, **117**(4), 398.
[87] K.A. Alsweify, C. Booker, E.I. Elghandour, F.A. Kolkailah and S.H. Farghaly, *Proc. of the 43rd Int. SAMPE Symp. and Exhibition*, SAMPE, Covina, 1998, **43**(1), 426 pp.
[88] S.W. Hudnut and D.D.L. Chung, *Carbon*, 1995, **33**(11), 1627.
[89] M. Segiet and D.D.L. Chung, *Composite Interfaces*, 2000, **7**(4) 257.
[90] C. Wang, Q. Peng, J. Liu, X. Sun, J. Zhu and Q. Cai, *Acta Metallurgica Sinica* (English edition), Series A: Physical Metallurgy and Materials Science, 1994, **7**(2), 114.
[91] X. Fu, X. Li and D.D.L. Chung, *J. Mater. Sci.*, 1998, **33**, 3601.
[92] Y. Wang and D.D.L. Chung, *Cem. Concr. Res.*, 1998, **28**(10), 455.
[93] Y. Xu and D.D.L. Chung, *Cem. Concr. Res.*, 1999, **29**(7), 1107.
[94] G.T. Olsen, A. Wolfenden and M.G. Hebsur, *J. Mater. Eng. and Perf.*, 2000, **9**(1), 116.
[95] S.W. Parks, M.S. project, Department of Mechanical and Aerospace Engineering, State University of New York at Buffalo, 1995.

13 Intrinsically smart structural composites

13.1 Introduction

Smart structures are important because of their relevance to hazard mitigation, structural vibration control, structural health monitoring, transportation engineering and thermal control. Research on smart structures has emphasized the incorporation of various devices in a structure for providing sensing, energy dissipation, actuation, control or other functions. Research on smart composites has emphasized the incorporation of a smart material in a matrix material for enhancing smartness or durability. Research on smart materials has emphasized the study of materials (e.g., piezoelectric materials) used for making the devices. However, relatively little attention has been given to the development of structural materials (e.g., concrete and composites) that are inherently able to provide some of the smart functions, so that the need for embedded or attached devices is reduced or eliminated, thereby lowering cost, enhancing durability, increasing the smart volume and minimizing mechanical property degradation (which usually occurs in the case of embedded devices).

Smart structures are those that have the ability to sense certain stimuli and are able to respond to them in an appropriate fashion, somewhat like a human being. Sensing is the most fundamental aspect of a smart structure. A structural composite which is itself a sensor is multifunctional.

This chapter focuses on structural composites for smart structures. It addresses cement–matrix and polymer–matrix composites. The smart functions addressed include strain sensing (for structural vibration control and traffic monitoring), damage sensing (both mechanical and thermal damage in relation to structural health monitoring), temperature sensing (for thermal control, hazard mitigation and structural performance control), thermoelectricity (for thermal control and energy saving) and vibration reduction (for structural vibration control). These functional abilities of structural composites have been shown in the laboratory. Applications in the field are forthcoming.

13.2 Cement–matrix composites for smart structures

Cement–matrix composites include concrete (containing coarse and fine aggregates), mortar (containing fine aggregate but no coarse aggregate) and cement paste (containing no aggregate, whether coarse or fine). Other fillers, called admixtures, can be added to the mix to improve the properties of the composite. Admixtures are discontinuous, so that they can be included in the mix. They can be particles, such as silica fume (a fine particulate) and latex (a polymer in the form of a dispersion). They can be short fibers, such as polymer, steel, glass or carbon fibers. They can be liquids such as methylcellulose aqueous solution, water reducing agent, defoamer, etc. Admixtures for rendering the composite smart while maintaining or even improving the structural properties are the focus of this section.

13.2.1 Background on cement–matrix composites

Cement–matrix composites (Section 2.3) for smart structures include those containing short carbon fibers (for sensing strain, damage and temperature and for thermal control), short steel fibers (for sensing temperature and for thermal control) and silica fume (for vibration reduction). This section provides background on cement–matrix composites, with emphasis on carbon fiber cement–matrix composites because of their dominance among intrinsically smart cement–matrix composites.

13.2.2 Cement–matrix composites for strain sensing

The electrical resistance of strain-sensing concrete (without embedded or attached sensors) changes reversibly with strain, such that the gage factor (fractional change in resistance per unit strain) is up to 700 under compression or tension [1–16]. The resistance (DC/AC) increases reversibly upon tension and decreases reversibly upon compression, owing to fiber pull-out upon microcrack opening (<1 μm) and the consequent increase in fiber–matrix contact resistivity. The concrete contains as low as 0.2 vol.% short carbon fibers, which are preferably those that have been surface treated. The fibers do not need to touch one another in the composite. The treatment improves wettability with water. The presence of a large aggregate decreases the gage factor, but the strain-sensing ability remains sufficient for practical use. Strain-sensing concrete works even when data acquisition is wireless. The applications include structural vibration control and traffic monitoring.

Figure 13.1 shows the fractional change in resistivity along the stress axis as well as the strain during repeated compressive loading at an increasing stress amplitude for carbon fiber latex cement paste at 28 days of curing. The strain varies linearly with stress up to the highest stress amplitude. The strain returns to zero at the end of each cycle of loading. Resistivity decreases upon loading in every cycle (due to fiber push-in) and increases upon unloading in every cycle (due to fiber pull-out). The resistivity has a net increase after the first cycle, due to damage. Little further damage occurs in subsequent cycles, as shown by the resistivity after unloading not increasing much after the first cycle. The greater the strain amplitude, the more is the resistivity decrease during loading, although the resistivity and strain are not linearly related. The effects of Figure 13.1 were similarly observed in carbon fiber silica-fume cement paste at 28 days of curing.

Figures 13.2 and 13.3 show the fractional changes in the longitudinal and transverse resistivities respectively for carbon fiber silica-fume cement paste at 28

Intrinsically smart structural composites 255

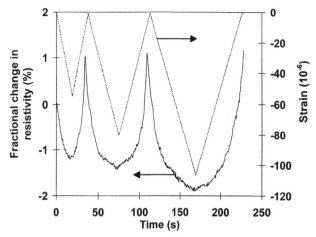

Figure 13.1 Variation of the fractional change in volume electrical resistivity with time and of the longitudinal strain (negative for compressive strain) with time during dynamic compressive loading at increasing stress amplitudes within the elastic regime for carbon fiber latex cement paste at 28 days of curing.

days of curing during repeated uniaxial tensile loading at increasing strain amplitudes. The strain essentially returns to zero at the end of each cycle, indicating elastic deformation. The longitudinal strain is positive (i.e., elongation); the transverse strain is negative (i.e., shrinkage due to the Poisson effect). Both longitudinal and transverse resistivities increase reversibly upon uniaxial tension. The reversibility of both strain and resistivity is more complete in the longitudinal direction than the transverse direction. The gage factor is 89 and −59 for the longitudinal and transverse resistances respectively.

Figures 13.4 and 13.5 show corresponding results for silica-fume cement paste. The strain is essentially totally reversible in both the longitudinal and transverse directions, but the resistivity is only partly reversible in both directions, in

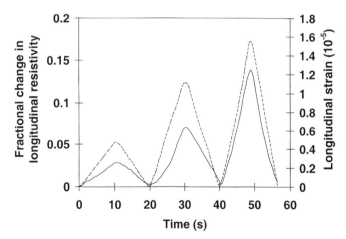

Figure 13.2 Variation of the fractional change in longitudinal electrical resistivity with time (solid curve) and of the longitudinal strain with time (dashed curve) during dynamic uniaxial tensile loading at increasing stress amplitudes within the elastic regime for carbon fiber silica-fume cement paste.

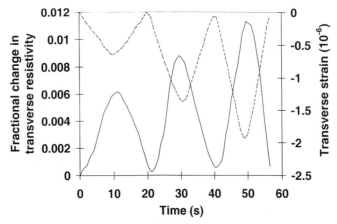

Figure 13.3 Variation of the fractional change in transverse electrical resistivity with time (solid curve) and of the transverse strain with time (dashed curve) during dynamic uniaxial tensile loading at increasing stress amplitudes within the elastic regime for carbon fiber silica-fume cement paste.

contrast to the reversibility of the resistivity when fibers are present (Figures 13.2 and 13.3). As in the case with fibers, both longitudinal and transverse resistivities increase upon uniaxial tension. However, the gage factor is only 7.2 and −7.1 for Figures 13.4 and 13.5 respectively.

Comparison of Figures 13.2 and 13.3 (with fibers) with Figures 13.4 and 13.5 (without fibers) shows that fibers greatly enhance the magnitude and reversibility of the resistivity effect. The gage factors are much smaller in magnitude when fibers are absent.

The increase in both longitudinal and transverse resistivities upon uniaxial tension for cement pastes, whether with or without fibers, is attributed to defect (e.g., microcrack) generation. In the presence of fibers, fiber bridging across

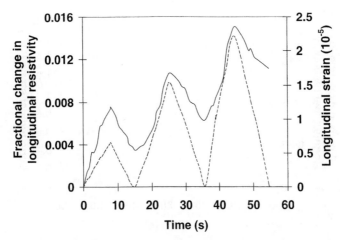

Figure 13.4 Variation of the fractional change in longitudinal electrical resistivity with time (solid curve) and of the longitudinal strain with time (dashed curve) during dynamic uniaxial tensile loading at increasing stress amplitudes within the elastic regime for silica-fume cement paste.

Intrinsically smart structural composites

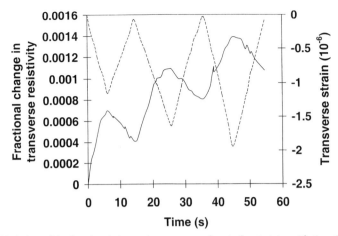

Figure 13.5 Variation of the fractional change in transverse electrical resistivity with time (solid curve) and of the transverse strain with time (dashed curve) during dynamic uniaxial tensile loading at increasing stress amplitudes within the elastic regime for silica-fume cement paste.

microcracks occurs and slight fiber pull-out occurs upon tension, thus enhancing the possibility of microcrack closing and causing more reversibility in the resistivity change. The fibers are much more electrically conductive than the cement matrix. The presence of fibers introduces interfaces between fibers and matrix. The degradation of the fiber–matrix interface due to fiber pull-out or other mechanisms is an additional type of defect generation which will increase the resistivity of the composite. Therefore, the presence of fibers greatly increases the gage factor.

The transverse resistivity increases upon uniaxial tension, even though the Poisson effect causes the transverse strain to be negative. This means that the effect of transverse resistivity increase overshadows the effect of the transverse shrinkage. The resistivity increase is a consequence of the uniaxial tension. In contrast, under uniaxial compression, the resistance in the stress direction decreases at 28 days of curing. Hence, the effects of uniaxial tension on the transverse resistivity and of uniaxial compression on the longitudinal resistivity are different; the gage factors are negative and positive for these cases respectively.

The similarity of the resistivity change in longitudinal and transverse directions under uniaxial tension suggests similarity for other directions as well. This means that the resistance can be measured in any direction in order to sense the occurrence of tensile loading. Although the gage factor is comparable in both longitudinal and transverse directions, the fractional change in resistance under uniaxial tension is much higher in the longitudinal direction than the transverse direction. Thus, the use of the longitudinal resistance for practical self-sensing is preferred.

13.2.3 Cement–matrix composites for damage sensing

Concrete, whether with or without admixtures, is capable of sensing major and minor damage – even damage during elastic deformation – due to the electrical resistivity increase that accompanies damage [2,6,17–19]. That both strain and damage can be sensed simultaneously through resistance measurement means that the strain/stress condition (during dynamic loading) under which damage occurs can be obtained, thus facilitating damage origin identification. Damage is indicated by a resistance increase, which is larger and less reversible when the stress amplitude is higher. The resistance increase can be a sudden increase during loading. It can also be a gradual shift of the baseline resistance.

Figure 13.6 [18] shows the fractional change in resistivity along the stress axis as well as the strain during repeated compressive loading at an increasing stress amplitude for plain cement paste at 28 days of curing. The strain varies linearly with stress up to the highest stress amplitude. The strain returns to zero at the end of each cycle of loading. During the first loading, resistivity increases due to damage initiation. During the subsequent unloading, resistivity continues to increase, probably due to opening of the microcracks generated during loading. During the second loading, resistivity decreases slightly as the stress increases up to the maximum stress of the first cycle (probably due to closing of the microcracks) and then increases as the stress increases beyond this value (probably due to the generation of additional microcracks). During unloading in the second cycle, resistivity increases significantly (probably due to opening of the microcracks). During the third loading, resistivity essentially does not change (or decreases very slightly) as the stress increases to the maximum stress of the third cycle (probably due to the balance between microcrack generation and microcrack closing). Subsequent unloading causes the resistivity to increase very significantly (probably due to opening of the microcracks).

Figure 13.7 shows the fractional change in resistance and strain during repeated compressive loading at increasing and decreasing stress amplitudes for carbon fiber (0.18 vol.%) concrete (with fine and coarse aggregates) at 28 days of curing. The highest stress amplitude is 60% of the compressive strength. A group of cycles in which the stress amplitude increases cycle by cycle and then decreases cycle by cycle back to the initial low stress amplitude is referred to as a "group". Figure 13.7 shows the results for three groups. The strain returns to zero at the end

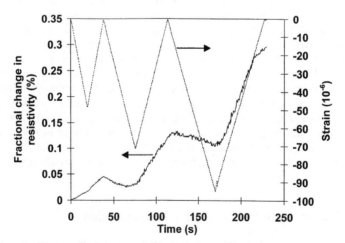

Figure 13.6 Variation of the fractional change in electrical resistivity with time and of the longitudinal strain (negative for compressive strain) with time during dynamic compressive loading at increasing stress amplitudes within the elastic regime for silica-fume cement paste at 28 days of curing.

of each cycle for any of the stress amplitudes, indicating elastic behavior. The resistance decreases upon loading in each cycle, as in Figure 13.1. An extra peak at the maximum stress of a cycle grows as the stress amplitude increases, resulting in two peaks per cycle. The original peak (strain induced) occurs at zero stress, while the extra peak (damage induced) occurs at the maximum stress. Hence, during loading from zero stress within a cycle, the resistance drops and then increases sharply, reaching the maximum resistance of the extra peak at the maximum stress of the cycle. Upon subsequent unloading, the resistance decreases and then increases as unloading continues, reaching the maximum resistance of the original peak at zero stress. In the part of this group where the stress amplitude decreases cycle by cycle, the extra peak diminishes and disappears, leaving the original peak as the sole peak. In the part of the second group where the stress amplitude increases cycle by cycle, the original peak (peak at zero stress) is the sole peak, except that the extra peak (peak at maximum stress) returns in a minor way (more minor than in the first group) as the stress amplitude increases. The extra peak grows as the stress amplitude increases, but, in the part of the second group in which the stress amplitude decreases cycle by cycle, it quickly diminishes and vanishes, as in the first group. Within each group, the amplitude of resistance variation increases as the stress amplitude increases and decreases as the stress amplitude subsequently decreases.

The greater the stress amplitude, the larger and the less reversible is the damage-induced resistance increase (the extra peak). If the stress amplitude has been experienced before, the damage-induced resistance increase (the extra peak) is small, as shown by comparing the result of the second group with that of the first group (Figure 13.7), unless the extent of damage is large (see Figure 13.8 for a highest stress amplitude of >90% the compressive strength). When the damage is extensive (as shown by a modulus decrease), damage-induced resistance increase occurs in every cycle, even at a decreasing stress amplitude, and it can overshadow the strain-induced resistance decrease (Figure 13.8). Hence, the damage-induced resistance increase occurs mainly during loading (even within the elastic regime), particularly at a stress above that in prior cycles, unless the

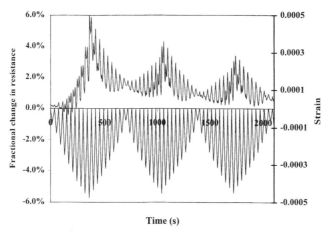

Figure 13.7 Fractional change in resistance and strain during repeated compressive loading at increasing and decreasing stress amplitudes, the highest of which was 60% of the compressive strength, for carbon fiber concrete at 28 days of curing.

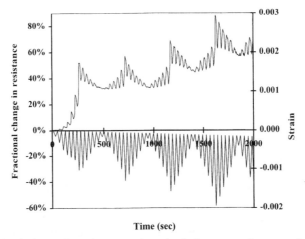

Figure 13.8 Fractional change in resistance and strain during repeated compressive loading at increasing and decreasing stress amplitudes, the highest of which was >90% of the compressive strength, for carbon fiber concrete at 28 days of curing.

stress amplitude is high and/or damage is extensive.

At a high stress amplitude, the damage-induced resistance increase cycle by cycle as the stress amplitude increases causes the baseline resistance to increase irreversibly (Figure 13.8). The baseline resistance in the regime of major damage (with a decrease in modulus) provides a measure of the extent of damage (i.e., condition monitoring). This measure works in the loaded or unloaded state. In contrast, the measure using the damage-induced resistance increase (Figure 13.7) works only during stress increase and indicates the occurrence of damage (whether minor or major) as well as the extent of damage.

13.2.4 Cement–matrix composites for temperature sensing

A thermistor is a thermometric device consisting of a material (typically a semiconductor, but in this case a cement paste) whose electrical resistivity decreases with rise in temperature. The carbon fiber concrete described in Section 13.2.2 for strain sensing is a thermistor owing to its resistivity decreasing reversibly with increasing temperature [20]; the sensitivity is comparable to that of semiconductor thermistors. (The effect of temperature will need to be compensated in using the concrete as a strain sensor, Section 13.2.2.)

Figure 13.9 [20] shows the current–voltage characteristic of carbon fiber (0.5% by weight of cement) silica-fume (15% by weight of cement) cement paste at 38°C during stepped heating. The characteristic is linear below 5 V and deviates positively from linearity beyond 5 V. The resistivity is obtained from the slope of the linear portion. The voltage at which the characteristic starts to deviate from linearity is referred to as the critical voltage.

Figure 13.10 shows a plot of the resistivity vs. temperature during heating and cooling for carbon fiber silica-fume cement paste. The resistivity decreases upon heating and the effect is quite reversible upon cooling. That the resistivity is slightly increased after a heating–cooling cycle is probably due to thermal

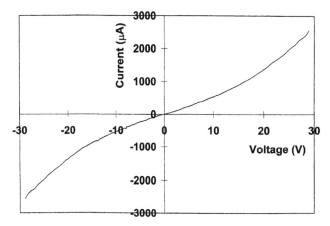

Figure 13.9 Current–voltage characteristic of carbon fiber silica-fume cement paste at 38°C during stepped heating.

degradation of the material. Figure 13.11 shows the Arrhenius plot of log conductivity (conductivity = 1/resistivity) vs. reciprocal absolute temperature. The slope of the plot gives the activation energy, which is 0.390 ± 0.014 and 0.412 ± 0.017 eV during heating and cooling respectively.

Results similar to those of carbon fiber silica-fume cement paste were obtained with carbon fiber (0.5% by weight of cement) latex (20% by weight of cement) cement paste, silica-fume cement paste, latex cement paste and plain cement paste. However, for all these four types of cement paste, (i) the resistivity is higher by about an order of magnitude, and (ii) the activation energy is lower by about an order of magnitude, as shown in Table 13.1. The critical voltage is higher when fibers are absent (Table 13.1).

The Seebeck [21–24] effect is a thermoelectric effect which is the basis for thermocouples for temperature measurement. This effect involves charge carriers

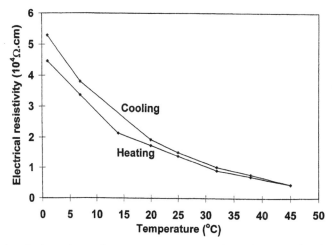

Figure 13.10 Plot of volume electrical resistivity vs. temperature during heating and cooling for carbon fiber silica-fume cement paste.

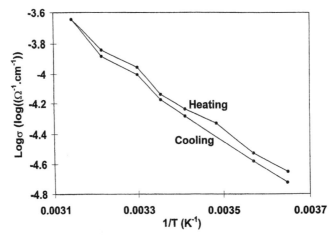

Figure 13.11 Arrhenius plot of log electrical conductivity vs. reciprocal absolute temperature for carbon fiber silica-fume cement paste.

moving from a hot point to a cold point within a material, thereby resulting in a voltage difference between the two points. The Seebeck coefficient is the negative of the voltage difference (hot minus cold) per unit temperature difference between the two points (hot minus cold). Negative carriers (electrons) make it more negative and positive carriers (holes) make it more positive.

The Seebeck effect in carbon fiber-reinforced cement paste involves electrons from the cement matrix [23] and holes from the fibers [21,22], such that the two contributions are equal at the percolation threshold, a fiber content between 0.5% and 1.0% by weight of cement [23]. The hole contribution increases monotonically with increasing fiber content below and above the percolation threshold [23].

Owing to the free electrons in a metal, cement containing metal fibers such as steel fibers is even more negative in thermoelectric power than cement without fiber [24]. The attainment of a very negative thermoelectric power is attractive, since a material with a positive thermoelectric power and a material with negative thermoelectric power are two very dissimilar materials, the junction of which is a thermocouple junction. (The greater the dissimilarity, the more sensitive is the thermocouple.)

Table 13.2 and Fig. 13.12 show the thermopower results. The absolute thermoelectric power is much more negative for all the steel fiber cement pastes compared to all the carbon fiber cement pastes. An increase of the steel fiber content from 0.5% to 1.0% by weight of cement makes the absolute thermoelectric

Table 13.1 Resistivity, critical voltage and activation energy of five types of cement paste.

Formulation	Resistivity at 20°C (Ω.cm)	Critical voltage at 20°C (V)	Activation energy (eV)	
			Heating	Cooling
Plain	$(4.87 \pm 0.37) \times 10^5$	10.80 ± 0.45	0.040 ± 0.006	0.122 ± 0.006
Silica fume	$(6.12 \pm 0.15) \times 10^5$	11.60 ± 0.37	0.035 ± 0.003	0.084 ± 0.004
Carbon fibers + silica fume	$(1.73 \pm 0.08) \times 10^4$	8.15 ± 0.34	0.390 ± 0.014	0.412 ± 0.017
Latex	$(6.99 \pm 0.12) \times 10^5$	11.80 ± 0.31	0.017 ± 0.001	0.025 ± 0.002
Carbon fibers + latex	$(9.64 \pm 0.08) \times 10^4$	8.76 ± 0.35	0.018 ± 0.001	0.027 ± 0.002

Intrinsically smart structural composites

Table 13.2 Volume electrical resistivity, Seebeck coefficient ($\mu V/°C$) with copper as the reference, and absolute thermoelectric power ($\mu V/°C$) of various cement pastes with steel fibers (S_f) or carbon fibers (C_f).

Cement paste	Volume fraction fibers	Resistivity ($\Omega \cdot cm$)	Seebeck coefficient	Absolute thermoelectric power	Seebeck coefficient	Absolute thermoelectric power
S_f (0.5[a])	0.10%	$(7.8 \pm 0.5) \times 10^4$	-51.0 ± 4.8	-53.3 ± 4.8	-45.3 ± 4.4	-47.6 ± 4.4
S_f (1.0[a])	0.20%	$(4.8 \pm 0.4) \times 10^4$	-56.8 ± 5.2	-59.1 ± 5.2	-53.7 ± 4.9	-56.0 ± 4.9
S_f (0.5[a]) + SF	0.10%	$(5.6 \pm 0.5) \times 10^4$	-54.8 ± 3.9	-57.1 ± 3.9	-52.9 ± 4.1	-55.2 ± 4.1
S_f (1.0[a]) + SF	0.20%	$(3.2 \pm 0.3) \times 10^4$	-66.2 ± 4.5	-68.5 ± 4.5	-65.6 ± 4.4	-67.9 ± 4.4
S_f (0.5[a]) + L	0.085%	$(1.4 \pm 0.1) \times 10^5$	-48.1 ± 3.2	-50.4 ± 3.2	-45.4 ± 2.9	-47.7 ± 2.9
S_f (1.0[a]) + L	0.17%	$(1.1 \pm 0.1) \times 10^5$	-55.4 ± 5.0	-57.7 ± 5.0	-54.2 ± 4.5	-56.5 ± 4.5
C_f (0.5[a]) + SF	0.48%	$(1.5 \pm 0.1) \times 10^4$	$+1.45 \pm 0.09$	-0.89 ± 0.09	$+1.45 \pm 0.09$	-0.89 ± 0.09
C_f (1.0[a]) + SF	0.95%	$(8.3 \pm 0.5) \times 10^2$	$+2.82 \pm 0.11$	$+0.48 \pm 0.11$	$+2.82 \pm 0.11$	$+0.48 \pm 0.11$
C_f (0.5[a]) + L	0.41%	$(9.7 \pm 0.6) \times 10^4$	$+1.20 \pm 0.05$	-1.14 ± 0.05	$+1.20 \pm 0.05$	-1.14 ± 0.05
C_f (1.0[a]) + L	0.82%	$(1.8 \pm 0.2) \times 10^3$	$+2.10 \pm 0.08$	-0.24 ± 0.08	$+2.10 \pm 0.08$	-0.24 ± 0.08

[a] % by weight of cement.
SF, silica fume; L, latex.

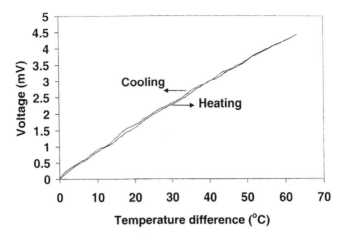

Figure 13.12 Variation of Seebeck voltage (with copper as the reference) vs. the temperature difference during heating and cooling for steel fiber silica-fume cement paste containing steel fibers in the proportion of 1.0% by weight of cement.

power more negative, whether silica fume (or latex) is present or not. An increase of the steel fiber content also increases the reversibility and linearity of the change in Seebeck voltage with the temperature difference between the hot and cold ends, as shown by comparing the values of the Seebeck coefficient obtained during heating and cooling in Table 13.2. The values obtained during heating and cooling are close for the pastes with the higher steel fiber content, but are not so close for the pastes with the lower steel fiber content. In contrast, for pastes with carbon fibers in place of steel fibers, the change in Seebeck voltage with the temperature difference is highly reversible for both carbon fiber contents of 0.5% and 1.0% by weight of cement, as shown in Table 13.2 by comparing the values of the Seebeck coefficient obtained during heating and cooling.

Table 13.2 shows that the volume electrical resistivity is much higher for steel fiber cement pastes than the corresponding carbon fiber cement pastes. This is attributed to the much lower volume fraction of fibers in the former (Table 13.2). An increase in the steel or carbon fiber content from 0.5% to 1.0% by weight of cement decreases the resistivity, although the decrease is more significant in the carbon fiber case than the steel fiber case. That the resistivity decrease is not large when the steel fiber content is increased from 0.5% to 1.0% by weight of cement and that the resistivity is still high at a steel fiber content of 1.0% by weight of cement suggests that a steel fiber content of 1.0% by weight of cement is below the percolation threshold.

Whether with or without silica fume (or latex), the change in Seebeck voltage with temperature is more reversible and linear at a steel fiber content of 1.0% by weight of cement than at a steel fiber content of 0.5% by weight of cement. This is attributed to the larger role of the cement matrix at the lower steel fiber content and the contribution of the cement matrix to the irreversibility and non-linearity. Irreversibility and non-linearity are particularly significant when the cement paste contains no fiber.

From the practical point of view, the steel fiber silica-fume cement paste containing steel fibers in an amount of 1.0% by weight of cement is particularly attractive for use in temperature sensing, as the absolute thermoelectric power is highest (−68 μV/°C) and the variation of Seebeck voltage with the temperature difference between the hot and cold ends is reversible and linear. The absolute thermoelectric power is as high in magnitude as those of commercial thermocouple materials.

Joints between concretes with different values of thermoelectric power, as made by multiple pouring, provide concrete thermocouples [25].

13.2.5 Cement–matrix composites for thermal control

Concretes that are inherently able to provide heating through Joule heating, provide temperature sensing (Section 13.2.4) or provide temperature stability through a high specific heat (high thermal mass) are highly desirable for thermal control of structures and energy saving in buildings. Concretes of low electrical resistivity [26–35] are useful for joule heating, concrete thermistors and thermocouples are useful for temperature sensing, and concretes of high specific heat [36–39] are useful for heat retention. These concretes involve the use of admixtures such as fibers and silica fume. For example, silica fume introduces interfaces which promote the specific heat [36]; short carbon fibers enhance the electrical conductivity [31] and render the concrete p-type [23]. (Plain concrete is n-type [23].)

Figure 13.13 [31] gives the volume electrical resistivity of composites at 7 days of curing. The resistivity decreases much with increasing fiber volume fraction,

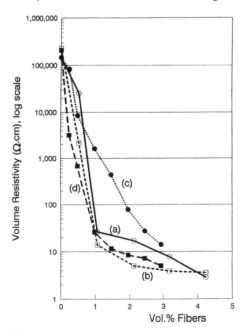

Figure 13.13 Variation of the volume electrical resistivity of cement–matrix composites with carbon fiber volume fraction: (a) without sand, with methylcellulose, without silica fume; (b) without sand, with methylcellulose, with silica fume; (c) with sand, with methylcellulose, without silica fume; (d) with sand, with methylcellulose, with silica fume.

whether a second filler (silica fume or sand) is present or not. When sand is absent, the addition of silica fume decreases the resistivity at all carbon fiber volume fractions except the highest volume fraction of 4.24%; the decrease is most significant at the lowest fiber volume fraction of 0.53%. When sand is present, the addition of silica fume similarly decreases the resistivity, such that the decrease is most significant at fiber volume fractions below 1%. When silica fume is absent, the addition of sand decreases the resistivity only when the fiber volume fraction is below about 0.5%; at high fiber volume fractions, the addition of sand even increases the resistivity owing to the porosity induced by the sand. Thus, the addition of a second filler (silica fume or sand) that is essentially non-conducting decreases the resistivity of the composite only at low volume fractions of the carbon fibers, and the maximum fiber volume fraction for the resistivity to decrease is larger when the particle size of the filler is smaller. The resistivity decrease is attributed to the improved fiber dispersion due to the presence of the second filler. Consistent with the improved fiber dispersion is the increased flexural toughness and strength due to the presence of the second filler.

Table 13.3 [37,40] shows the specific heat of cement pastes. Specific heat is significantly increased by the addition of silica fume. It is further increased by the further addition of methylcellulose and defoamer. It is still further increased by the addition of carbon fibers.

Table 13.3 [37,40] also shows the thermal conductivity. It is significantly decreased by the addition of silica fume. The further addition of methylcellulose and defoamer or the still further addition of fibers has little effect on thermal conductivity.

Sand is a much more common component in concrete than silica fume. It is different from silica fume in its relatively large particle size and negligible reactivity with cement. Sand gives effects that are opposite from those of silica fume, i.e., sand addition decreases the specific heat and increases thermal conductivity [39].

Table 13.3 [39] shows the thermal behavior of cement pastes and mortars. Comparison of the results on cement paste without silica fume and those on mortar without silica fume shows that sand addition decreases specific heat by 13% and increases thermal conductivity by 9%. Comparison of the results on cement paste with silica fume and those on mortar with silica fume shows that sand addition decreases specific heat by 11% and increases thermal conductivity

Table 13.3 Thermal behavior of cement pastes and mortars.

	Cement paste		Mortar	
	Without silica fume[a]	With silica fume[a]	Without silica fume[a]	With silica fume[a]
Density (g/cm^3, ± 0.02)	2.01	1.73	2.04	2.20
Specific heat (J/g.K, ± 0.001)	0.736	0.788	0.642	0.705
Thermal diffusivity (mm^2/s, ± 0.03)	0.36	0.24	0.44	0.35
Thermal conductivity[b] (W/m.K, ± 0.03)	0.53	0.33	0.58	0.54

[a]Silane treated.
[b]Product of density, specific heat and thermal diffusivity.

by 64%. That sand addition has more effect on thermal conductivity when silica fume is present than when it is absent is due to the low value of the thermal conductivity of cement paste with silica fume (Table 13.3).

Comparison of the results on cement paste without silica fume and on cement paste with silica fume shows that silica fume addition increases the specific heat by 7% and decreases thermal conductivity by 38%. Comparison of the results on mortar without silica fume and on mortar with silica fume shows that silica fume addition increases the specific heat by 10% and decreases thermal conductivity by 6%. Hence, the effects of silica fume addition on mortar and cement paste are in the same direction. That the effect of silica fume on thermal conductivity is much less for mortar than for cement paste is mainly due to the fact that silica fume addition increases the density of mortar but decreases the density of cement paste (Table 13.3). That the fractional increase in specific heat due to silica fume addition is higher for mortar than cement paste is attributed to the low value of the specific heat of mortar without silica fume (Table 13.3).

Comparison of the results on cement paste with silica fume and those on mortar without silica fume shows that sand addition gives a lower specific heat than silica fume addition and a higher thermal conductivity than silica fume addition. Since sand has a much larger particle size than silica fume, sand results in much less interface area than silica fume, although the interface may be more diffuse for silica fume than for sand. The low interface area in the sand case is believed to be responsible for the low specific heat and higher thermal conductivity, as slippage at the interface contributes to the specific heat and the interface acts as a thermal barrier.

Silica fume addition increases the specific heat of cement paste by 7%, whereas sand addition decreases it by 13%. Silica fume addition decreases the thermal conductivity of cement paste by 38%, whereas sand addition increases it by 22%. Hence, silica fume addition and sand addition have opposite effects. The cause is believed to be mainly associated with the low interface area for the sand case and the high interface area for the silica fume case, as explained in the last paragraph. The high reactivity of silica fume compared to sand may contribute to the observed difference between silica fume addition and sand addition, although this contribution is believed to be minor, as the reactivity should have tightened up the interface, thus decreasing the specific heat (in contrast to the observed effects). The decrease in specific heat and increase in thermal conductivity upon sand addition are believed to be due to the higher level of homogeneity within a sand particle than within cement paste.

13.2.6 Cement–matrix composites for vibration reduction

Vibration reduction requires a high damping capacity and a high stiffness. Viscoelastic materials such as rubber have a high damping capacity but a low stiffness. Concretes having both high damping capacity (two or more orders higher than conventional concrete) and high stiffness (Table 12.1) [41] can be obtained by using surface-treated silica fume as an admixture in the concrete. Steel-reinforced concretes having improved damping capacity and stiffness can be obtained by surface treating the steel (say, by sand blasting) prior to incorporating the steel in the concrete (Table 13.4) [42], or by using silica fume in the concrete [41]. Owing to its small particle size, silica fume in concrete

Table 13.4 Loss tangent, storage modulus and loss modulus of mortars with and without steel reinforcement.

Property	Sample type	Frequency		
		0.2 Hz	0.5 Hz	1.0 Hz
Loss tangent	A	$<10^{-4}$	$<10^{-4}$	$<10^{-4}$
	B	$(2.73 \pm 0.19) \times 10^{-2}$	$(1.56 \pm 0.08) \times 10^{-2}$	$(7.20 \pm 0.37) \times 10^{-3}$
	C	$(3.32 \pm 0.15) \times 10^{-2}$	$(1.98 \pm 0.17) \times 10^{-2}$	$(1.07 \pm 0.09) \times 10^{-2}$
	D	$(3.65 \pm 0.27) \times 10^{-2}$	$(2.50 \pm 0.22) \times 10^{-2}$	$(1.24 \pm 0.16) \times 10^{-2}$
Storage modulus (GPa)	A	20.2 ± 3.5	27.5 ± 4.3	25.8 ± 3.7
	B	44.2 ± 4.8	47.7 ± 5.3	44.4 ± 5.0
	C	36.9 ± 4.3	41.0 ± 3.9	38.4 ± 3.0
	D	46.0 ± 4.0	51.2 ± 6.4	49.3 ± 5.8
Loss modulus (GPa)	A	$<10^{-3}$	$<10^{-3}$	$<10^{-3}$
	B	1.21 ± 0.22	0.74 ± 0.12	0.32 ± 0.05
	C	1.23 ± 0.20	0.81 ± 0.15	0.41 ± 0.07
	D	1.68 ± 0.27	1.28 ± 0.27	0.61 ± 0.51

Note on sample type designations: A, no rebar; B, as-received steel rebar; C, ozone-treated steel rebar; D, sandblasted steel rebar.

introduces interfaces which enhance damping. Sand blasting of a steel rebar increases the interface area between steel and concrete, thereby enhancing damping. Carbon fiber addition has relatively small effect on the damping capacity and stiffness [43].

13.3 Polymer–matrix composites for smart structures

Polymer–matrix composites (Section 2.2) for structural applications typically contain continuous fibers such as carbon, polymer and glass fibers, as continuous fibers tend to be more effective than short fibers as a reinforcement. Polymer–matrix composites with continuous carbon fibers are used for aerospace, automobile and civil structures. (In contrast, continuous fibers are too expensive for reinforcing concrete.) Due to the fact that carbon fibers are electrically conducting, whereas polymer and glass fibers are not, carbon fiber composites are predominant among polymer–matrix composites that are intrinsically smart.

13.3.1 Background on polymer–matrix composites

Polymer–matrix composites containing continuous carbon fibers are important structural materials owing to their high tensile strength, high tensile modulus and low density. They are used for lightweight structures such as satellites, aircraft, automobiles, bicycles, ships, submarines, sporting goods, wheelchairs, armor and rotating machinery (such as turbine blades and helicopter rotors). Owing to the recent emphasis on repair of civil infrastructural systems, composites are beginning to be used for the repair of concrete structures and for bridges, even though they are much more expensive than concrete. As the price of carbon fibers has been dropping steadily during the last two decades, the spectrum of applications has been widening tremendously.

Continuous carbon fibers used are primarily based either on polyacrylonitrile (PAN) or mesophase pitch. Mesophase-pitch-based carbon fibers, if heat treated to high temperatures exceeding 2500°C, can be graphitized and attain very high values of tensile modulus and thermal conductivity (in-plane), in addition to improved oxidation resistance. The high thermal conductivity is attractive for thermal management, which is particularly important for electronics (e.g., heat sinks). However, graphitized fibers tend to be relatively low in strength, due to the ease of shear between the graphite layers, and they are very expensive. On the other hand, PAN-based fibers cannot be graphitized, although they compete well with mesophase-pitch-based fibers which have not been graphitized, in that both materials exhibit reasonably high values of both strength and modulus and are not very expensive. These fibers are the most widely used among carbon fibers. The fabrication of both pitch-based and PAN-based carbon fibers involves stabilization (infusibilization) and then carbonization (conversion from hydrocarbon molecules to a carbon network). Graphitization optionally follows carbonization.

Owing to the importance of carbon fiber polymer–matrix composites for structural applications, much investigation has been made on the mechanical behavior of these materials. Much less work has been done in studying the electrical behavior [44–51]. On the other hand, due to the fact that carbon fibers are much more conducting than the polymer matrix, the electrical behavior gives much information on the microstructure, such as the degree of fiber alignment, the number of fiber–fiber contacts, the amount of delamination and the extent of fiber breakage. Such information is not only useful for scientific understanding of the properties of the composite, but is also valuable for the purpose of rendering the composite able to sense its strain, damage and temperature in real time via electrical measurement. In other words, the strain, damage and temperature affect the electrical behavior, such as the electrical resistance, which thus serves to indicate strain, damage and temperature. In this way, the composite is self-sensing, i.e., intrinsically smart, without the need for attached or embedded sensors (such as optical fibers, acoustic sensors and piezoelectric sensors), which raise the cost, reduce the durability and, in the case of embedded sensors, weaken the structure.

Carbon fibers are electrically conducting, while the polymer matrix is electrically insulating (except for the rare situation in which the polymer is an electrically conducting one [52]). The continuous fibers in a composite laminate are in the form of layers called laminae. Each lamina comprises many bundles (called tows) of fibers in a polymer matrix. Each tow consists of thousands of fibers. There may or may not be twist in a tow. Each fiber has a diameter typically ranging from 7 to 12 μm. The tows within a lamina are typically oriented in the same direction, but tows in different laminae may or may not be in the same direction. A laminate with tows in all the laminae oriented in the same direction is said to unidirectional. A laminate with tows in adjacent laminae oriented at an angle of 90° is said to be crossply. In general, an angle of 45° and other angles may also be involved for the various laminae, as desired for attaining the mechanical properties required for the laminate in various directions in the plane of the laminate.

Within a lamina with tows in the same direction, the electrical conductivity is highest in the fiber direction. In the transverse direction in the plane of the lamina, the conductivity is not zero, even though the polymer matrix is insulating. This is because there are contacts between fibers of adjacent tows [53]. In other words, a fraction of the fibers of one tow touch a fraction of the fibers of an adjacent tow here and there along the length of the fibers. These contacts result from the fact that fibers are not perfectly straight or parallel (even though the lamina is said to be unidirectional), and that the flow of the polymer matrix (or resin) during composite fabrication can cause a fiber to be not completely covered by the polymer or resin (even though, prior to composite fabrication, each fiber may be completely covered by the polymer or resin, as in the case of a prepreg, i.e., a fiber sheet impregnated with the polymer or resin). Fiber waviness is known as marcelling. Thus, the transverse conductivity gives information on the number of fiber–fiber contacts in the plane of the lamina.

For similar reasons, the contacts between fibers of adjacent laminae cause the conductivity in the through-thickness direction (direction perpendicular to the plane of the laminate) to be non-zero. Thus, the through-thickness conductivity gives information on the number of fiber–fiber contacts between adjacent laminae.

Matrix cracking between the tows of a lamina decreases the number of fiber–fiber contacts in the plane of the lamina, thus decreasing the transverse conductivity. Similarly, matrix cracking between adjacent laminae (as in delamination [54]) decreases the number of fiber–fiber contacts between adjacent laminae, thus decreasing the through-thickness conductivity. This means that the transverse and through-thickness conductivities can indicate damage in the form of matrix cracking.

Fiber damage (as distinct from fiber fracture) decreases the conductivity of a fiber, thereby decreasing the longitudinal conductivity (conductivity in the fiber direction). However, owing to the brittleness of carbon fibers, the decrease in conductivity due to fiber damage prior to fiber fracture is rather small [55].

Fiber fracture causes a much larger decrease in the longitudinal conductivity of a lamina than fiber damage. If there is only one fiber, a broken fiber results in an open circuit, i.e., zero conductivity. However, a lamina has a large number of fibers and adjacent fibers can make contact here and there. Therefore, the portions of a broken fiber still contribute to the longitudinal conductivity of the lamina. As a result, the decrease in conductivity due to fiber fracture is less than what it would be if a broken fiber did not contribute to the conductivity. Nevertheless, the effect of fiber fracture on the longitudinal conductivity is significant, so that the longitudinal conductivity can indicate damage in the form of fiber fracture [56].

The through-thickness volume resistance of a laminate is the sum of the volume resistance of each of the laminae in the through-thickness direction and the contact resistance of each of the interfaces between adjacent laminae (i.e., the interlaminar interface). For example, a laminate with eight laminae has eight volume resistances and seven contact resistances, all in the through-thickness direction. Thus, to study the interlaminar interface, it is better to measure the contact resistance between two laminae rather than the through-thickness volume resistance of the entire laminate.

Measurement of the contact resistance between laminae can be made by allowing two laminae (strips) to contact at a junction and using the two ends of each strip for making four electrical contacts [57]. An end of the top strip and an end of the bottom strip serve as contacts for passing current. The other end of the top strip and of the bottom strip serve as contacts for voltage measurement. The fibers in the two strips can be in the same direction or in different directions. This method is a form of the four-probe method of electrical resistance. The configuration is illustrated in Figure 13.14 for crossply and unidirectional laminates. To make sure that the volume resistance within a lamina in the through-thickness direction does not contribute to the measured resistance, the fibers at each end of a lamina strip should be electrically shorted together by using silver paint or other conducting media. The measured resistance is the contact resistance of the junction. This resistance, multiplied by the area of the junction, gives the contact resistivity, which is independent of the area of the junction and just depends on the nature of the interlaminar interface. The unit of contact resistivity is $\Omega.m^2$, whereas that of volume resistivity is $\Omega.m$.

The structure of the interlaminar interface tends to be more prone to change than the structure within a lamina. For example, damage in the form of delamination is much more common than damage in the form of fiber fracture. Moreover, the structure of the interlaminar interface is affected by the interlaminar stress (whether thermal stress or curing stress), which is particularly significant when the laminae are not unidirectional (as the anisotropy within each lamina enhances the interlaminar stress). The structure of the interlaminar interface also depends on the extent of consolidation of the laminae during composite fabrication. The contact resistance provides a sensitive probe of the structure of the interlaminar interface.

Measurement of volume resistivity in the through-thickness direction can be conducted by using the four-probe method, in which each of the two current

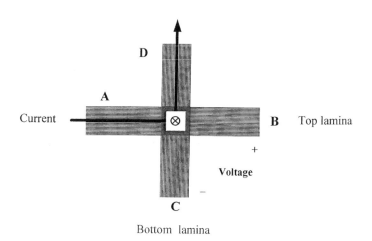

Figure 13.14 Specimen configuration for measurement of contact electrical resistivity between laminae: (a) crossply laminae; (b) unidirectional laminae.

contacts is in the form of a conductor loop (made by silver paint, for example) on each of the two outer surfaces of the laminate in the plane of the laminate and each of the two voltage contacts is in the form of a conductor dot within the loop [54]. An alternative method is to have four of the laminae in the laminate extra long so as to extend out for the purpose of serving as electrical leads [58]. The two outer leads are for current contacts and the two inner leads are for voltage contacts. The use of a thin metal wire inserted at an end into the interlaminar space during composite fabrication in order to serve as an electrical contact is not recommended, because the quality of the electrical contact between the metal wire and carbon fibers is hard to control and the wire is intrusive to the composite. The alternative method is less convenient than that involving loops and dots, but it approaches more closely the ideal four-probe method.

In order to attain zero conductivity in the through-thickness direction of a laminate, it is necessary to use an insulating layer between two adjacent laminae [59]. The insulating layer can be a piece of writing paper. Tissue paper is ineffective in preventing contact between fibers of adjacent laminae, owing to its porosity. The attainment of zero conductivity in the through-thickness direction allows the laminate to serve as a capacitor. This means that the structural composite stores energy by serving as a capacitor.

13.3.2 Polymer–matrix composites for strain/stress sensing

Smart structures which can monitor their own strain (which relates to stress) are valuable for structural vibration control. Self-monitoring of strain (reversible) has been achieved in carbon fiber epoxy–matrix composites without the use of embedded or attached sensors [58,60–63], as the volume electrical resistance of the composite in the through-thickness or longitudinal direction changes reversibly with longitudinal strain (gage factor up to 40) due to change in the degree of fiber alignment. Tension in the fiber direction of the composite increases the degree of fiber alignment, thereby decreasing the chance for fibers of adjacent laminae to touch one another. As a consequence, the through-thickness resistance increases while the longitudinal resistance decreases.

Figure 13.15 [61] shows the change in longitudinal volume resistance during cyclic longitudinal tension in the elastic regime for a unidirectional continuous carbon fiber epoxy–matrix composite with eight fiber layers (laminae). The stress amplitude is equal to 14% of the breaking stress. The strain returns to zero at the end of each cycle. Because of the small strains involved, the fractional resistance change $\Delta R/R_0$ is essentially equal to the fractional change in resistivity. The longitudinal $\Delta R/R_0$ decreases upon loading and increases upon unloading in every cycle, such that R irreversibly decreases slightly after the first cycle (i.e., $\Delta R/R_0$ does not return to 0 at the end of the first cycle). At higher stress amplitudes, the effect is similar, except that both the reversible and irreversible parts of $\Delta R/R_0$ are larger.

Figure 13.16 [61] shows the change in through-thickness volume resistance during cyclic longitudinal tension in the elastic regime for the same composite. The stress amplitude is equal to 14% of the breaking stress. The through-thickness $\Delta R/R_0$ increases upon loading and decreases upon unloading in every cycle, such that R irreversibly decreases slightly after the first cycle (i.e., $\Delta R/R_0$ does not return to 0 at the end of the first cycle). Upon

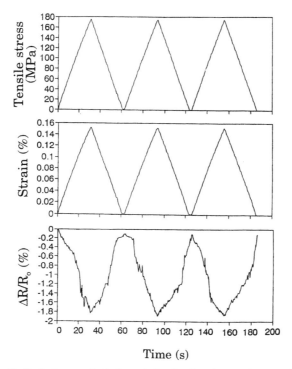

Figure 13.15 Longitudinal stress and strain and fractional resistance increase ($\Delta R/R_0$) obtained simultaneously during cyclic tension at a stress amplitude equal to 14% of the breaking stress for continuous fiber epoxy–matrix composite.

increasing the stress amplitude, the effect is similar, except that the reversible part of $\Delta R/R_0$ is larger.

The strain sensitivity (gage factor) is defined as the reversible part of $\Delta R/R_0$ divided by the longitudinal strain amplitude. It is negative (from −18 to −12) for the longitudinal $\Delta R/R_0$ and positive (17–24) for the through-thickness $\Delta R/R_0$. The magnitudes are comparable for the longitudinal and through-thickness strain sensitivities. As a result, whether the longitudinal R or the through-thickness R is preferred for strain sensing just depends on the convenience of electrical contact application for the geometry of the particular smart structure.

Figure 13.17 [61] shows the compressive stress, strain and longitudinal $\Delta R/R_0$ obtained simultaneously during cyclic compression at stress amplitudes equal to 14% of the breaking stress for a similar composite having 24 rather than eight fiber layers. The longitudinal $\Delta R/R_0$ increases upon compressive loading and decreases upon unloading in every cycle, such that resistance R irreversibly increases very slightly after the first cycle. The magnitude of the gage factor is lower in compression (−1.2) than in tension (from −18 to −12).

A dimensional change without any resistivity change would have caused longitudinal R to increase during tensile loading and decrease during compressive loading. In contrast, the longitudinal R decreases upon tensile loading and increases upon compressive loading. In particular, the magnitude of $\Delta R/R_0$ under tension is 7–11 times that of $\Delta R/R_0$ calculated by assuming that $\Delta R/R_0$ is only due to dimensional change and not to any resistivity change. Hence

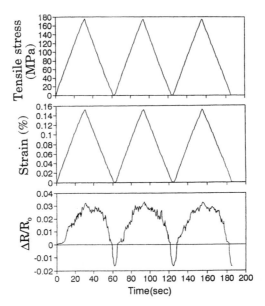

Figure 13.16 Longitudinal stress and strain and the through-thickness $\Delta R/R_0$ obtained simultaneously during cyclic tension at a stress amplitude equal to 14% of the breaking stress for continuous fiber epoxy-matrix composite..

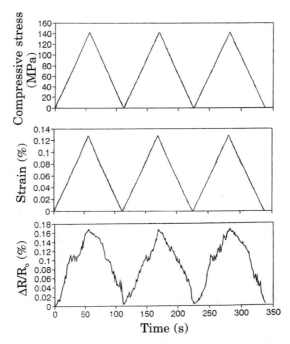

Figure 13.17 Longitudinal stress, strain and $\Delta R/R_0$ obtained simultaneously during cyclic compression (longitudinal) at a stress amplitude equal to 14% of the breaking stress for continuous fiber epoxy–matrix composite.

the contribution of $\Delta R/R_0$ from the dimensional change is negligible compared to that from the resistivity change.

The irreversible behavior, although small compared to the reversible behavior, is such that R (longitudinal or through-thickness) under tension is irreversibly decreased after the first cycle. This behavior is attributed to the irreversible disturbance to the fiber arrangement at the end of the first cycle, such that the fiber arrangement becomes less neat. A less neat fiber arrangement means more chance for the adjacent fiber layers to touch one another.

Instead of measuring the resistivity of a volume of composite material, one can measure the resistivity of an interface in the composite. The contact resistivity of the interlaminar interface decreases upon compression in the direction perpendicular to the interface [64]. The effect is quite reversible and can be used for stress sensing.

13.3.3 Polymer–matrix composites for damage sensing

Self-monitoring of damage (whether due to stress or temperature, under static or dynamic conditions) has been achieved in continuous carbon fiber polymer–matrix composites, as the electrical resistance of the composite changes with damage [54,60,65–78]. Minor damage in the form of slight matrix damage and/or disturbance to the fiber arrangement is indicated by the longitudinal and through-thickness resistance decreasing irreversibly due to increase in the number of contacts between fibers, as shown after one loading cycle in Figures 13.15–13.17. More significant damage in the form of delamination or interlaminar interface degradation is indicated by the through-thickness resistance (or more exactly the contact resistivity of the interlaminar interface) increasing due to decrease in the number of contacts between fibers of different laminae. Major damage in the form of fiber breakage is indicated by the longitudinal resistance increasing irreversibly.

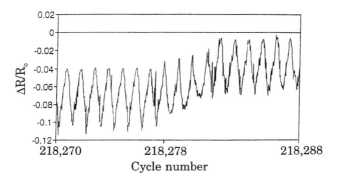

Figure 13.18 Variation of longitudinal $\Delta R/R_0$ with cycle number during tension–tension fatigue testing for a carbon fiber epoxy-matrix composite. Each cycle of reversible decrease in resistance is due to strain. The irreversible increase in resistance at around cycle no. 218,281 is due to damage in the form of fiber breakage.

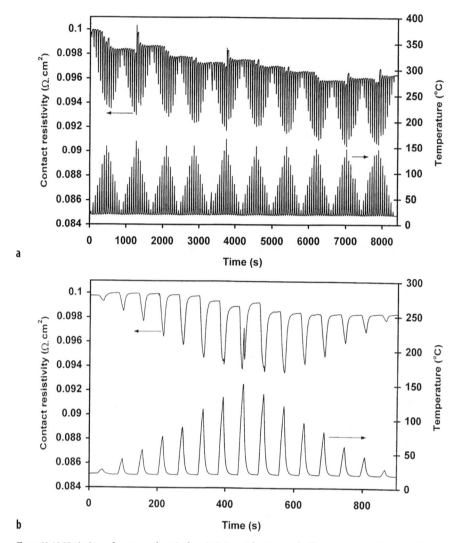

Figure 13.19 Variation of contact electrical resistivity with time and of temperature with time during thermal cycling of a carbon fiber epoxy-matrix composite. Part (**b**) shows the magnified view of the first 900 s of part (**a**).

During mechanical fatigue, delamination was observed to begin at 30% of the fatigue life, whereas fiber breakage was observed to begin at 50% of the fatigue life. Figure 13.18 [65] shows an irreversible resistance increase occurring at about 50% of the fatigue life during tension–tension fatigue testing of a unidirectional continuous carbon fiber epoxy-matrix composite. The resistance and stress are in the fiber direction. The reversible changes in resistance are due to strain, which causes the resistance to decrease reversibly in each cycle, as in Figure 13.15.

Figure 13.19 shows the variation in contact resistivity of the interlaminar interface with temperature during thermal cycling. The temperature is repeatedly increased to various levels. A group of cycles in which the temperature amplitude increases cycle by cycle and then decreases cycle by cycle back to the initial low temperature amplitude is referred to as a "group". Figure 13.19(a) shows the results of the first 10 groups, while Figure 13.19(b) shows the first group only. The contact resistivity decreases upon heating in every cycle of every group. At the highest temperature (150°C) of a group, a spike of resistivity increase occurs, as shown in Figure 13.19(b). It is attributed to damage at the interlaminar interface. In addition, the baseline resistivity (i.e., the top envelope) gradually and irreversibly shifts downward as cycling progresses, as shown in Figure 13.19(a). The baseline decrease is probably due to matrix damage within a lamina and the resulting decrease in modulus and hence decrease in residual stress.

Self-monitoring of damage due to stress has also been achieved in continuous glass fiber polymer–matrix composites containing an electrically conductive component such as carbon particles or continuous carbon fibers [79–86].

13.3.4 Polymer–matrix composites for temperature sensing

Continuous carbon fiber epoxy–matrix composites provide temperature sensing by serving as thermistors [57,87] and thermocouples [88].

The thermistor function stems from the reversible decrease of contact electrical resistivity at the interface between fiber layers (laminae) upon heating. Figure 13.20 shows the variation of contact resistivity ρ_c with temperature during reheating and subsequent cooling, both at 0.15°C/min, for carbon fiber epoxy–matrix composites cured at 0 and 0.33 MPa. The corresponding Arrhenius

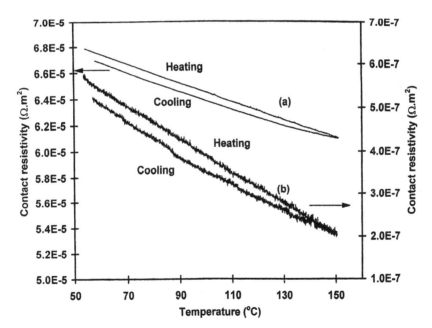

Figure 13.20 Variation of contact electrical resistivity with temperature during heating and cooling of carbon fiber epoxy–matrix composites at 0.15°C/min (**a**) for composite made without any curing pressure and (**b**) for composite made with a curing pressure 0.33 MPa.

plots of log contact conductivity (inverse of contact resistivity) vs. inverse absolute temperature during heating are shown in Figure 13.21. From the slope (negative) of the Arrhenius plot, which is quite linear, the activation energy can be calculated. The linearity of the Arrhenius plot means that the activation energy does not change throughout the temperature variation. This activation energy is the energy for electron jumping from one lamina to the other. Electronic excitation across this energy enables conduction in the through-thickness direction.

The activation energies, thicknesses and room temperature contact resistivities for samples made at different curing pressures and composite configurations are shown in Table 13.5. All the activation energies were calculated based on the data at 75–125°C. In this temperature regime, the temperature change was very linear and well controlled. From Table 13.5 it can be seen that, for the same composite configuration (crossply), the higher the curing pressure, the smaller the composite thickness (because of more epoxy being squeezed out), the lower the contact resistivity, and the higher was the activation energy. A smaller composite thickness corresponds to a higher fiber volume fraction in the composite. During curing and subsequent cooling the matrix shrinks, so a longitudinal compressive stress will develop in the fibers. For carbon fibers, the modulus in the longitudinal direction is much higher than that in the transverse direction. Moreover, the carbon fibers are continuous in the longitudinal direction. Thus, the overall shrinkage in the longitudinal direction tends to be less than that in the transverse direction. Therefore, there will be a residual interlaminar stress in the two crossply layers in a given direction. This stress accentuates the barrier for the electrons to jump from one lamina to the other. After curing and subsequent

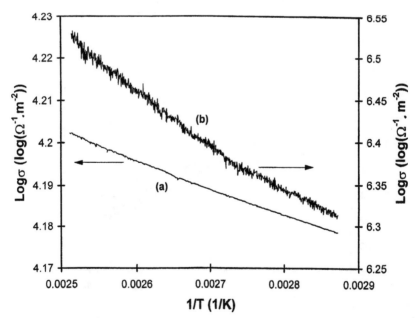

Figure 13.21 21Arrhenius plot of log contact conductivity vs. inverse absolute temperature during heating of carbon fiber epoxy–matrix composites at 0.15°C/min (a) for composite made without any curing pressure and (b) for composite made with curing pressure 0.33 MPa.

cooling, heating will decrease the thermal stress. Both the thermal stress and the curing stress contribute to the residual interlaminar stress. Therefore, the higher the curing pressure, the larger the fiber volume fraction, the greater the residual interlaminar stress, and the higher the activation energy, as shown in Table 13.5. Besides the residual stress, thermal expansion can also affect contact resistance by changing the contact area. However, calculation shows that the contribution of thermal expansion is less than one-tenth of the observed change in contact resistance with temperature.

The electron jump primarily occurs at points where there is direct contact between fibers of the adjacent laminae. Direct contact is possible because of the flow of the epoxy resin during composite fabrication and the slight waviness of the fibers, as explained in Reference 54 in relation to the through-thickness volume resistivity of a carbon fiber epoxy–matrix composite.

The curing pressure for the sample in the unidirectional composite configuration was higher than that of any of the crossply samples (Table 13.5). Consequently, the thickness was the lowest. As a result, the fiber volume fraction was the highest. However, the contact resistivity of the unidirectional sample was the second highest rather than being the lowest, and its activation energy was the lowest rather than the highest. The low activation energy is consistent with the fact that there was no CTE or curing shrinkage mismatch between the two unidirectional laminae and, as a result, no interlaminar stress between the laminae. This low value supports the notion that interlaminar stress is important in affecting the activation energy. The high contact resistivity for the unidirectional case can be explained in the following way. In the crossply samples, the pressure during curing forced the fibers of the two laminae to press on to one another and hence contact tightly. In the unidirectional sample, the fibers of one of the laminae just sank into the other lamina at the junction, so pressure helped relatively little in the contact between fibers of adjacent laminae. Moreover, in the crossply situation, every fiber at the lamina–lamina interface contacted many fibers of the other lamina, while, in the unidirectional situation, each fiber had little chance to contact the fibers of the other lamina. Therefore, the number of contact points between the two laminae was less for the unidirectional sample than the crossply samples.

The thermocouple function stems from the use of n-type and p-type carbon fibers (as obtained by intercalation) in different laminae. The thermocouple sensitivity and linearity are as good as or better than those of commercial thermocouples. By using two laminae that are crossply, a two-dimensional array of thermistors or thermocouple junctions are obtained, thus allowing temperature distribution sensing.

Table 13.6 shows the Seebeck coefficient and the absolute thermoelectric power of carbon fibers and the thermocouple sensitivity of epoxy–matrix composite junctions. A positive value of the absolute thermoelectric power indicates p-type behavior; a negative value indicates n-type behavior. Pristine P-25 is slightly n-type; pristine T-300 is strongly n-type. A junction comprising pristine P-25 and pristine T-300 has a positive thermocouple sensitivity that is close to the difference of the Seebeck coefficients (or the absolute thermoelectric powers) of T-300 and P-25, whether the junction is unidirectional or crossply. Pristine P-100 and pristine P-120 are both slightly n-type. Intercalation with sodium causes P-100 and P-120 to become strongly n-type. Intercalation with bromine causes P-100 and P-120 to become strongly p-type. A junction comprising bromine-

Table 13.6 Seebeck coefficient [64].

Seebeck coefficient (μV/°C) and absolute thermoelectric power (μV/°C) of carbon fibers and thermocouple sensitivity (μV/°C) of epoxy–matrix composite junctions. All junctions are unidirectional unless specified as crossply. The temperature range is 20–110°C

	Seebeck coefficient with copper as the reference (μV/°C)	Absolute thermoelectric power (μV/°C)	Thermocouple sensitivity (μV/°C)
P-25[a]	+0.8	−1.5	
T-300[a]	−5.0	−7.3	
P-25[a] + T-300[a]			+5.5
P-25[a] + T-300[a] (crossply)			+5.4
P-100[a]	−1.7	−4.0	
P-120[a]	−3.2	−5.5	
P-100 (Na)	−48	−50	
P-100 (Br$_2$)	+43	+41	
P-100 (Br$_2$) + P-100 (Na)			+82
P-120 (Na)	−42	−44	
P-120 (Br$_2$)	+38	+36	
P-120 (Br$_2$) + P-120 (Na)			+74

[a]Pristine (i.e., not intercalated).

intercalated P-100 and sodium-intercalated P-100 has a positive thermocouple sensitivity that is close to the sum of the magnitudes of the absolute thermoelectric powers of the bromine-intercalated P-100 and the sodium-intercalated P-100. Similarly, a junction comprising bromine-intercalated P-120 and sodium-intercalated P-120 has a positive thermocouple sensitivity that is close to the sum of the magnitudes of the absolute thermoelectric powers of the bromine-intercalated P-120 and the sodium-intercalated P-120. Figure 13.22 shows the linear relationship of the measured voltage with the temperature difference between hot and cold points for the junction comprising bromine-intercalated P-100 and sodium-intercalated P-100.

A junction comprising n-type and p-type partners has a thermocouple sensitivity that is close to the sum of the magnitudes of the absolute thermoelectric powers of the two partners. This is because the electrons in the n-type partner as well as the holes in the p-type partner move away from the hot point toward the corresponding cold point. As a result, the overall effect on the voltage difference between the two cold ends is additive.

By using junctions comprising strongly n-type and strongly p-type partners, a thermocouple sensitivity as high as +82 μV/°C was attained. Semiconductors are known to exhibit much higher values of Seebeck coefficient than metals, but the need to have thermocouples in the form of long wires makes metals the main materials for thermocouples. Intercalated carbon fibers exhibit much higher values of Seebeck coefficient than metals. Yet, unlike semiconductors, their fiber form and fiber composite form make them convenient for practical use as thermocouples.

The thermocouple sensitivity of the carbon fiber epoxy–matrix composite junctions is independent of the extent of curing and is the same for unidirectional and crossply junctions. This is consistent with the fact that the thermocouple effect hinges on the difference in the bulk properties of the two partners, and is not an interfacial phenomenon. This behavior means that the interlaminar interfaces in a fibrous composite serve as thermocouple junctions

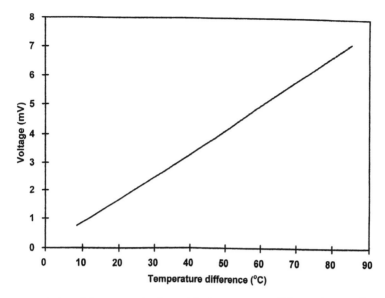

Figure 13.22 Variation of the measured voltage with the temperature difference between hot and cold points for the epoxy-matrix composite junction comprising bromine-intercalated P-100 and sodium-intercalated P-100 carbon fibers.

in the same way, irrespective of the lay-up configuration of the dissimilar fibers in the laminate. As a structural composite typically has fibers in multiple directions, this behavior facilitates the use of a structural composite as a thermocouple array.

It is important to note that the thermocouple junctions do not require any bonding agent other than the epoxy, which serves as the matrix of the composite but not as an electrical contact medium (since it is not conducting). In spite of the presence of epoxy matrix in the junction area, direct contact occurs between a fraction of the fibers of a lamina and a fraction of the fibers of the other lamina, thus resulting in a conduction path in a direction perpendicular to the junction.

13.3.5 Polymer-matrix composites for vibration reduction

Polymer-matrix composites containing continuous carbon fibers and exhibiting both high damping capacity and high stiffness have been attained by using interlayers in the form of discontinuous 0.1 μm diameter carbon filaments [89]. The damping enhancement is due to the large area of the interface between filaments and matrix. Viscoelastic constrained interlayers, although common, degrade the strength and modulus and become less effective as the temperature increases [90]. Surface treatment of the filaments helps, at least in the case of a thermoplastic matrix [90]. Other techniques for damping enhancement are matrix modification [91,92], fiber coating [92] and filament winding angle optimization [93].

13.4 Conclusion

Intrinsically smart structural composites for strain sensing, damage sensing, temperature sensing, thermal control and vibration reduction are attractive for smart structures. They include cement–matrix and polymer–matrix composites, particularly cement–matrix composites containing short carbon fibers and polymer–matrix composites containing continuous carbon fibers. The electrical conductivity of the fibers enables the DC electrical resistivity of the composites to change in response to strain, damage or temperature, thereby allowing sensing. In addition, conduction enables the Seebeck effect, which is particularly large in cement–matrix composites containing short steel fibers and in polymer–matrix composites containing intercalated continuous carbon fibers. By using the interfaces in composites to enhance damping, cement–matrix and polymer–matrix composites having both enhanced damping capacity and increased stiffness are obtained. By using composite interfaces, cement–matrix composites with increased specific heat for thermal control are also obtained.

Review questions

1. Carbon fiber-reinforced concrete is useful for pothole repair. Why?
2. Describe the scientific origin for the ability of carbon fiber (short)-reinforced cement to function as a strain sensor.
3. Describe the scientific origin for the ability of a carbon fiber (continuous) polymer–matrix composite to function as a strain sensor.
4. Describe how a carbon fiber (continuous) polymer–matrix composite can be a two-dimensional array of thermistors.
5. Describe how delamination in a carbon fiber (continuous) polymer–matrix composite can be sensed by electrical resistance measurement.
6. How does the curing pressure of a carbon fiber (continuous) epoxy–matrix composite affect the activation energy of electrical conduction in the through-thickness direction?
7. Why is the interlaminar interface of a crossply composite a more sensitive thermistor than that of a unidirectional composite?
8. How does the volume electrical resistivity of a carbon fiber (short) cement paste change upon uniaxial compression for the resistivity (a) in the longitudinal (stress) direction and (b) in the transverse direction?

References

[1] X. Fu, W. Lu and D.D.L. Chung, *Carbon*, 1998, **36**(9), 1337–1345.
[2] P. Chen and D.D.L. Chung, *Smart Mater. Struct.*, 1992, **2**, 22–30.
[3] P. Chen and D.D.L. Chung, *Composites*, Part B, 1996, **27B**, 11–23.
[4] P. Chen and D.D.L. Chung, *J. Am. Ceram. Soc.*, 1995, **78**(3), 816–818.
[5] D.D.L. Chung, *Smart Mater. Struct.*, 1995, **4**, 59–61.
[6] P. Chen and D.D.L. Chung, *ACI Mater. J.*, 1996, **93**(4), 341–350.
[7] X. Fu and D.D.L. Chung, *Cem. Concr. Res.*, 1996, **26**(1), 15–20.
[8] X. Fu, E. Ma, D.D.L. Chung and W. A. Anderson, *Cem. Concr. Res.*, 1997, **27**(6), 845–852.
[9] X. Fu and D.D.L. Chung, *Cem. Concr. Res.*, 1997, **27**(9), 1313–1318.
[10] X. Fu, W. Lu and D.D.L. Chung, *Cem. Concr. Res.*, 1998, **28**(2), 183–187.
[11] S. Wen and D.D.L. Chung, *Cem. Concr. Res.*, 2000, **30**(8), 1289–1294.

[12] Z. Shi and D.D.L. Chung, *Cem. Concr. Res.*, 1999, **29**(3), 435–439.
[13] Q. Mao, B. Zhao, D. Sheng and Z. Li, *J. Wuhan U. Tech.*, Mater. Sci. Ed., 1996, **11**(3), 41–45.
[14] Q. Mao, B. Zhao, D. Shen and Z. Li, *Fuhe Cailiao Xuebao/Acta Materiae Compositae Sinica*, 1996, **13**(4), 8–11.
[15] M. Sun, Q. Mao and Z. Li, *J. Wuhan U. Tech.*, Mater. Sci. Ed., 1998, **13**(4), 58–61.
[16] B. Zhao, Z. Li and D. Wu, *J. Wuhan Univ. Tech.*, Mater. Sci. Ed., 1995, **10**(4), 52–56.
[17] D. Bontea, D.D.L. Chung and G.C. Lee, *Cem. Concr. Res.*, 2000, **30**(4), 651–659.
[18] S. Wen and D.D.L. Chung, *Cem. Concr. Res.*, 2001, **31**(2), 297–301.
[19] J. Lee and G. Batson, *Materials for the New Millennium*, Proc. 4th Mater. Eng. Conf., ASCE, New York, 1996, **2**, 887–896.
[20] S. Wen and D.D.L. Chung, *Cem. Concr. Res.*, 1999, **29**(6), 961–965.
[21] M. Sun, Z. Li, Q. Mao and D. Shen, *Cem. Concr. Res.*, 1998, **28**(4), 549–554.
[22] M. Sun, Z. Li, Q. Mao and D. Shen, *Cem. Concr. Res.*, 1998, **28**(12), 1707–1712.
[23] S. Wen and D.D.L. Chung, *Cem. Concr. Res.*, 1999, **29**(12), 1989–1993.
[24] S. Wen and D.D.L. Chung, *Cem. Concr. Res.*, 2000, **30**(4), 661–664.
[25] S. Wen and D.D.L. Chung, *Cem. Concr. Res.*, 2001, **31**(3), 507–510.
[26] X. Fu and D.D.L. Chung, *Cem. Concr. Res.*, 1995, **25**(4), 689–694.
[27] J. Hou and D.D.L. Chung, *Cem. Concr. Res.*, 1997, **27**(5), 649–656.
[28] G. G. Clemena, *Materials Performance*, 1988, **27**(3), 19–25.
[29] R. J. Brousseau and G. B. Pye, *ACI Mater. J.*, 1997, **94**(4), 306–310.
[30] P. Chen and D.D.L. Chung, *Smart Mater. Struct.*, 1993, **2**, 181–188.
[31] P. Chen and D.D.L. Chung, *J. Electron. Mater.*, 1995, **24**(1), 47–51.
[32] X. Wang, Y. Wang and Z. Jin, *Fuhe Cailiao Xuebao/Acta Materiae Compositae Sinica* 1998, **15**(3), 75–80.
[33] N. Banthia, S. Djeridane and M. Pigeon, *Cem. Concr. Res.*, 1992, **22**(5), 804–814 (1992).
[34] P. Xie, P. Gu and J. J. Beaudoin, *J. Mater. Sci.*, 1996, **31**(15), 4093–4097.
[35] Z. Shui, J. Li, F. Huang and D. Yang, *J. Wuhan Univ. Tech.*, Mater. Sci. Ed., 1995, **10**(4), 37–41.
[36] X. Fu and D.D.L. Chung, *ACI Mater. J.*, 1999, **96**(4), 455–461.
[37] Y. Xu and D.D.L. Chung, *Cem. Concr. Res.*, 1999, **29**(7), 1117–1121.
[38] Y. Shinozaki, *Adv. Mater.: Looking Ahead to the 21st Century*, Proc. 22nd National SAMPE Tech. Conf., SAMPE, Covina, CA, 1999, **2**, 986–997.
[39] Y. Xu and D.D.L. Chung, *Cem. Concr. Res.*, 2000, **30**(7), 1175–1178.
[40] Y. Xu and D.D.L. Chung, *ACI Mater. J.*, 2000, **97**(3), 333–342.
[41] Y. Wang and D.D.L. Chung, *Cem. Concr. Res.*, 1998, **28**(10), 1353–1356.
[42] S. Wen and D.D.L. Chung, *Cem. Concr. Res.*, 2000, **30**(2), 327–330.
[43] Y. Xu and D.D.L. Chung, *Cem. Con. Res.*, 1999, **29**(7), 1107–1109.
[44] W.F.A. Davies, *J. Phys.*, D: App. Phys., 1974, **7**, 120–130.
[45] W.J. Gadja, Report RADC-TR-78-158, A059029, 1978.
[46] P. Li, W. Strieder and T. Joy, *J. Comp. Mater.*, 1982, **16**, 53–64.
[47] T. Choi, P. Ajmera and W. Strieder, *J. Comp. Mater.*, 1980, **14**, 130–141.
[48] T. Joy and W. Strieder, *J. Comp. Mater.*, 1979, **13**, 72–78.
[49] K.W. Tse and C.A. Moyer, *Mater. Sci. and Eng.*, 1981, **49**, 41–46.
[50] V. Volpe, *J. Comp. Mater.*, 1980, **14**, 189–198.
[51] V.G. Shevchenko, A.T. Ponomarenko and N.S. Enikolopyan, *Int. J. Appl. Electromagnetics in Materials*, 1994, **5**(4), 267–277.
[52] X.B. Chen and D. Billaud, *Ext. Abstr. Program – 20th Bienn. Conf. Carbon*, 1991, 274–275.
[53] X. Wang and D.D.L. Chung, *Composite Interfaces*, 1998, **5**(3), 191–199.
[54] X. Wang and D.D.L. Chung, *Polymer Composites*, 1997, **18**(6), 692–700.
[55] X. Wang and D.D.L. Chung, *Carbon*, 1997, **35**(5), 706–709.
[56] X. Wang and D.D.L. Chung, *J. Mater. Res.*, 1999, **14**(11), 4224–4229.
[57] S. Wang and D.D.L. Chung, *Composite Interfaces*, 1999, **6**(6), 497–506.
[58] S. Wang and D.D.L. Chung *Polymer Composites*, 2001, **22**(1), 42–46.
[59] X. Luo and D.D.L. Chung, *Composites Sci. Tech.*, 2001, **61**, 885–888.
[60] N. Muto, H. Yanagida, T. Nakatsuji, M. Sugita, Y. Ohtsuka and Y. Arai, *Smart Mater. Struct.*, 1992, **1**, 324–329.
[61] X. Wang, X. Fu and D.D.L. Chung, *J. Mater. Res.*, 1999, **14**(3), 790–802.
[62] X. Wang and D.D.L. Chung, *Composites: Part B*, 1998, **29B**(1), 63–73.
[63] P.E. Irving and C. Thiogarajan, *Smart Mater. Struct.*, 1998, **7**, 456–466.
[64] S. Wang, D.P. Kowalik and D.D.L. Chung, *J. Adh.*, 2002, **78**, 189–200.
[65] X. Wang, S. Wang and D.D.L. Chung, *J. Mater. Sci.*, 1999, **34**(11), 2703–2714.
[66] S. Wang and D.D.L. Chung, *Polymer Composites*, 2001, **9**(2), 135–140.

[67] N. Muto, H. Yanagida, M. Miyayama, T. Nakatsuji, M. Sugita and Y. Ohtsuka, *J. Ceramic Soc. Japan*, 1992, **100**(4), 585–588.
[68] N. Muto, H. Yanagida, T. Nakatsuji, M. Sugita, Y. Ohtsuka, Y. Arai and C. Saito, *Adv. Composite Mater.*, 1995, **4**(4), 297–308.
[69] R. Prabhakaran, *Experimental Techniques*, 1990, **14**(1), 16–20.
[70] M. Sugita, H. Yanagida and N. Muto, *Smart Mater. Struct.*, 1995, **4**(1A), A52–A57.
[71] A.S. Kaddour, F.A.R. Al-Salehi, S.T.S. Al-Hassani and M.J. Hinton, *Composites Sci. Tech.*, 1994, **51**, 377–385.
[72] O. Ceysson, M. Salvia and L. Vincent, *Scripta Materialia*, 1996, **34**(8), 1273–1280.
[73] K. Schulte and Ch. Baron, *Composites Sci. Tech.*, 1989, **36**, 63–76.
[74] K. Schulte, *J. Physique IV*, Colloque C7, 1993, **3**, 1629–1636.
[75] J.C. Abry, S. Bochard, A. Chateauminois, M. Salvia and G. Giraud, *Composites Sci. Tech.*, 1999, **59**(6), 925–935.
[76] A. Tedoroki, H. Kobayashi and K. Matuura, *JSME Int. J. Series A – Solid Mechanics Strength of Materials*, 1995, **38**(4), 524–530.
[77] S. Hayes, D. Brooks, T. Liu, S. Vickers and G.F. Fernando, *Proc. SPIE Int. Soc. for Optical Engineering*, Smart Structures and Materials 1996: Smart Sensing, Processing, and Instrumentation, SPIE, Bellingham, WA, 1996, **2718**, 376–384.
[78] S.Wang, Z. Mei and D.D.L. Chung, *Int. J. Adh. Adh.*, 2001, **21**(ER6), 465–471.
[79] Y. Okuhara, S.-G. Shin, H. Matsubara, H. Yanagida and N. Takeda, *Proc. SPIE – Int. Soc. for Optical Engineering*, 2001, **4328**, 314–322.
[80] H. Matsubara, S.-G. Shin, Y. Okuhara, H. Nomura and H. Yanagida, *Proc. SPIE – Int. Soc. for Optical Engineering*, 2001, **4234**, 36–43.
[81] N. Muto, Y. Arai, S.G. Shin, H. Matsubara, H. Yanagida, M. Sugita and T. Nakatsuji, *Compos. Sci. and Tech.*, 2001, **61**(6), 875–883.
[82] H. Yanagida, *Materials and Design*, 2000, **21**(6), 507–509.
[83] Y. Okuhara, S.-G. Shin, H. Matsubara, H. Yanagida and N. Takeda, *Proc. SPIE – Int. Soc. for Optical Engineering*, 2000, **3986**, 191–198.
[84] H. Nishimura, T. Sugiyama, Y. Okuhara, S.-G. Shin, H. Matsubara and H. Yanagida, *Proc. SPIE – Int. Soc. for Optical Engineering*, 2000, **3985**, 335–342.
[85] H. Yanagida, *Ceramic Eng. and Sci. Proc.*, 1998, **2**, 567–577.
[86] M. Sugita, H. Yanagida and N. Muto, *Smart Mater. and Struct.*, 1995, **4**(1), A52–A57.
[87] S. Wang and D.D.L. Chung, *Composites: Part B*, 1999, **30**(6), 591–601.
[88] S. Wang and D.D.L. Chung, *Composite Interfaces*, 1999, **6**(6), 519–530.
[89] S.W. Hudnut and D.D.L. Chung, *Carbon*, 1995, **33**(11), 1627–1631.
[90] M. Segiet and D.D.L. Chung, *Composite Interfaces*, 2000, **7**(4), 257–276.
[91] B. Benchekchou, M. Coni, H.V.C. Howarth and R.G. White, *Composites: Part B*, 1998, **29B**, 809–817.
[92] I.C. Finegan and R.F. Gibson, *Composite Structures*, 1999, **44**(2), 89–98.
[93] H.L. Wettergren, *J. Composite Mater.*, 1998, **32**(7), 652–663.

Index

A

Absolute thermoelectric power, 110, 112, 117, 118, 262, 279
Absorption, 92
Absorption coefficient, 176
Absorptivity, 171
AC loss, 129
Acceptance angle, 174
Activated carbon fibers, 12
Activated carbons, 12
Activation energy, 277
Admixtures, 26, 254
Aging, 144
Aging rate, 144
Air void content, 26
AlN, see also aluminum nitride, 61, 154
Aluminum nitride, see also AlN, 58, 62
Aluminum-matrix composites, 57, 59
Angle of refraction, 172
Anode, 213
Antiferromagnetic material, 198
Attenuation loss, 176
Auger electron, 183
Austenite, 246
Avalanche breakdown, 133

B

Bag molding, 22
Battery, 10, 219
Beryllium-matrix composites, 60
Bimorph, 150
Binder, 40, 57, 225
Bioactive, 233
Bioactive glass, 236, 237
Bioactivity, 238
Biocompatible, 233, 236, 238
Biocompatible material, 13
Biodegradable polymer, 11, 234
Biomedical applications, 12, 233, 235, 237, 238
Bipolar plates, 222, 226
Bohr magneton, 191
Borosilicate glass, 61
Butterfly curve, 143

C

Calendering, 22
Capacitance, 126, 149

Capacitor, 131, 154, 223
Carbon black, 224
Carbon fibers, 15, 16, 19, 20, , 26, 27, 28, 32, 108, 113, 249, 254, 260, 262, 269
Carbon filaments, 27
Carbon nanotubes, 3
Carbon yield, 29, 30
Carbon-carbon composites, see also carbon-matrix composites, 16, 28, 32, 33, 34, 38, 39, 60, 238
Carbonization, 29, 31, 32, 38, 61
Carbonization-impregnation cycles, 29
Carbon-matrix composites, 2, 16, 28, 60, 94, 238
Carrier multiplication, 133
Cathode, 213
Cathodic protection, 5
Cement paste, 253, 266
Cemented fixation, 233
Ceramic-matrix composites, 2, 17, 25, 26, 46, 61, 66, 84, 94, 105, 108, 154, 156, 204, 237, 248, 253, 254, 255, 260, 265, 267
Ceramics, 1, 7, 9, 56, 248
Char yield, 29, 33
Charge density, 125
Chemical vapor deposition, 31, 43
Chemical vapor infiltration, see also CVI, 28, 61
Chip carrier, 74
Cladding, 174, 175
Closure domains, 200
Coefficient of thermal expansion, 62
Coercive field, 143
Coherence, 182
Coherence length, 182
Coherent, 181
Colossal magnetoresistance, 203
Commingling, 23, 24
1-3 composite, 149
Composite materials, 2, 15
Composition cell, 215
III-V compound semiconductors, 180
Compression molding, 22, 23
Concentration cell, 217
Concrete, 2, 5, 25, 26, 248, 253, 255, 265
Conduction, 8
Connectivities, 149
Consolidation, 24
Controlled resistivity materials, 84
Convection, 8

285

Copper-matrix composites, 59
Corrosion resistance, 5, 25
Critical angle, 172
CTE, 60, 61, 153
Curie temperature, 140, 141, 195
Curing, 24
Current collector, 11, 223
CVI, 28, 31, 32

D

Damage sensing, 255
Damping capacity, 245, 248, 267
Dark current, 183
Decibel, 176
Depoling, 144
Diamagnetic material, 199
Diamond, 3, 7, 8, 9, 56, 60, 62
Diamond-like carbon, 3
Die attach, 75
Dielectric applications, 125, 152
Dielectric breakdown, 133
Dielectric constant, 61, 62, 73, 125, 127
Dielectric loss, 129
Dielectric loss angle, 129
Dielectric strength, 133
Dielectrics, 6, 7
Die-less forming, 24
Diffusion bonding, 41
Dipole friction, 128, 132
Dipole moment, 128, 134, 142
Direct gap semiconductor, 180
Direct piezoelectric effect, 133, 135
Dissipation factor, 129
Domain boundaries, 142
Dry cell, 10, 220
Drying shrinkage, 5, 26, 46

E

Easy-plane anisotropy, 201
Elecrical applications, 6
Electric double-layer capacitor, 223
Electric field grading, 156
Electric susceptibility, 128
Electrical insulation, 153
Electrical interconnections, 6
Electrical resistivity, 39, 112
Electrochemical applications, 9
Electrochemical capacitor, 223
Electrochemical reaction, 9
Electrode, 9, 11, 223
Electroluminescence, 178
Electrolyte, 9, 11, 213, 226
Electromagnetic applications, 91
Electromagnetic interference, see also EMI, 79
Electromagnetic interference shielding, see also EMI shielding, 6, 73, 91
Electromagnetic observability, 91
Electromagnetic radiation, 167
Electromagnetic windows, 157

Electromechanical coupling coefficient, 139
Electromechanical coupling factor, 139
Electronic applications, 6
Electronic devices, 6
Electronic materials, 7
Electronic packaging, 7
Electronic pollution, 12
Electrostatic discharge, 83
EMI, 79
EMI gaskets, 94
EMI shielding, 83, 84, 94
EMI shielding gaskets, 83
Encapsulation, 75
Energy loss, 132
Environmental applications, 11
Epoxy, 18, 20, 21, 62, 78, 106, 249, 276
Evanescent-wave sensor, 177
Exchange interaction, 193
Extrusion, 22

F

Ferrimagnetic material, 195
Ferroelectric, 140, 141, 143, 145, 148, 155
Ferroelectric domains, 140, 142
Ferromagnetic, 204, 245
Ferromagnetic domain, 193
Ferromagnetic material, 193
Fiber-optic sensor, 177
Filament windings, 23, 24
Flexible graphite, 3, 66
Fluid permeability, 5
Fluorescence, 171
Freeze-thaw durability, 5, 27
Frequency, 168
Frequency bandwidth, 181
Fuel cell, 10, 221, 222, 226
Fullerenes, 3
Functionally gradient electrode-electrolyte structure, 227

G

Gage factor, 254, 255
Galvanic cell, 10, 215
Galvanic couple, 59, 215
Gauss's law, 126
Giant magnetoresistance, 202
Giant magnetostriction, 201
Glass, 46, 62, 185
Glass-ceramic, 238
Glass-matrix composites, 154, 205
Graded index fiber, 175
Graphite, 2, 225
Graphitization, 3, 28, 31, 38, 39, 60

H

Hard magnet, 198, 199
Heat retention, 66
Heat sinks, 7, 76

Heat transfer, 8
Heating element, 77
High-temperature consolidation, 31
Hot isostatic pressure, 28
Hot pressing, 28, 31, 41
Hydrogen-oxygen fuel cell, 222
Hydroxyapatite, 237, 238

I
Implants, 13, 233
Impregnation, 23, 29, 31
Impregnation carbonization, 28
Indirect gap semiconductor, 180
Infiltration, 41
Inhibitors, 34, 36
Injection molding, 22
Intercalation, 3, 105, 106
Interconnection, 74
Interface engineering, 19
Interlaminar interface, 271
Interlaminar shear strength, 19
Interlayer dielectric, 75
Intermetallic compounds, 5, 64
Internal-sensing sensor, 177
Isothermal method, 31

J
Joints, 5
Joule heating, 265

L
Lasers, 178, 179, 185
Laves phases, 201
Le Chatelier's principle, 216
Leachability, 235
Lead storage battery, 10
Leadless chip carriers, 74
Light detector, 182, 184
Light throughput, 177
Light-emitting diode, 178, 179
Liquid metal infiltration, 57
Liquid metal transfer agent technique, 43
Liquid phase impregnation, see also LPI, 28
Lithium-ion cell, 221
Loss factor, 129
Loss tangent, 129, 245, 249
Low observability, 205
LPI, 32
Luminescence, 169

M
Macropores, 12
Magnetic applications, 6, 191, 203, 204
Magnetic moment, 191, 197
Magnetocrystalline anisotropy constant, 201
Magnetoresistance, 202
Magnetostriction, 199

Magnetostrictive constant, 200
Magnetotagging, 206
Martensite, 246
Martensitic transformation, 246
Matched die molding, 22
Matched-die forming, 24
Mechanical properties, 38
Mesophase, 29, 32
Mesopores, 12
Metal-matrix composites, 2, 5, 39, 40, 41, 57, 203, 238
Metals, 1, 56
Microelectronics, 7, 73
Micro-micropores, 12
Micropores, 12
Microwave switching, 156
Monochromatic, 181
Mortar, 2, 253, 266
Multifunctional, 4
Multilayer, 205
Multimode optical fiber, 175
Multiple reflections, 92

N
Negative electrode, 213
Negative magnetostriction, 200
Network, 154
Nickel ferrite, 197
Nickel-cadmium battery, 220
Non-destructive evaluation, 206
Numerical aperture, 174

O
Open-circuit voltage, 217
Optical applications, 6
Optical detector, 182
Optical fiber, 6, 173, 175, 177, 178, 184
Optical fiber sensor, 177
Optical waveguides, 184
Optoelectronic devices, 6
Organometallic, 45
Organometallic compounds, 44
Overall cell potential, 219
Oxidation, 214
Oxidation protection, 33
Oxidation treatments, 19
Oxygen concentration cell, 217, 218

P
Pack cementation, 34, 35, 37
Paraelectric, 142, 145
Parallel-plate capacitor, 126, 149
Paramagnetic material, 194
Particle tagging, 206
PEG, see also polyethylene glycol, 64, 65
Peltier effect, 101, 104
Percolation, 80
Percolation threshold, 81

Permeability, 193
Permittivity of free space, 126
Perovskite structure, 140
Phase-change materials, 65
Phonons, 8
Phosphorescence, 171
Photoconduction, 170, 182
Photoconductivity, 182
Photon energy, 168
Photoresponse, 183
Photovoltage, 182
Photovoltaic device, 184
Photovoltaic effect, 182
Piezoceramic, 147
Piezoelectric, 133, 134, 135, 140, 155
Piezoelectric aging, 144
Piezoelectric charge coefficient, 135
Piezoelectric constant, 137
Piezoelectric coupling coefficient, 135
Piezoelectric devices, 6
Piezoelectric voltage coefficient, 137
Piezopolymer, 147
Pin-inserting-type packages, 74
Pitch, 30, 32
Plasma spraying, 42, 58
PMN, 147, 154
PMN-PT, 147, 154
Pn junction, 178, 184
Polarization, 126, 127, 134, 136, 142, 145, 151
Polarization reversal, 143
Poling, 143
Polyethylene glycol, see also PEG, 64
Polymer concrete, 154
Polymer-matrix composites, 2, 5, 7, 8, 17, 21, 25, 40, 62, 73, 78, 79, 93, 94, 103, 105, 153, 154, 155, 203, 235, 236, 268, 276, 281
Polymers, 1, 7, 9, 66, 185, 247
Population inversion, 180
Pores, 12
Positive electrode, 213
Positive magnetostriction, 200
Powder metallurgy, 58
Preform, 22, 40, 57
Prepregs, 24
Pressure gradient method, 31
Primary cell, 221
Printed circuit board, 74
Prostheses, 235
Pseudoelasticity, 246
PT, 147
Pultrusion, 22, 236
PVDF, 148, 149, 155, 156
Pyroelectric, 151, 155
Pyroelectric coefficient, 151
Pyrolytic carbon, 233
Pyrolytic graphite, 61
PZT, 141, 145, 146, 147, 149

Q

Quartz, 135, 140

R

Radiation, 8
Reaction sintering, 34, 35
Reduction, 214
Reflection, 91
Reflectivity, 171
Refraction, 171, 172
Refractive index, 171, 173, 185
Relative dielectric constant, 126, 127
Relative permeability, 194
Relaxor ferroelectric, 146, 147
Remanence, 194
Remanent magnetization, 194
Remanent polarization, 143
Repair, 5
Resin transfer molding, 22
Resistance heating, 77
Resorbable materials, 234
Reverse piezoelectric effect, 133, 140

S

Sacrificial anode, 5
Saturation magnetization, 194
Saturation polarization, 142
Scaffolds, 233
Screen printing, 79
Sealant, 34, 35
Secondary cell, 221
Seebeck coefficient, 101, 262, 279
Seebeck effect, 101, 104, 261
Seebeck voltage, 101
Self-polarization, 140
Self-sensing, 4
Semiconductor, 170, 180
Semiconductor detector, 183
Semiconductor laser, 180, 181
Semiconductors, 2, 7, 171, 182, 183
Shape-memory alloys, 245
SiC, see also silicon carbide, 61
SiC conversion coating, 34
Silica fume, 26, 66, 113, 248, 254, 263, 266, 267
Silicon carbide, see also SiC, 58, 59
Single-mode optical fiber, 175
Skin depth, 93
Skin effect, 83, 93, 94
Slurry casting, 42
Slurry infiltration, 47
Smart structures, 253, 268
Snell's law, 172
Sodium silicate, 64
Soft magnet, 199
Solar cell, 184
Solder, 7, 63
Solid electrolytes, 157
Solution coating, 43
Solution coating method, 44
Specific heat, 266
Spin scattering, 202
Spin-orbit coupling, 199
Spontaneous emission, 179

Standard electrode potential, 218
Standard electromotive force series, 218
Standard hydrogen reference half-cell, 218
Static dielectric constant, 129
Stealth, 205
Steel fiber, 108, 112, 254, 262
Stepped index fiber, 174
Stimulated emission, 179
Stir casting, 57
Storage modulus, 245
Strain sensing, 254
Strain/stress sensing, 272
Stress shielding, 233
Stress-graphitization, 31
Structural health monitoring, 4
Substrate, 6, 74
Supercapacitor, 223
Superelasticity, 246
Superexchange interaction, 195
Surface treatment, 20, 33, 38
Surface-mounting-type packages, 74

T
Tan d, 132, 133, 245
Temperature gradient method, 31
Temperature sensing, 260, 276
Thermal applications, 8, 9, 55
Thermal conductivity, 39, 47, 55, 56, 58, 59, 60, 66, 73, 79, 266
Thermal contact conductance, 64, 65
Thermal control, 265
Thermal expansion, 39, 47, 73
Thermal expansion coefficient, 81
Thermal fatigue, 7
Thermal fluids, 63
Thermal greases, 63
Thermal insulation, 66
Thermal interface materials, 63, 64, 65, 74, 76
Thermal pastes, 64
ThermalGraph, 61
Thermistor, 260
Thermocouple, 110, 111, 113, 114, 279
Thermocouple junction, 106, 108, 114
Thermoelectric applications, 101
Thermoelectric devices, 6
Thermoelectric effect, 261

Thermoelectric energy conversion, 103
Thermoelectric figure of merit, 101, 102
Thermoelectric power, 101
Thermoforming, 22, 24
Thermoplastics, 1, 17, 19, 21, 23, 24, 25, 62
Thermopower, 101, 106, 262
Thermosets, 1, 17, 21, 24, 62
Thick-film, 7
Thromboresistant, 233
Time domain reflectometry, 177
Titanium, 237
Total internal reflection, 172
Tow, 17
TPG, 61
Transmission-gap sensor, 177
Transverse waves, 168
Turbostratic carbon, 3

U
Ultracapacitor, 223

V
Vibration damping, 4, 6, 245, 247, 248
Vibration reduction, 267, 281
Viscoelastic material, 6, 267
Viscous glass consolidation, 47
Voltage sensitivity, 149

W
Wavelength, 167
Workability, 26

X
x-ray, 183
x-ray spectrometry, 183
x-ray spectroscopy, 183

Z
z-axis anisotropic electrical conductor film, 75, 81
z-axis conductor, 62